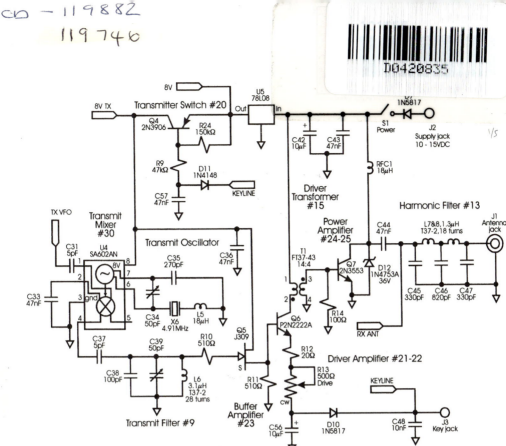

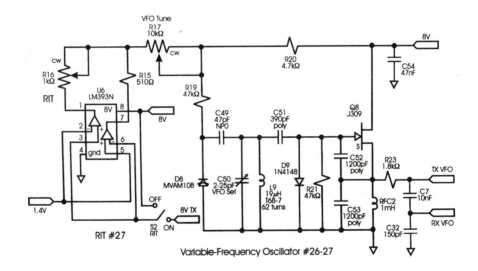

THE ELECTRONICS OF RADIO

This innovative book provides a stimulating introduction to analog electronics by analyzing the design and construction of a radio transceiver. Essential theoretical background is given at each step, along with carefully designed laboratory and homework exercises. This structured approach ensures a good grasp of basic electronics as well as an excellent foundation in wireless communications systems.

The author begins with a thorough description of basic electronic components and simple circuits. He then goes on to describe the key elements of radio electronics, including filters, amplifiers, oscillators, mixers, and antennas. In the laboratory exercises, he leads the reader through the design, construction, and testing of a popular radio transceiver (the NorCal 40A), thereby illustrating and reinforcing the theoretical material. A diskette containing the widely known circuit simulation software, *Puff*, is included in the book.

This is the first book to deal with elementary electronics in the context of radio. It can be used as a textbook for introductory analog electronics courses, or for more advanced undergraduate classes on radio-frequency electronics. It will also be of great interest to electronics hobbyists, radio enthusiasts, and to anyone who wants to find out more about the field of wireless communications.

David Rutledge is Professor of Electrical Engineering at the California Institute of Technology. He attended Williams College and the University of Cambridge and received his Ph.D. from the University of California at Berkeley. He is a recipient of the IEEE Microwave Prize and the Distinguished Educator Award of the Microwave Theory and Techniques Society. He is a Fellow of the IEEE and the holder of an Amateur Extra Class radio license with the call sign KN6EK.

THE ELECTRONICS
OF RADIO

DAVID B. RUTLEDGE

California Institute of Technology

Illustrations by Dale Yee

CAMBRIDGE
UNIVERSITY PRESS

PUBLISHED BY THE PRESS SYNDICATE OF THE UNIVERSITY OF CAMBRIDGE
The Pitt Building, Trumpington Street, Cambridge, United Kingdom

CAMBRIDGE UNIVERSITY PRESS
The Edinburgh Building, Cambridge CB2 2RU, UK   http://www.cup.cam.ac.uk
40 West 20th Street, New York, NY 10011-4211, USA   http://www.cup.org
10 Stamford Road, Oakleigh, Melbourne 3166, Australia

First published 1999

Printed in the United States of America

Typeset in Stone Serif 9.25/13 pt., Eurostile, and Optima in LaTeX 2_ε [TB]

A catalog record for this book is available from the British Library.

Library of Congress Cataloging-in-Publication-Data
Rutledge, David B., 1952–
        The electronics of radio / David B. Rutledge.
            p.   cm.
        ISBN 0-521-64136-5 (hc.). – ISBN 0-521-64645-6 (pbk.)
        1. Radio circuits – Design and construction.   2. Radio – Transmitter
    -receivers – Design and construction.   3. Electronics.   I. Title.
    TK6560.R83   1999
    621.384 – dc21   98-39967
                                                                     CIP
ISBN 0 521 64136 5 hardback
ISBN 0 521 64645 6 paperback

To my children
Robb, Kate, and Alan

The first rule of tinkering is to keep all the pieces.

– Aldo Leopold

Contents

Preface

A modern electrical-engineering textbook is formidable. One thousand pages of matrices and theorems and problems sap enthusiasm from the hardiest students. Even after wading through this massive amount of material, students may be no closer to designing or building electronic circuits. A delightful contrast to these books is Paul Nahin's *The Science of Radio*. Nahin, who is also a historian of great skill, approaches the mathematics of communications engineering in top-down fashion, by telling a history of early radio and introducing the mathematics only when ("just in time") he needs it for his story. However, in one sense, Professor Nahin only tells half the story, and we would like to tell the rest of it. The mathematics of communications, although beautiful, is limited – engineering products must be built. Today's electrical-engineering students have usually not built stereos or tinkered with cars, and this means that they do not know the smoke and smell of construction or the excitement of electronic circuits coming to life. Many universities encourage this trend, with exercises where students switch components in and out of a circuit, never even heating up the soldering iron.

This is an introduction to electronics based on the progressive construction of a radio transceiver, the NorCal 40A, through thirty-nine exercises. At Caltech, beginning electrical-engineering students complete one problem as homework for each lecture. These exercises may also be useful for students in radio engineering classes. Radio amateurs who want to learn more about the transceivers they build have also found the material helpful.

The approach is not traditional. The reader will not find Laplace transforms or matrix-circuit solutions. On the other hand, Philips's SA602AN double-balanced mixer and oscillator looms large in our story. In addition, as the students progress through the material, they will become adept at working with complex numbers and learn about Fourier series. Our experience is that it is valuable for students to learn to put knobs on without destroying the screw heads and to learn to completely strip the enamel from magnet wire before soldering it into the circuit. One benefit of this approach is that when the transceivers are finished, students can do quite sophisticated tests on complete systems with only modest equipment.

The first chapter introduces the fundamental ideas in radio, and the second discusses basic components in circuits. We tackle phasors in the third chapter and begin the construction of the transceiver, which proceeds through much of the rest of the book. I have included a chapter on transmission lines because of the connections to filters and acoustics. The focus is on material and measurements that show how the transceiver works. The discussions of power amplifiers and oscillators as nonlinear circuits are more serious than those usually found in introductory electronics textbooks. I conclude with a chapter on antennas and propagation.

There are problems at the end of each chapter. After introductory exercises, the problems take a student through the transceiver construction. They include background and construction notes. The parts with numerical answers and plots that students do for homework are distinguished by boldface letters, **A**, **B**, and so on. Appendix A by Kent Potter gives a list of the supplies and equipment we use in the measurements. Appendix B explains Fourier series. Appendix C has the instructions for the circuit-simulator program *Puff* that is included and is used in the problems. Appendix D has a set of data sheets for the parts in the transceiver.

It is a pleasure to acknowledge the many ideas of longtime friend and colleague Kent Potter, laboratory engineer at Caltech. I wish to thank William Bridges, Carl F. Braun Professor of Engineering, for introducing me to the community of amateurs. Paul Nahin contributed the epigraph. The teaching assistants for the 96–97 class, Lon Christensen, Kai-Wai Chiu, John Davis, and Jonathan Little, were exceptional and made many perceptive suggestions. I would like to thank Alwin Chi for proofreading and indexing, and Connie Rodriguez for handling the correspondence. Bob Dyer at Wilderness Radio and Wayne Burdick, the designer of the NorCal 40A, have been extremely helpful. I also appreciate the suggestions from reviewers. Finally, this book would never have been written without the encouragement of my wife Dale and my editor, Philip Meyler. We have worked hard to fix errors, but many surely remain. Please let me know about them at rutledge@caltech.edu.

1

The Wireless World

On Sunday, April 14, 1912, shortly before midnight, the *RMS Titanic* struck an iceberg off the coast of Newfoundland. The radio operator, John Phillips, repeatedly transmitted the distress call CQD in Morse Code. He also sent the newly established signal $\overline{SOS}$. Fifty-eight miles away, the *Carpathia* received the messages, and steamed toward the sinking liner. The *Carpathia* pulled 705 survivors out of their lifeboats. Phillips continued transmitting until power failed. He and the other passengers could have been saved if more lifeboats had been available, or if the *California*, which was so close that it could be seen from the deck of the *Titanic*, had had a radio operator on duty. However, this dramatic rescue established the power of wireless communication. Always before, ships out of sight of land and each other were cut off from the rest of the world. Now the veil was lifted. Since the *Titanic* disaster, wireless communications have expanded beyond the dreams of radio pioneers. Billions of people around the world receive radio and television broadcasts every day. Millions use cellular telephones and pagers and receive television programs from satellites. Thousands of ships and airplanes communicate by radio over great distances and navigate by the radio-navigation systems LORAN and GPS.

The enormous increase of wireless communications is tied to the growth of electronics in general, and computers in particular. Often people distinguish between digital and analog electronics. By digital electronics, we mean the circuits that deal with binary levels of voltage, the ones and zeros that are the syllables of computers and calculators. In analog electronics, we deal with voltages and currents that vary continuously. Most systems today actually use a mixture of analog and digital electronics. It is analog electronics that is the focus of this book. We will study the design, construction, and testing of radio circuits and systems. The approach that we take is the progressive construction of a transceiver, the NorCal 40A.

The NorCal 40A transceiver was designed by Wayne Burdick. It is named after the Northern California QRP club. The club is a group of radio amateurs that develop new transceiver designs. QRP is a radio signal indicating a low-power station. Wayne originally designed the NorCal 40 transceiver as a club project, but it proved so popular that the club could not keep up with orders. An improved version, the NorCal 40A, is now available from the Wilderness Radio Company. The NorCal 40A operates in the frequency range from 7.00 MHz to 7.04 MHz, which is the 40-meter amateur band. The NorCal 40A is well suited for learning electronics. It includes an exceptional variety of interesting analog circuits ranging from audio frequencies to radio frequencies and operates at power levels that vary from a picowatt, or 10^{-12} watts, at the receiver input, to two watts at the transmitter output. Because the transmitter was designed for Morse Code, it produces superb sine-wave signals that are extremely useful in receiver testing. In contrast to cellular telephone circuits, the frequencies are low enough that the signals can be conveniently observed on an oscilloscope, and the level of integration is low, so that the detailed behavior of the circuits can be probed. Because of its history as a club project, the circuits are bullet proof, the components are inexpensive, and the design is open.

1.1 Kirchhoff's Laws

The key quantities in electronics are voltage and current. We will study formulas for voltage and current at two levels. At the bottom, we have *components*, such as resistors, capacitors, and inductors, which have particular relationships between voltage and current. We will study these formulas in the next chapter. We connect components to form a *circuit*. In a circuit, two fundamental laws of physics – conservation of energy and charge – govern voltages and currents. Conservation of energy gives us *Kirchhoff's voltage law*, which applies to a loop of components. Conservation of charge gives *Kirchhoff's current law*, which applies to a junction between components. These laws are named for the German physicist Gustav Kirchhoff. Kirchhoff's laws, together with the current–voltage relations for individual components, allow us to predict what circuits will do.

We can define *voltage* by relating electric charge and potential energy. When a charge moves from one place to another, its potential energy changes. The potential-energy change comes from the force exerted by an electric field. We write the charge as Q and the potential-energy change as E. The units of charge are coulombs, with the abbreviation C, and the units of energy are joules, abbreviated J. See Table 1.1 for a list of units and prefixes that we will use in this book. We can write the voltage V as the ratio of the energy change and charge:

$$V = E/Q. \tag{1.1}$$

The units of voltage are volts, abbreviated as V. We can think of the voltage as the potential energy per unit charge. Often the voltage is called the potential, and that may help make the connection to potential energy clearer. In this connection, it is precise to speak of a voltage *difference*, because the voltage depends on the potential-energy difference between the beginning and ending points. Also, the sign of the voltage depends on the order of the points, and we will often add plus and minus signs to indicate the plus and minus terminals. However, it is typical to consider all the voltages relative to a single reference point, the *ground*, which in an electronic circuit is often the metal box. We will often talk about the voltage at some point in a circuit, and by this, we mean the voltage relative to the reference

Table 1.1. Units (a) and Prefixes (b), and Their Abbreviations.

Quantity	Symbol	Unit	Abbr.	Prefix	Multiplier	Abbr.
voltage	V	volt	V	atto	10^{-18}	a
charge	Q	coulomb	C	femto	10^{-15}	f
energy	E	joule	J	pico	10^{-12}	p
current	I	ampere	A	nano	10^{-9}	n
time	t	second	s	micro	10^{-6}	μ
resistance	R	ohm	Ω	milli	10^{-3}	m
conductance	G	siemen	S	kilo	10^{3}	k
power	P	watt	W	mega	10^{6}	M
capacitance	C	farad	F	giga	10^{9}	G
inductance	L	henry	H	tera	10^{12}	T
length	l	meter	m	peta	10^{15}	P
frequency	f	hertz	Hz	exa	10^{18}	E
absolute temperature	T	kelvin	K			
temperature, Celsius	T	degree	°C			
pressure	P	pascal	Pa			

(a)   (b)

Note: We use the MKS (meter-kilogram-second) system, so that the expressions we derive will be combinations of these units. A few dimensions are also given in centimeters (cm). It is important to distinguish between upper and lower case. For example, one could mistake milli (m) for mega (M), or kilo (k) for kelvin (K). We use both the absolute temperature scale and the Celsius scale. The unit has the same size in each scale, but the zero is different. For absolute temperature, the zero is absolute zero. For the Celsius scale, the zero is the ice point of water, which is 273.15 kelvins on the absolute scale.

ground. We shall also speak of the voltage drop *across* a resistor or capacitor, and this is the voltage difference between the connecting wires.

We can apply our voltage definition to a circuit with components connected in *parallel* (Figure 1.1a). A parallel connection forces the voltage across each component to be the same because we calculate the voltage across each component at the same points:

$$V_1 = V_2. \tag{1.2}$$

It is more interesting to consider the voltages for components connected in a loop (Figure 1.1b). Consider a charge at point P. When we move the charge around the loop, the potential energy changes at each component. If we return to P, the potential energy returns to its initial value. We say that the potential energy is *path independent*. This is true as long as the magnetic-field changes inside the loop are not important. We will consider the effect of magnetic fields when we study inductors. This means that the sum of the voltages around the loop is zero:

$$0 = \sum_i V_i, \tag{1.3}$$

where i is an index for the components. This is *Kirchhoff's voltage law*, and it is a statement of the conservation of energy. You have to be careful about the

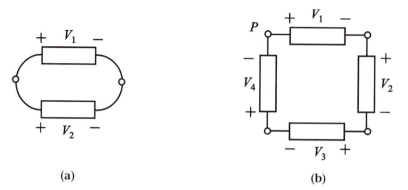

(a)   (b)

Figure 1.1. Parallel connection (a) and connecting components to form a loop (b). Kirchhoff's voltage law applies to a loop.

signs of voltages when you apply the formula. You might test yourself on your understanding of Kirchhoff's law by considering the parallel circuit in Figure 1.1a as a simple loop.

The *current* is the flow of charge past some point. The units are coulombs per second, or amperes (A). Often this is shortened to "amps." We will talk about the current *through* a wire or a component. Like the voltage, we will need to be careful about signs, and we will add arrows to indicate the positive direction. We can apply our current definition to a series connection (Figure 1.2a). If the current does not leak out along the way, the current in both components must be the same, and we write

$$I_1 = I_2,$$ (1.4)

where I is the traditional letter for current. For a junction between several components (Figure 1.2b), if charge does not pile up at the junction, positive current in

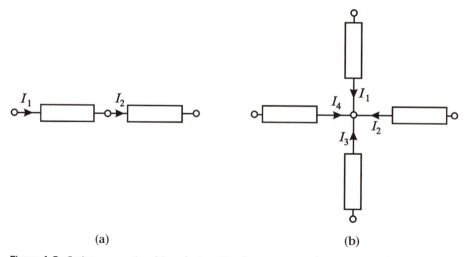

(a)   (b)

Figure 1.2. Series connection (a) and a junction between several components (b). Kirchhoff's current law applies to a junction between components.

some components must be offset by negative current in others, and we can write

$$0 = \sum_i I_i. \tag{1.5}$$

In words, the net current at a junction must be zero. This is *Kirchhoff's current law*, and it is a statement of conservation of charge. Test yourself on Kirchhoff's current law by considering the series circuit in Figure 1.2a as a simple junction.

Since the voltage is energy per charge, and the current is the rate of charge flow, we can find the power, which is the rate at which work is done, by multiplying voltage and current. We write the power P in watts (W) as

$$P = VI. \tag{1.6}$$

In a radio transmitter, power radiates into the air from an antenna as radio waves. In an audio system, power radiates from a speaker as sound. It may also go into heat, and we will have to watch carefully to see how hot our components get.

1.2 Frequency

In radio engineering, we commonly use voltages and currents that are cosine functions of time. For example, we might write a voltage $V(t)$ and a current $I(t)$ as

$$V(t) = V_p \cos(2\pi f t), \tag{1.7}$$
$$I(t) = I_p \cos(2\pi f t). \tag{1.8}$$

In these expressions f is the *frequency*. The units of frequency are hertz (Hz), after the German physicist Heinrich Hertz who first demonstrated radio waves. The frequency is the number of cycles of the cosine that are completed in one second. Often the zero of the time scale will not be important. In fact, if we shift the time scale by a quarter of a period, the cosine becomes a sine. This means that we could use sine functions if we wanted instead of cosine functions. In these formulas, V_p and I_p are called *peak amplitudes*. In the laboratory, we will often see cosine voltages on an oscilloscope, where the zero of the voltage scale may not be clear. For this reason, people usually measure the peak-to-peak voltage, V_{pp}, shown in Figure 1.3, rather than the peak voltage. The peak-to-peak voltage is twice the

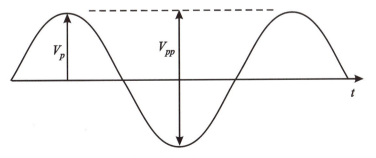

Figure 1.3. Peak and peak-to-peak voltages.

peak voltage. It is important to keep track of whether one is talking about peak voltage, which is easier to use in mathematical formulas, or peak-to-peak voltage, which is easier to measure on an oscilloscope.

We can calculate power by multiplying voltage and current. We write

$$P(t) = V(t)I(t) = V_p I_p \cos^2(2\pi f t). \tag{1.9}$$

The power is not constant as it is for a steady voltage and current. We can separate the power into a steady part and a variable part by writing

$$P(t) = \frac{V_p I_p}{2} + \left(\frac{V_p I_p}{2} \right) \cos(4\pi f t). \tag{1.10}$$

The last term has zero average value, and so we can write the average power P_a as

$$P_a = \frac{V_p I_p}{2}. \tag{1.11}$$

The factor of 2 in the denominator takes into account the fact that the voltage and current vary over a cycle. This formula is for peak amplitudes. In terms of peak-to-peak amplitudes, the average power becomes

$$P_a = \frac{V_{pp} I_{pp}}{8}. \tag{1.12}$$

These formulas depend on the fact that the voltage and current are synchronized so that they are both cosines. This is the situation for a resistor or a properly adjusted antenna. However, for a capacitor or an inductor, one waveform is a cosine and the other is a sine. The power becomes

$$P(t) = V_p I_p \sin(2\pi f t) \cos(2\pi f t) = \left(\frac{V_p I_p}{2} \right) \sin(4\pi f t). \tag{1.13}$$

The average power is zero. This means that an inductor or capacitor does not consume power. However, we will see in the next chapter that they can store energy.

Radio waves travel at the speed of light, and this means that we can write a simple relation between the frequency f and the wavelength λ (the Greek letter *lambda*) as

$$f\lambda = c, \tag{1.14}$$

where c is the velocity of light, given exactly as

$$c = 299,792,458 \text{ m/s}. \tag{1.15}$$

For our purposes, 3.00×10^8 m/s is close enough. It is useful to be able to convert between wavelength and frequency quickly. To do this, divide the quantity that you know into 300:

$$f \text{ (MHz)} = 300/\lambda \text{ (m)} \quad \text{or} \quad \lambda \text{ (m)} = 300/f \text{ (MHz)}. \tag{1.16}$$

One reason that the wavelength is important is that it determines the antenna size. Transmitting antennas need to be at least an eighth wavelength long to be efficient. If the antenna is much shorter than this, most of the power will heat up the wire rather than radiate as radio waves. For example, consider an AM station transmitting at 1 MHz, where the wavelength is 300 m. A large tower must be used for the transmitting antenna. This wavelength would not be suitable for portable telephones. In contrast, cellular phones use frequencies near 1 GHz, where the wavelength is 30 cm, and the antennas need only be a few centimeters long.

At any time, there are thousands of radio services in use around the world. Radios avoid interfering with each other by transmitting on different frequencies. Television stations do the same thing, but the different frequencies are called channels. In cellular phones the process is more complicated. In some systems, different frequencies are assigned. These frequencies may change as a car moves from the area covered by one antenna to another. In other cellular phone systems, the phones may share a range of frequencies with other users but transmit with different codes that allow them to be distinguished. The frequency bands that can be used for different applications are assigned by the Federal Communications Commission (FCC) in the United States and the corresponding communications authorities in other countries. Radio waves propagate long distances, and for this reason, the different communications authorities cooperate to help keep transmitters in different countries from interfering with each other. One factor in assigning frequencies is that waves at different frequencies travel different distances. For example, frequencies below 30 MHz can propagate around the world by reflecting off the ionosphere. Another factor is that some services like TV need large channel spacings. Each TV channel takes up 6.5 MHz. In contrast, AM stations are 10 kHz apart, and stations communicating by Morse Code may be only 500 Hz apart. Table 1.2 gives the names for the radio bands.

Fundamentally, radios are limited by unwanted signals that are at the same frequency as the signal we want. Sometimes these signals are produced by other

Table 1.2. Naming Radio Bands.

VLF (*very low frequency*, 3–30 kHz, or 100–10 km) – submarine communication (24 kHz)

long wave (30–300 kHz, or 10–1 km) – LORAN navigation system (100 kHz)

medium wave (300 kHz–3 MHz, or 1 km–100 m) – AM radio (500–1,600 kHz)

HF (*high frequency* or *short wave*, 3 MHz–30 MHz, or 100–10 m) – international broadcasting, air and ship communication, amateur communication

VHF (*very high frequency*, 30 MHz–300 MHz, or 10–1 m) – television (channels 2–6: 54–88 MHz), channels 7–13: 174–216 MHz) and FM radio (88–108 MHz)

UHF (*ultra high frequency*, 300 MHz–1 GHz, or 1 m–30 cm) – television (channels 14–69: 470–806 MHz), cellular telephone (824–894 MHz)

microwaves (1–30 GHz, or 30–1 cm) – GPS (1.575 GHz), PCS (personal communications services, 1.85–2.2 GHz), ovens (2.45 GHz), satellite TV (C-band: 3.7–4.2 GHz and Ku band: 10.7–12.75 GHz)

millimeter waves (30–300 GHz, or 10–1 mm) – This is the frontier and many applications are under development: car radar (76 GHz), computer networks inside buildings (60 GHz), aircraft landing in fog (94 GHz).

submillimeter waves (frequencies greater than 300 GHz, wavelengths less than 1 mm) – These frequencies are strongly absorbed by water vapor, and this limits them to scientific applications such as radio astronomy and fusion-plasma diagnostics.

Note: The frequency and wavelength ranges are identified, together with the prominent applications.

transmitters at the same frequency, and we call this *interference*. In addition, there are natural radio waves that are called *noise*. Noise appears at all frequencies, and the noise at the receiver frequency is treated just the same way as the signal we want. We hear the noise from a speaker as *hiss*. In a television receiver, noise makes speckles on the screen that we call *snow*. There are many sources of noise. For example, at the frequency used by the NorCal 40A, 7 MHz, the most important noise source is lightning in tropical thunderstorms. The transistors in a transceiver also produce noise, but typically the power is a thousand times less than the lightning noise. However, if the antenna is small, the lightning noise is not received well, and the transistor noise can dominate. At higher frequencies the noise may be quite different. For example, in the UHF TV channels between 400 and 800 MHz, radio waves from the center of the Milky Way dominate. At 12 GHz, which is used for satellite TV transmissions, the most important noise may be radio waves from the earth's surface.

In the simplest form, a radio communications system consists of a *transmitter*, with its antenna to launch the radio waves, and a *receiver*, with its antenna to receive them. In TV and radio broadcasting, that is all there is, but for two-way communications we need a transmitter and a receiver at each end. Typically these are combined in one box to make a *transceiver*. In the NorCal 40A transceiver, the transmitter and receiver share an oscillator circuit that determines the operating frequency, and they share the antenna.

1.3 Modulation

A cosine waveform by itself does not tell us much. Usually we would like to send a message, which might be an audio signal or a digital signal made up of 1s and 0s from a computer that might represent almost anything. We can include the message in our transmitted signal by varying the amplitude or frequency. This is called *modulation*. In *amplitude modulation*, or AM, we vary the amplitude. This is used in AM radio stations. For example, if we have an audio signal $a(t)$, we get an amplitude-modulated signal by using it as the amplitude:

$$V(t) = a(t)\cos(2\pi f t). \tag{1.17}$$

The signal $a(t)$ is called the *modulating waveform*. Usually the voltage is not transmitted in this form. We will see when we build an AM detector that it is easy to recover the audio voltage if a cosine is added:

$$V(t) = V_c \cos(2\pi f t) + a(t)\cos(2\pi f t). \tag{1.18}$$

Here $V_c \cos(2\pi f t)$ is the *carrier*. AM radio stations use this kind of signal. An AM modulated waveform is shown in Figure 1.4a.

We can also modulate by varying the frequency. This is *frequency modulation*, or FM. We can write an FM signal in the form

$$V(t) = V_c \cos(2\pi (f_c + a(t))t). \tag{1.19}$$

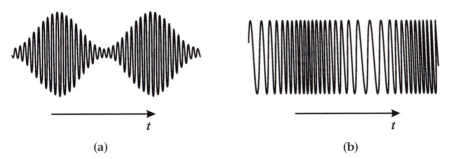

(a)	**(b)**

Figure 1.4. AM waveform (a) and FM waveform (b). In each case, the modulating signal is a cosine function. In AM, the amplitude changes, but the frequency stays the same. In FM, the frequency varies, but the amplitude is constant.

An FM waveform is shown in Figure 1.4b. FM signals usually require wider chan-nels than AM signals, but FM systems are usually better at rejecting noise and interference. For this reason FM stations often program music, which is quite sen-sitive to noise. In contrast, AM stations often emphasize talk shows and news, which are less bothered by noise. Television uses both AM and FM – AM for the picture, and FM for the audio.

For digital signals, a variety of modulation types are used. The oldest is Morse Code, which may properly be considered digital modulation. In Morse Code, the letters are sent as a series of short and long pulses (*dits* and *dahs*), and different lengths of spaces are used to separate letters and words. This could be considered a form of AM. Morse Code is traditionally sent by hand and decoded by ear. A different approach that is better suited for machine reception is to use one frequency for a 1 and another frequency for a 0. We write

$$V(t) = \begin{cases} V_p \cos(2\pi f_1 t) & \text{for a 1,} \\ V_p \cos(2\pi f_0 t) & \text{for a 0.} \end{cases} \qquad (1.20)$$

You will also hear these called *mark* and *space* frequencies. This approach is called *frequency-shift keying*, or FSK, and it is a form of frequency modulation. An FSK signal is shown in Figure 1.5a. Another approach is to keep the frequency the same, but to change the sign of the voltage. This is called *phase-shift keying* or

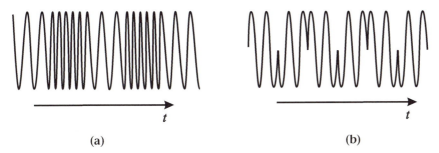

(a)	**(b)**

Figure 1.5. An FSK waveform (a) and a PSK waveform (b). The modulating signal in both cases is a series of alternating 1s and 0s.

PSK (Figure 1.5b):

$$V(t) = \begin{cases} +V_p \cos(2\pi f t) & \text{for a 1,} \\ -V_p \cos(2\pi f t) & \text{for a 0.} \end{cases} \tag{1.21}$$

Some of the most sophisticated modulation approaches are found in computer modems, where a combination of amplitude shifts and phase shifts are used.

1.4 Amplifiers

The radio signals that are received by an antenna are often weak, with typical power levels in the range of one picowatt. To make the level convenient for listening, a huge increase in the power is required, in the range of a factor of a billion. The device that does this is called an *amplifier*, and it is key to all electronic systems. Early amplifiers were made with vacuum tubes, but most amplifiers now are made with transistors, which are smaller and lighter. Vacuum tubes can handle much more power than transistors, however, so they are still found in high-power transmitters. We will study several different amplifiers in the NorCal 40A. An amplifier is characterized by the *gain*, which is the ratio of output power to input power. We write the gain G as

$$G = P/P_i, \tag{1.22}$$

where P is the output power and P_i is the input power. Figure 1.6a shows the symbol for an amplifier that is used in circuit diagrams. Amplifiers also appear in transmitters, because the signals are initially generated in low-power oscillators and therefore must be amplified.

1.5 Decibels

In radio systems, power varies by enormous factors. International broadcasters may transmit as much as a megawatt. Receivers work with signals that may be

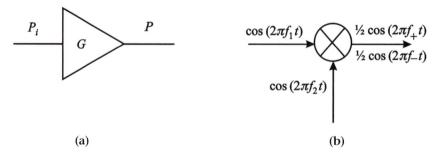

(a)   (b)

Figure 1.6. (a) Symbol for an amplifier that appears in circuit diagrams. The symbol is a triangle that indicates the direction of power flow. (b) Operation of a mixer. The symbol for a mixer is a circle with a large × inside it to indicate multiplication. The inputs and outputs are distinguished by arrows.

Table 1.3. Comparing the Gain Ratio with the Gain in dB.

G, dB	0	1	2	3	4	5	6	7	8	9	10
G, ratio	1	1.3	1.6	2.0	2.5	3.2	4.0	5.0	6.3	8	10

Note: It is an interesting exercise to construct this table with pencil and paper, using the fact that 3 dB corresponds to a gain ratio of 2, 10 dB corresponds to 10, and doubling and halving the gain in dB are equivalent to squaring and taking the square root of the gain ratio.

less than a femtowatt. In order to deal with such a wide range of numbers, it is convenient to use a logarithmic scale for comparing power levels. It is traditional to use a base-10 logarithm and multiply the result by 10. The units are *decibels*, but people usually say "dB." For an amplifier, we can write the gain in dB as

$$G = 10 \log(P/P_i) \, \text{dB}. \tag{1.23}$$

We will write "log" for the base-10 logarithm, to distinguish it from "ln," the natural logarithm. We will use the symbol G for both the gain ratio and the gain in dB; so it is important to include the dB units to distinguish them. For example, if P is twice P_i, then

$$G = 10 \log(P/P_i) = 10 \log(2) = 3.0 \, \text{dB}. \tag{1.24}$$

We say that the amplifier has a gain of 3 dB. Because engineers use dB as commonly as numerical ratios, it is a good idea to learn the conversions. These are given in Table 1.3. Decibels are a relative measure based on the ratio of two powers. However, dBs are also used for absolute powers. For this we use a reference power level in place of P_i. It is common to use 1 watt or 1 milliwatt. For example, 4 W is written as 6 dBw. We add the "w" to dB to indicate that 1 W is the reference level. A femtowatt is written as -120 dBm, with "m" standing for milliwatt.

1.6 Mixers

One thing that you might notice from looking at the frequencies used by the different services is that they are much higher than the ones we hear, which are in the low kilohertz range. In a receiver, we need to be able to convert from a radio frequency down to an audio frequency. In a transmitter, we must go the other way. The device that shifts frequency is called a *mixer*. A mixer effectively multiplies two signals. The output contains two different frequencies that are the sum and difference of the original frequencies. This process is called *heterodyning*. To see how this works, let the two inputs be $\cos(2\pi f_1 t)$ and $\cos(2\pi f_2 t)$. The product is given by

$$V(t) = \cos(2\pi f_1 t) \cos(2\pi f_2 t) = (1/2) \cos(2\pi f_+ t) + (1/2) \cos(2\pi f_- t), \tag{1.25}$$

where f_+ is the *sum frequency*, given by

$$f_+ = f_1 + f_2, \tag{1.26}$$

and f_- is the *difference frequency*, given by

$$f_- = |f_1 - f_2|. \tag{1.27}$$

Now we can identify two frequency components in the output. The sum-frequency component, V_+, is given by

$$V_+ = (1/2)\cos(2\pi f_+ t). \tag{1.28}$$

The other term is the difference-frequency voltage, V_-, given by

$$V_- = (1/2)\cos(2\pi f_- t). \tag{1.29}$$

The operation of a mixer is shown schematically in Figure 1.6b. In a receiver, one frequency might come from the antenna and the other from an oscillator inside the receiver. In practice, we may use either the sum frequency or the difference frequency, depending on whether we want to shift the frequency up or down.

1.7 Filters

The extra frequency that is generated by a mixer is a problem, and we use a device called a *filter* to remove it. A filter lets the frequencies we want through, while blocking the others. For example, if we wanted the difference frequency, but not the sum frequency, we would use a filter with a characteristic like that shown in Figure 1.7a. This is called a *low-pass* filter. In the figure, the output power P is plotted as a function of frequency. At low frequencies, P is at a maximum level P_m, but at high frequencies, the output power is greatly reduced. The lower frequency region where P is large is called the *pass band*. The higher frequencies that are rejected fall into the *stop band*. The response of a filter falls as we move into the

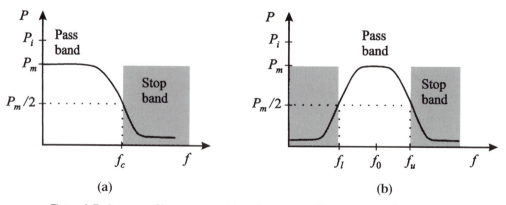

Figure 1.7. Low-pass filter response (a), and band-pass filter response (b). The plots show the output power versus frequency.

stop band, and so people also call this the *roll-off* region. Traditionally the boundary between the pass band and the stop band is taken as the frequency where the output power is reduced to half the maximum value. This half-power frequency is called the *cut-off frequency*, and we write it as f_c. We also refer to it as the 3-dB frequency, because 3 dB corresponds to a 2:1 power ratio.

We characterize a filter by two different numbers. In the pass band we specify the *loss*, which is the ratio of P_i to P_m. We write the loss factor L as

$$L = P_i/P_m. \tag{1.30}$$

We can think of the loss as the inverse of the gain that we defined for amplifiers in Equation 1.22. Often the loss is specified in dB, and we write it as

$$L = 10 \log(P_i/P_m) \, \text{dB}. \tag{1.31}$$

For example, if the maximum output power is half the input power, we say that the loss is 3 dB. The loss in dB is the negative of the gain in dB. This means that for our example, we could in principle talk about a gain of -3 dB instead of a loss of 3 dB. However, people typically use gain when the gain is positive and loss when the loss is positive. In addition, we specify a *rejection factor*, which shows how well the filter blocks a signal at a particular frequency in the stop band compared to the power in the pass band. We write the rejection factor R as

$$R = P_m/P, \tag{1.32}$$

where P is the output power in the stop band. We can write the rejection in dB as

$$R = 10 \log(P_m/P) \, \text{dB}. \tag{1.33}$$

For example, if the output power at a particular frequency in the stop band is a million times lower than the power in the pass band, we would say that the filter rejection is 60 dB.

Figure 1.7b shows a band-pass filter that rejects frequencies above and below a particular operating frequency f_0. In this case we need to consider two 3-dB frequencies, f_u and f_l. The bandwidth of the filter is the difference between the upper and lower 3-dB frequencies. We write the bandwidth as

$$\Delta f = f_u - f_l, \tag{1.34}$$

where Δ is the Greek capital *delta*. You might consider what the response curves for *high-pass* filters and *band-stop* filters would look like.

1.8 Direct-Conversion Receivers

Now we consider a heterodyne receiver that converts a radio signal to an audio frequency that we can hear. It is called a *direct-conversion* receiver because there is one mixer (Figure 1.8). The antenna picks up a signal, which is the mixer input. This signal is traditionally called the *RF* signal, with RF standing for radio

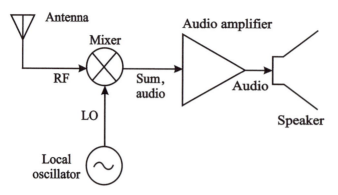

Figure 1.8. Direct-conversion receiver.

frequency. The other input to the mixer is from an oscillator inside the receiver. This oscillator is called the *local oscillator*, or LO for short. The output of the mixer will have a component at both the sum frequency and the difference frequency. The LO frequency is chosen so that the difference frequency is convenient for hearing, and it passes through an amplifier and loudspeaker, and then to our ears. For example, for receiving Morse Code, many operators like a frequency in the range of 600 Hz. For this we could use an LO frequency that is 600 Hz above the RF signal. Typically the sum frequency is out of the range of the amplifier, speaker, and our ears, and so it is ignored. To tune the receiver to pick out a radio station that we want to hear, we adjust the LO frequency until it sounds right. In fact, we will hear the station twice, once when the LO is 600 Hz above the RF frequency, and again when it is 600 Hz below.

Direct-conversion receivers are simple, but they have a fundamental problem: Signals the same distance above and below the LO produce an audio output at the same frequency. The frequency that we do not want to receive is called the *image*. This is shown in Figure 1.9. We can write the relations for the frequencies as

$$f_{rf} = f_{lo} - f_a, \tag{1.35}$$
$$f_i = f_{lo} + f_a, \tag{1.36}$$

where f_i is the image frequency and f_a is the audio frequency. This means that there is a problem if we have a station at the RF frequency and at the image. We hear

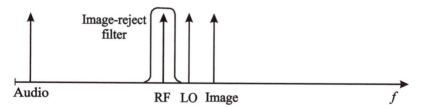

Figure 1.9. Image-frequency problem for a direct-conversion receiver. An RF band-pass filter is needed to remove the image.

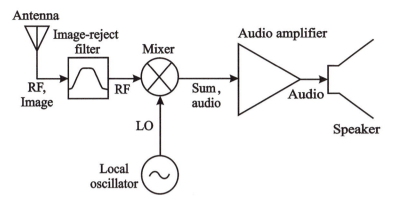

Figure 1.10. Adding an image-reject band-pass filter to a direct-conversion receiver to make a single-signal receiver.

both stations at the same time! This is a major limitation for a direct-conversion receiver.

There are several possible solutions. One is to add a second receiver and shift the oscillator waveform by a quarter of a cycle. It turns out that it is possible to cancel the image by splitting and combining with the correct shifts. This is complicated because it requires a second receiver. Sometimes channel assignments leave gaps, and the image can be put in one of the gaps. The fundamental solution is to put a band-pass filter in front of the mixer to reject the image (Figure 1.10). This requires a band-pass filter with an extremely narrow bandwidth. For example, in the 40-meter band that the NorCal 40A uses, the operating frequency is 7 MHz. The difference between the RF frequency and the image is only one part in 6,000. Surprisingly, it turns out to be quite practical to make a filter like this, with quartz crystals like those used in a wristwatch. However, there is a catch. The pass band in a crystal filter is fixed, and this means that we cannot tune to different stations.

1.9 Superheterodyne Receivers

We can solve the problem of a fixed filter pass band by adding a variable oscillator, another mixer, and another image-reject filter in front (Figure 1.11). This is the *superheterodyne* receiver, arguably the most important invention in the history of communications. The "superhet" is the classic receiver design that has been used in some form in the vast majority of radio and television receivers. The superheterodyne receiver was invented by an American, Howard Armstrong, during the First World War. He wanted to develop a receiver that could be used to intercept German radio transmissions. Armstrong was a brilliant engineer who also invented FM, but he was a tragic figure who spent many years fighting legal battles over patents. He finally committed suicide by jumping off a building in 1954.

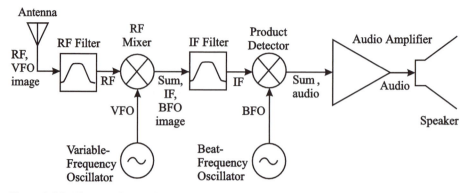

Figure 1.11. The superheterodyne receiver. An additional filter, mixer, and oscillator are added to remove the image.

A superheterodyne receiver is a complicated system, and we need to introduce terminology for the different filters, mixers, and oscillators. The first oscillator can tune in frequency, and it is called the *Variable-Frequency Oscillator*, or VFO. The second, fixed oscillator is the *Beat-Frequency Oscillator*, or BFO. The first mixer is the *RF Mixer*, and the second mixer is the *Product Detector*. The input frequency for the Product Detector is the *Intermediate Frequency*, or IF. The *RF Filter* rejects the image of the VFO, and the *IF Filter* rejects the image of the BFO.

Figure 1.12 shows the relationships among the frequencies in a superheterodyne receiver. These are complicated. We can write

$$f_{rf} = f_{if} + f_{vfo}, \tag{1.37}$$

$$f_{vi} = f_{if} - f_{vfo}, \tag{1.38}$$

$$f_{if} = f_{bfo} - f_a, \tag{1.39}$$

$$f_{bi} = f_{bfo} + f_a, \tag{1.40}$$

where f_{vi} is the VFO image and f_{bi} is the BFO image. We tune the receiver by changing the VFO frequency. The RF-Filter pass band must be broad enough to allow the full range of RF frequencies. In addition, the RF Filter must stop the VFO image. From the figure, we can see that the VFO image is far away from the RF frequency, and this makes it easy for the RF Filter to block it.

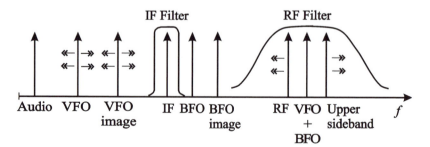

Figure 1.12. Frequencies for a superheterodyne receiver. The variable frequencies are indicated with horizontal arrows.

The BFO-image frequency f_{bi} is blocked by the IF Filter. The frequency that produces a signal at the image frequency f_{bi} is called the *upper sideband*, or USB, and it is given by

$$f_{usb} = f_{vfo} + f_{bi}. \tag{1.41}$$

If we substitute for f_{bi} from Equation 1.40, we find

$$f_{usb} = f_{vfo} + f_{bfo} + f_a. \tag{1.42}$$

The *lower sideband*, or LSB, is given by

$$f_{lsb} = f_{vfo} + f_{bfo} - f_a. \tag{1.43}$$

The lower-sideband frequency is the same as our RF frequency, and so we call this is a *lower-sideband receiver*. Equivalently, we could make an *upper-sideband receiver* by shifting the IF-Filter pass band above the frequency of the BFO so that the upper sideband would pass through, and the lower sideband would be rejected.

1.10 The NorCal 40A

The NorCal 40A transceiver includes a superheterodyne receiver and a transmitter. Figure 1.13 shows a block diagram for the transceiver, with the transmitter on the left, and the receiver on the right. There are two new audio components in the receiver that appear near the bottom. These are the *Automatic Gain Control*, or AGC, and the *AGC Detector*. The purpose of the AGC is to help the receiver adapt to signals with varying power levels. If a signal is strong, the AGC Detector will cause the AGC to reduce the audio signal before it gets to the Audio Amplifier. However, if the signal is weak, the AGC will let the audio signal through to the Audio Amplifier with little loss. This helps to keep the output sound level in a comfortable range for listening.

On the transmitter side, we start at the bottom of the figure. To transmit, a telegraph key activates the *Transmit Oscillator*, which produces a cosine with a frequency near 4.9 MHz. The output of the Transmit Oscillator mixes with the VFO, which is near 2.1 MHz, to give a sum frequency at 7.0 MHz. We write this as

$$f_+ = f_t + f_{vfo} = f_{rf} \approx 7.0\,\text{MHz}, \tag{1.44}$$

where f_t is the frequency of the Transmit Oscillator. We use the same VFO as for the receiver, so that we can tune both the receiver and transmitter together. The difference frequency is given by

$$f_- = f_t - f_{vfo} \approx 2.8\,\text{MHz}. \tag{1.45}$$

Since we do not want this frequency in the output, the Transmit Mixer is followed by the *Transmit Filter*, which passes the sum frequency and removes the difference frequency. At this point, the signal power is low, only about 10 μW. Next there is

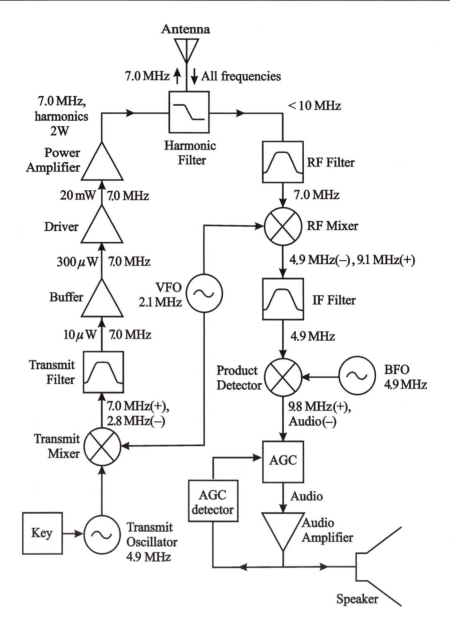

Figure 1.13. Block diagram for the NorCal 40A. Sum frequencies are noted by + signs, and difference frequencies by − signs. Adapted from Appendix C of the *NorCal 40A Assembly and Operating Manual*, by Wayne Burdick, published by Wilderness Radio. Used by permission.

a series of three amplifiers, the *Buffer*, the *Driver*, and the *Power Amplifier*, that raise the power level successively to 300 μW, 20 mW, and finally to 2 W. The output of the Power Amplifier contains the RF frequency, 7.0 MHz, but there is also power at the harmonic frequencies, 14 MHz and 21 MHz. Because these frequencies interfere with other operators, it is important to remove them. This is done with the *Harmonic Filter*, which is a low-pass filter with a 3-dB frequency of 10 MHz.

Inside the front cover is a complete schematic of the NorCal 40A. As you work through the problems, it is helpful to refer to this figure for component values and to see how the different circuits you build go together.

FURTHER READING

Paul Nahin's book, *The Science of Radio*, published by the American Institute of Physics, is a history of early radio, in addition to being a textbook on communications theory. Ken Burns's documentary, *Empire of the Air*, is a fascinating story of Howard Armstrong and the other pioneers of radio. It is available on video cassette through PBS Home Video.

Voltage, current, Kirchhoff's laws, and power all have simple definitions in this chapter, but the underlying ideas in electricity and magnetism are quite subtle and present problems for very fast circuits. My favorite book on this topic is *Electricity and Magnetism*, by the late Edward Purcell, published by McGraw-Hill. We will also explore these issues when we discuss transmission lines in Chapter 4. For broader coverage of radio engineering at a more advanced level, I recommend *Radio-Frequency Electronics*, by Jon Hagen, published by Cambridge University Press.

2

Components

Resistors, capacitors, and inductors are the basic components that make up electrical circuits. For each, we will give a formula that relates voltage and current that is based on an underlying physical law. We will see what happens when we apply a voltage or current source to circuits with these components. In addition, we consider circuits with a semiconductor device, the diode. Analyzing circuits requires a lot of algebra. The expressions get so complicated that it is difficult to see what is going on. For this reason, it is important to express the results in a simple form that can be understood, and it is worthwhile to try different approaches to find the simplest path to a solution and to make sure that you understand it.

2.1 Resistors

In a resistor, the voltage is proportional to the current. This is Ohm's law. We call the ratio of voltage and current the *resistance* and write it as

$$R = V/I. \tag{2.1}$$

The units of resistance are ohms. The abbreviation for ohms is Ω (the Greek capital letter *omega*). Figure 2.1a gives the circuit symbol for a resistor. In a series connection of resistors (Figure 2.1b), the voltage across the entire combination is the sum of the voltages across the individual resistors. Kirchhoff's current law tells us that the current in each resistor is the same. This means that the resistance of a series connection is the sum of the individual resistances:

$$R = \sum_i R_i, \tag{2.2}$$

where i is an index for the resistors.

The inverse of the resistance is called the *conductance*, with units of siemens (S). The conductance G is written as

$$G = I/V. \tag{2.3}$$

In a parallel connection of resistors (Figure 2.1c), the total current is the sum of the individual currents. Kirchhoff's voltage law tells us that voltage across each

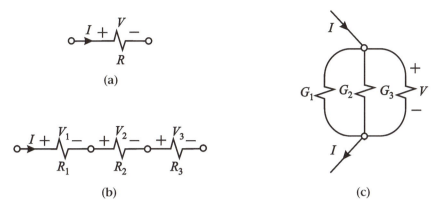

Figure 2.1. Circuit symbol for a resistor (a), series connection (b), and parallel connection (c).

resistor is the same. This means that we can write the conductance of a parallel connection as the sum of the individual conductances:

$$G = \sum_i G_i. \tag{2.4}$$

Since most people think in terms of resistances rather than conductances, we can substitute to find the equivalent formula and get

$$R = \frac{1}{\sum_i 1/R_i}. \tag{2.5}$$

This formula is easy to compute on a calculator, but it is a little complicated to write, and we will use the shorthand $\|$ to indicate a parallel connection. For example, for three resistors connected in parallel, we write

$$R_1 \| R_2 \| R_3 = \frac{1}{1/R_1 + 1/R_2 + 1/R_3} \tag{2.6}$$

to avoid writing out the formula. Circuits can often be simplified by repeatedly applying the parallel and series formulas. In more complicated cases, Kirchhoff's laws can be used to set up matrix equations for a solution.

The power dissipated in a resistor can be written in several ways. We can write

$$P(t) = V(t)I(t) = V^2(t)/R = I^2(t)R. \tag{2.7}$$

In terms of conductance, we can write

$$P(t) = V^2(t)G = I^2(t)/G. \tag{2.8}$$

These formulas mean that power is proportional to the square of a voltage or current. For cosine signals, we can rewrite Equation 1.11 to find the average power P_a as

$$P_a = \frac{V_p I_p}{2} = \frac{V_p^2}{2R} = \frac{I_p^2 R}{2}, \tag{2.9}$$

Color	Digit	Multiplier
Silver		0.01
Gold		0.1
Black	0	1
Brown	1	10
Red	2	100
Orange	3	1,000
Yellow	4	10,000
Green	5	100,000
Blue	6	1,000,000
Violet	7	
Grey	8	
White	9	

1st digit
2nd digit
Multiplier
Tolerance
(5% gold
10% silver)

(a)   (b)

Figure 2.2. Axial lead resistor (a), and the resistor color code (b). Silver and gold bands are not used for numbers, only for multipliers, while violet, grey, and white are used only for numbers. The color sequence from red to violet is the same as that of the visible spectrum, and this may help you remember it.

where V_p and I_p are peak values. In terms of the peak-to-peak values that we measure in the lab, we have

$$P_a = \frac{V_{pp}I_{pp}}{8} = \frac{V_{pp}^2}{8R} = \frac{I_{pp}^2 R}{8}. \tag{2.10}$$

Resistors come in a variety of sizes and configurations. The ones we use in the NorCal 40A are called axial-lead resistors (Figure 2.2a), because the leads extend along the axis of the resistor. It is traditional to indicate the resistance with three color bands. The first two bands give the first two digits of the resistance, while the third band indicates a multiplier. The colors and their digits and multipliers are given in Figure 2.2b. For example, a 100-Ω resistor would have the following bands: brown, black, brown. There is a fourth band that indicates the tolerance. We use 5% resistors with a gold band. The actual paint colors you see vary somewhat, so that it is not always easy to read the code, and you may have to check the resistance with a meter.

2.2 Sources

To provide the power for electronic circuits, we can connect a battery or a plug-in power supply. These *sources* are specified by the voltage that they deliver, such as 1.5 V for AA batteries or 12 V for an adapter. However, one should think of this as a nominal voltage, because the actual voltage depends on the current, generally dropping as more current is drawn (Figure 2.3a). It is as if there is a resistor inside

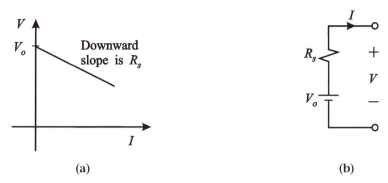

Figure 2.3. Voltage versus current for a source (a), and circuit model with an ideal voltage source V_o and a resistor R_s (b).

the battery or adapter. As the current increases, the voltage drops because the voltage across the resistor increases. Often, this is really what is happening, and you may feel a battery or adapter getting hot if you draw a lot of current. We will represent our source by a circuit *model* that has two parts (Figure 2.3b). The first is an *ideal voltage source* V_0. An ideal voltage source maintains the same voltage regardless of the current. The second is a *source resistance* R_s that provides the drop in voltage as the current increases.

We can relate the circuit model to our voltage plot to find the values of V_o and R_s. When the circuit has no components attached, no current is drawn. This is the *open-circuit* condition. Because there is no current, there is no voltage drop across R_s, and the output voltage is the same as that of the ideal source. We call V_o the open-circuit voltage, and it is just the y intercept of the voltage plot. The downward slope of the plot is the voltage drop per unit current, which is just the resistance R_s. With these values, the circuit model has the same relation between voltage and current as the real source. This circuit model with an ideal voltage source and a series resistor is called a *Thevenin* equivalent circuit.

Alternatively, we could use a different circuit model called the *Norton* equivalent circuit (Figure 2.4a). The Norton circuit also has two parts: an *ideal current source* I_s that maintains the same current regardless of what is connected to it

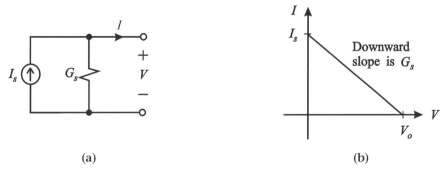

Figure 2.4. Norton equivalent circuit consisting of an ideal current source and a parallel resistor (a). Finding I_s and G_s from a plot of the current (b).

and a parallel resistor with a conductance G_s. We can find the components of the Norton equivalent circuit from a plot of the current (Figure 2.4b). We get I_s by considering the current when the voltage is zero. We call this the *short-circuit* condition, because it corresponds to putting a short circuit across the output. If the voltage is zero, then the current in the source conductance G_s is zero and the entire current I_s is delivered to the load. This means that I_s is the y intercept on our current plot. Next we find G_s by letting the output voltage of the circuit model increase. As the output voltage increases, the output current drops because some of the current begins to flow through G_s. The slope is just the negative of the conductance G_s. Because the Norton equivalent circuit also produces the same voltage and current as the real source, we could use it in place of the Thevenin. Usually we will choose the one that gives the simpler algebra.

If we compare the Norton circuit with the Thevenin circuit, the only difference in using the slope to calculate R_s and G_s is that we swap the current and voltage axes. This means that

$$R_s = 1/G_s \qquad (2.11)$$

and that the resistors are really one and the same. We could just as well use R_s in the Norton equivalent as G_s, or we could use G_s instead of R_s in the Thevenin equivalent. From the graph, we can also calculate R_s in terms of the open-circuit voltage V_o and the short-circuit current I_s from Figure 2.4b as

$$R_s = V_o/I_s. \qquad (2.12)$$

Thus if we know the Thevenin components, we can calculate the Norton components and vice versa. As a practical matter, many batteries and power adapters will not tolerate a short circuit, so that you may only be able to measure the voltage and current over a small range. To find I_s you may have to extrapolate out to the axes.

Next consider what happens if we connect an ideal voltage source V to a Thevenin source (Figure 2.5a). We use Kirchhoff's laws to write the voltage V

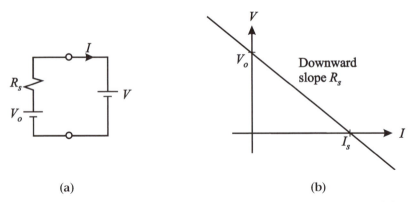

(a)   (b)

Figure 2.5. Connecting an ideal voltage source to a Thevenin source (a), and the extended voltage–current plot (b).

as

$$V = V_o - R_s I. \tag{2.13}$$

This is the equation of a straight line, with voltage intercept V_o and downward slope R_s (Figure 2.5b). If the load voltage V is negative, then the current is greater than the short-circuit current. However, if V is greater than V_o, then the current turns negative. This happens when we charge a battery. A battery charger has a larger voltage than the open-circuit voltage of the battery. For example, for "12-volt" lead-acid batteries, the open-circuit voltage for a battery that is 70% discharged is 12 V. We charge the batteries at 13.8 V. Initially, a large current flows into the battery, but this decreases as the battery charges. The open-circuit voltage when the battery is fully charged is 12.8 V.

2.3 Dividers

Now we are ready to analyze a circuit with a Thevenin source and a resistor R_l, which is called a *load* (Figure 2.6a). First we redraw the circuit in Figure 2.6b to emphasize that the two resistors are in series. We can write the output voltage V in terms of the current I as

$$V = IR_l. \tag{2.14}$$

We use Kirchhoff's laws to write the supply voltage V_o in a similar fashion as

$$V_o = I(R_l + R_s). \tag{2.15}$$

We can divide these two equations to find the output voltage:

$$\frac{V}{V_o} = \frac{R_l}{R_l + R_s} = \frac{1}{1 + R_s/R_l}. \tag{2.16}$$

The input voltage V_o divides proportionally between the load resistor R_l and the source resistor R_s, and for this reason we call this circuit a *voltage divider*. The larger the load resistor, the larger the output voltage. I have written two equivalent expressions, $R_l/(R_l + R_s)$ and $1/(1 + R_s/R_l)$. The first expression is easier to

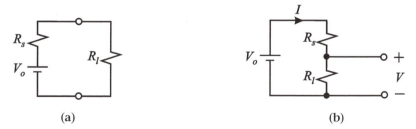

Figure 2.6. Voltage-divider circuit with a Thevenin source and a load resistor (a). Redrawn circuit for analysis (b).

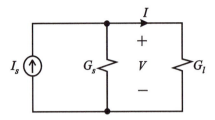

Figure 2.7. Current-divider circuit with a Norton source and a load resistor.

understand, because it can be interpreted as a ratio of load resistance to total circuit resistance. The second expression is convenient for a calculator because each resistance value need only be entered once.

Now consider the *current-divider* circuit, which has a Norton source and a load resistor with a conductance G_l (Figure 2.7). We can write the load current I in terms of the voltage V as

$$I = VG_l. \tag{2.17}$$

Using Kirchhoff's laws, we can write the supply current I_s as

$$I_s = V(G_l + G_s). \tag{2.18}$$

We divide these equations to find the output current:

$$\frac{I}{I_s} = \frac{G_l}{G_l + G_s} = \frac{R_s}{R_s + R_l}. \tag{2.19}$$

For the current divider, the larger the load conductance, the larger the output current. In terms of resistances, the smaller the load resistance, the larger the output current. Because voltage dividers and current dividers will arise repeatedly, it is a good idea to memorize these formulas.

2.4 Look-Back Resistance

So far we have considered a Thevenin or Norton equivalent circuit as a model for a real source. We make a plot of voltage and current to find the parameters of the model. We can also use a Thevenin or Norton source to simplify a section of a circuit diagram. The idea is to replace a complicated source circuit with a simpler equivalent. If the equivalent circuit has the same relationship between voltage and current, then the equivalent circuit will produce the same voltage and current in a load as the original source circuit, regardless of what the load is. This result is called *Thevenin's theorem*. To make the substitution, we need to find the components of the Thevenin equivalent circuit. One approach is to calculate the open-circuit voltage V_o and the short-circuit current I_s. Then we can take the ratio to find R_s.

There is another way to find R_s from the circuit diagram. If there is a voltage or current source in the circuit, then V_o and I_s are both proportional to the value of the source. This means that the slope of the voltage–current plot does not change

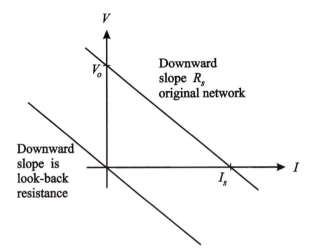

Figure 2.8. Voltage and current for a source network. The slope of the plot is the source resistance R_s. Also shown is a plot of voltage and current when the internal source is turned off. This slope is the look-back resistance. The slope of each plot is the same, so that the look-back resistance gives us R_s.

as we reduce the source to zero (Figure 2.8). We call the resistance when the internal source is zero the *look-back resistance*, because people talk about looking back into the circuit to find its resistance. The look-back resistance is equal to R_s.

As an example, we find the Thevenin equivalent circuit for a voltage-divider circuit (Figure 2.9a). We can write the open-circuit voltage V_o using the voltage-divider

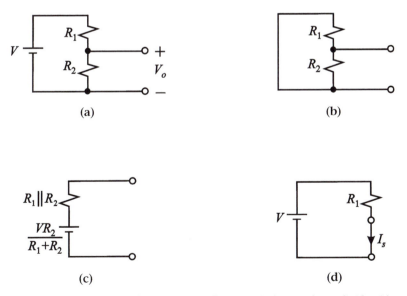

Figure 2.9. Finding the Thevenin equivalent circuit for a voltage divider (a). Circuit with source turned off for calculating the look-back resistance R_s (b), and the components of the Thevenin equivalent circuit (c). Calculating the short-circuit current I_s (d).

formula (Equation 2.16):

$$V_o = \frac{V R_2}{R_1 + R_2}. \tag{2.20}$$

To find the look-back resistance R_s, we set the ideal voltage source V to zero (Figure 2.9b). An ideal voltage source set to zero is no different from a short circuit, because its voltage is zero, regardless of the current. If we replace V by a short circuit, we have only the two resistors in parallel, which gives a look-back resistance of

$$R_s = R_1 \parallel R_2. \tag{2.21}$$

This gives us the Thevenin equivalent circuit shown in Figure 2.9c. This circuit is equivalent in that it will produce the same voltage and current in a load that the divider would, but it is simpler. As a check we can calculate the short-circuit current I_s by shorting out R_2 (Figure 2.9d). We can write

$$I_s = V/R_1. \tag{2.22}$$

If we divide V_o by I_s, we get R_s again.

However, the Thevenin equivalent is not identical to the original divider because the voltages and currents *inside* the divider and those of the Thevenin equivalent circuit are not the same. They could not be, because the Thevenin equivalent has fewer elements. It only produces the same voltage and current in an *outside* load. As an exercise, you should find the Thevenin and Norton equivalent for a current-divider circuit. For this you will need to set a current source to zero to find the look-back resistance. A current source set to zero acts just like an open circuit, since no current flows, regardless of the voltage. This means that the current source should be replaced by an open circuit for the calculation.

2.5 Capacitors

In a capacitor, voltage is proportional to charge. Fundamentally, this arises from the relation between charge and electric field expressed in Gauss's law. This is in contrast to a resistor, where voltage is proportional to current. We write

$$C = \frac{Q}{V}, \tag{2.23}$$

where C is the capacitance, with units of farads (F). We can think of the charge as the time integral of the current:

$$Q(t) = \int_0^t I(t)\,dt. \tag{2.24}$$

We have to be careful how we interpret this expression, because t is doing double duty mathematically. It is the argument of Q, and in this role, it appears in $Q(t)$

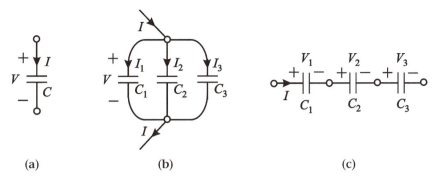

Figure 2.10. Circuit symbol for a capacitor (a), parallel connection (b), and series connection (c).

and as the limit of integration $\int_0^t$. It also appears in the integrand as the argument of I and the differential dt. Often people are disturbed by this, and for this reason they change the variable in the integrand to something else, like u. This makes the mathematical grammar correct but obscures the physics, because time really is the variable in both cases. We will say "ain't" in order to keep the physics. We can rewrite Equation 2.23 as

$$C = \frac{\int_0^t I(t)\,dt}{V}. \tag{2.25}$$

The circuit symbol for a capacitor represents a pair of plates (Figure 2.10a). If we have capacitors in parallel (Figure 2.10b), Kirchhoff's laws tell us that the voltage for each capacitor is the same, and the total current is the sum of the individual capacitor currents. The integral of the total current is also the sum of the integrals. This means that the capacitance of a parallel connection is the sum of the individual capacitances:

$$C = \sum_i C_i. \tag{2.26}$$

The series connection shown in Figure 2.10c is trickier, but we can follow the same logic we used for conductances. If we invert Equation 2.25, we have

$$\frac{1}{C} = \frac{V}{\int_0^t I\,dt}. \tag{2.27}$$

In a series circuit, the current in each capacitor is the same, and the current integral for each capacitor is the same. However, the total voltage is the sum of the individual voltages. Thus we can write

$$\frac{1}{C} = \sum_i \frac{1}{C_i}. \tag{2.28}$$

This is similar to the expression for the resistance of parallel resistors.

Earlier we studied resistor divider circuits. We can also make divider circuits with capacitors. Figure 2.11 shows a capacitive voltage divider. We can write the

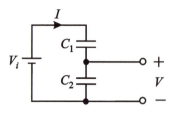

Figure 2.11. Capacitive voltage-divider circuit.

output voltage V as

$$V = \frac{\int_0^t I \, dt}{C_2} \tag{2.29}$$

and the input voltage V_i as

$$V_i = \frac{\int_0^t I \, dt}{C_1} + \frac{\int_0^t I \, dt}{C_2}. \tag{2.30}$$

If we divide these two expressions and simplify, we get

$$\frac{V}{V_i} = \frac{C_1}{C_1 + C_2}. \tag{2.31}$$

This formula says that the output voltage becomes larger if C_1 is made larger. This is the opposite of the resistive potential divider, where the voltage becomes larger if the load resistor is made larger.

2.6 Energy Storage in Capacitors

Capacitors store energy rather than dissipate it as heat like resistors. We can calculate the stored energy, starting with a capacitor with no charge or voltage on it at time $t = 0$. When a current flows into the capacitor, the voltage across it increases. The power going into the capacitor is given by $P(t) = V(t)I(t)$. We can write the energy $E(t)$ stored in the capacitor as an integral of the power $P(t)$. We write the stored energy $E(t)$ as

$$E(t) = \int_0^t P(t) \, dt = \int_0^t V(t)I(t) \, dt. \tag{2.32}$$

Now we rewrite Equation 2.25 in differential form as

$$I = CV', \tag{2.33}$$

where the prime denotes a time derivative. We substitute for current in the previous formula to get

$$E = \int_0^t VCV' \, dt. \tag{2.34}$$

If we use V as the variable of integration instead of t, we can write

$$dV = V' dt \tag{2.35}$$

and we get

$$E = C \int_0^V V \, dV = \frac{CV^2}{2}.$$ (2.36)

The fact that capacitors store energy rather than dissipate it means that ordinarily capacitors do not get hot. In practice, capacitors have some resistance, and large currents can cause them to heat up. The energy stored in a capacitor can discharge dangerously quickly if the output terminals are shorted together. It is important to remember this when working with large high-voltage capacitors. Even with the circuit off, the capacitors may be charged to a high voltage and can deliver a lethal shock.

2.7 RC Circuits

If we connect a resistor to a charged capacitor (Figure 2.12a), the capacitor will discharge through the resistor. This dissipates the energy stored in the capacitor as heat in the resistor. This is often used in high-voltage circuits to reduce the capacitor voltage to a safe level when the circuit is turned off. We call the resistor a *bleeder resistor*. We write the current I in two ways:

$$I = V/R = -CV'.$$ (2.37)

There is a minus sign because the current arrow points *out* from the capacitor. We could change the direction of the arrow and make it positive, but this would make the resistor current negative. We are stuck with one minus sign, no matter how we point the arrow. We can rewrite this as a differential equation

$$RCV' + V = 0.$$ (2.38)

The quantity RC has dimensions of time and is called the time constant. We will write it as τ (the Greek letter *tau*). We write

$$\tau = RC.$$ (2.39)

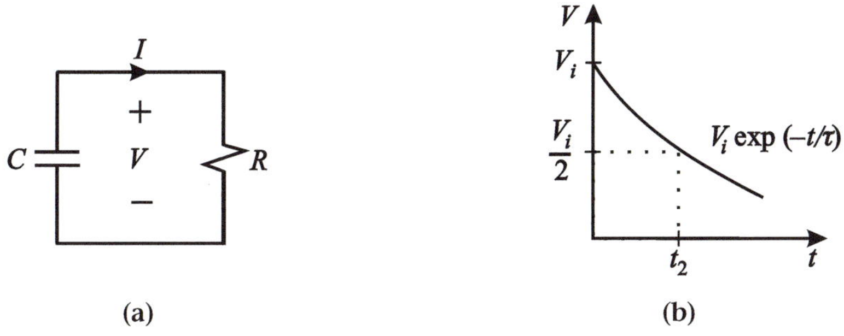

Figure 2.12. Charged capacitor with a bleeder resistor (a) and the decaying voltage (b).

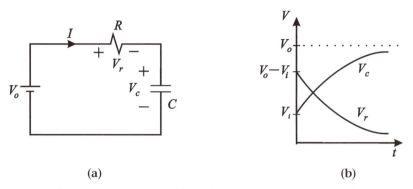

(a)   **(b)**

Figure 2.13. RC network with an added voltage source V_o (a). Resistor and capacitor voltages (b).

The solution is given by

$$V(t) = V_i \exp(-t/\tau), \tag{2.40}$$

where V_i is the initial voltage, and $\exp(x)$ is the exponential function e^x. The voltage decays exponentially with time (Figure 2.12b). It reaches half its initial level at time t_2, given by

$$t_2 = \tau \ln 2 = 0.69\tau. \tag{2.41}$$

In the lab, t_2 is convenient to measure, and this formula lets us work backward to find τ.

In the lab we study an RC network that is repeatedly charged and discharged by a function generator. We can understand the behavior if we add a voltage source V_o to our RC circuit (Figure 2.13a). The math is easier if we solve for the resistor voltage, which decays to zero. We use Kirchhoff's voltage law to write

$$V_o = V_c + V_r, \tag{2.42}$$

where V_c is the capacitor voltage and V_r is the resistor voltage. This formula lets us find V_c if we know V_r. The current I is written as

$$I = V_r/R = CV_c'. \tag{2.43}$$

Since V_o is fixed, Equation 2.42 tells us that $V_c' = -V_r'$, and we write

$$I = V_r/R = -CV_r'. \tag{2.44}$$

This gives us

$$\tau V_r' + V_r = 0. \tag{2.45}$$

The solution is a voltage that decays from its initial value with the same time constant τ as before. The initial value of V_r is given by $V_o - V_i$, where V_i is the initial voltage on the capacitor. We plot the capacitor voltage as $V_o - V_r$ (Figure 2.13b). This means that the capacitor voltage charges exponentially from V_i to V_o with the same time constant τ.

2.8 Diodes

Diodes are devices that let current pass more easily in one direction than the other. They do not obey a simple linear relation between voltage and current like resistors, or voltage and charge like capacitors, and so we say diodes are *nonlinear*. Figure 2.14 shows the schematic symbol for a diode and a representative plot of the current as a function of voltage. We call these plots I–V curves. For positive voltages, the diode conducts well (we say the diode is "on") if the voltage exceeds a small threshold that we call the *forward voltage*. For a silicon diode such as the 1N4148 in our transceiver, the forward voltage is 0.6 V. The power is usually low when the diode is conducting in the forward direction, because the voltage is low. For negative voltages the current is quite small (we say the diode is "off"). The reverse current for the 1N4148 is only a few nanoamps. Since the current is small, the power is also small. When we get to a sufficiently negative large voltage, 75 V for the 1N4148, the diode breaks down, and the current increases rapidly. In the breakdown region, both the voltage and current are large, and thus the power dissipated is large, and we have to be careful not to destroy the diode. Usually we avoid the breakdown region and operate with the diode either on or off. One way to think of a diode is that it limits a positive voltage and blocks a negative current.

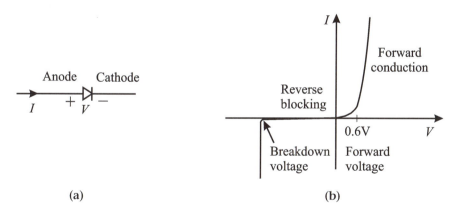

(a)   (b)

Figure 2.14. Circuit symbol and terminal names for a diode (a). We need names for each terminal, because unlike a resistor or capacitor, they are distinct. The names cathode and anode come from the days of vacuum-tube diodes. In a vacuum-tube diode the cathode emits electrons, and the anode collects them. Because electrons have a negative charge, however, we will usually say that current flows from the anode to the cathode, even though the electron flow is in the opposite direction. The diodes we use are made of silicon, but the names for the terminals are the same. Usually diode manufacturers mark the cathode end with a black stripe. In (b), a current plot for a diode is shown.

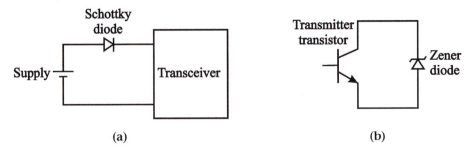

Figure 2.15. Protecting a transceiver from a negative voltage with a Schottky diode (a), and protecting a transistor from a high positive voltage with a Zener diode (b). The Zener symbol has short tags on the cathode bar that indicate that the diode conducts in both directions.

We can consider a diode as a kind of self-activated switch. When the voltage or current is positive, the switch is on. When the voltage or current is negative, the switch is off.

You should take note of the diode part number: 1N4148. There are thousands of different kinds of diodes and transistors, and there are standard registration and numbering systems to help us keep them straight. "1N" denotes a diode, and "2N" a transistor. The 1N and 2N designations are not trademarks, and they do not belong to a single manufacturer. Many different manufacturers make a 1N4148. For the complete specifications for this diode and other parts that appear in the transceiver, see Appendix D.

We use four different kinds of diodes in the NorCal 40A. We will only discuss how they act in a circuit, leaving the details of the operation of the diodes to a book on solid-state devices. In addition to the 1N4148 silicon diode, we use a Schottky diode, the 1N5817, that is made out of a contact between metal and silicon. Schottky diodes have a low forward voltage, only about 0.2 V, and this reduces the power dissipation if the current is large. We use the Schottky diode to prevent a negative power-supply voltage from being applied to the radio, which could damage some of the circuits (Figure 2.15a). If the voltage is positive, the current passes through the diode with only a small forward voltage drop. However, if the voltage is negative, the diode turns off, and current does not flow to the radio. In addition, we use a Zener diode that is fabricated to have a controlled breakdown voltage and to allow a reasonable amount of breakdown current to flow safely. A 1N4753A, 36-V, Zener appears across the output of our transmitter transistor (Figure 2.15b). This restricts the peak transistor voltage to 36 V to prevent damage to the transistor. The fourth type of diode we use is a varactor diode. A varactor operates with negative voltages, where the diode acts like a capacitor with a capacitance that is controlled by the voltage. We use Motorola's MVAM108 varactor to control the frequency of the transceiver.

Probably the single most important application of diodes is in changing the sinusoidal AC wall voltage to the steady DC voltage that is required in most circuits. AC is short for *alternating current*, and DC stands for *direct current*. This is called *rectification*. A rectifier circuit is shown in Figure 2.16a. The diode turns off whenever the voltage is negative, so that the negative parts of the waveform are removed (Figure 2.16b). This is called a *half-wave* rectifier.

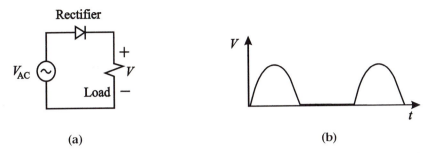

(a)   **(b)**

Figure 2.16. Rectifying an AC supply voltage with a diode to produce DC (a), and the voltage waveform (b).

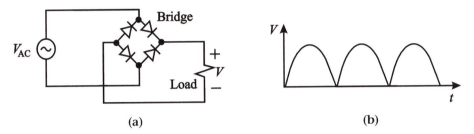

(a)   **(b)**

Figure 2.17. Bridge-rectifier circuit (a), and the full-wave rectified waveform (b).

Figure 2.17a shows a circuit that flips the negative parts of the AC waveform, rather than removes them. It has four diodes in a ring (Figure 2.17b). This arrangement is called a *bridge*. You should notice that the DC waveform is still pretty bumpy, and in most circuits we would have to smooth the voltage out with a capacitor before we could use it.

2.9 Inductors

In an inductor, the voltage is proportional to the time derivative of the current. The circuit symbol for an inductor is a coil (Figure 2.18a). The proportionality constant is called the inductance, with units of henries (H). The inductance is traditionally written as L, and we write

$$V = LI'. \tag{2.46}$$

The voltage arises from the fact that currents produce magnetic fields, and time-varying magnetic fields produce voltages through Faraday's law. We consider inductors in series (Figure 2.18b), and rewrite the inductance formula to get

$$L = \frac{V}{I'}. \tag{2.47}$$

In a series connection the current for each inductor is the same. This means that the denominator, I', is the same for each inductor. The total voltage is the sum of the individual inductor voltages. Therefore the total inductance is the sum of the

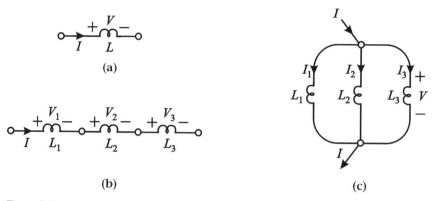

Figure 2.18. Circuit symbol for an inductor (a), series connection (b), and parallel connection (c).

inductances:

$$L = \sum_i L_i. \tag{2.48}$$

This is similar to the expression for series resistors. For the parallel connection (Figure 2.18c), we can start by inverting Equation 2.47 to get

$$\frac{1}{L} = \frac{I'}{V}. \tag{2.49}$$

In a parallel circuit, the voltage for each inductor is the same, and so the denominator does not change. The total current is the sum of the individual currents, and the derivative of the total current is the sum of the individual derivatives. This means that we can write

$$\frac{1}{L} = \sum_i \frac{1}{L_i}. \tag{2.50}$$

This is similar to the expression for parallel resistors.

2.10 Energy Storage in Inductors

Like capacitors, inductors store energy rather than dissipate it as heat. As in capacitors, there is some resistance, and if there are large currents, inductors will heat up. Again we calculate the energy $E(t)$ stored in the inductor as the integral of the power $P(t) = V(t)I(t)$. We write the stored energy $E(t)$ as

$$E(t) = \int_0^t P(t)\, dt = \int_0^t V(t)I(t)\, dt. \tag{2.51}$$

If we substitute for V from Equation 2.46, we get

$$E = \int_0^t LI'I\, dt. \tag{2.52}$$

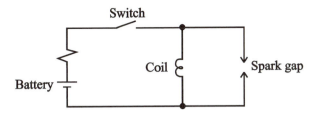

Figure 2.19. Inductor ignition system for cars. This circuit is simplified. In addition there would be a transformer to make sure that the spark voltage is bigger than the switch voltage. Otherwise the switch itself will arc.

We use I as the integration variable rather than t and get

$$E = L \int_0^I I \, dI = \frac{LI^2}{2}. \tag{2.53}$$

Notice that energy storage in an inductor is associated with the current rather than the voltage as in a capacitor. We said that for capacitors there is a danger in shorting the terminals because the current can be very large and the energy will be dissipated quickly in the short. In contrast, in an inductor, the danger is in trying to stop the current. For example, when a switch is opened in a circuit with an inductor that is carrying current, there will be a large voltage that can make an arc.

This feature is used in car ignition systems to make a spark inside the cylinders to ignite the gasoline. It is impressive because it takes 12 V from a battery and produces 10 kV to make the spark. Figure 2.19 shows a simplified ignition circuit. First a battery and resistor are connected by a switch to establish a current in the inductor. Then the switch is opened. When this happens, the current in the inductor drops quickly, causing a large voltage. When the voltage is high enough, there will be an arc across the spark gap, firing the spark plug.

2.11 RL Circuits

If we connect a resistor to an inductor carrying current (Figure 2.20a), the current will decay as the energy stored in the inductor is dissipated as heat in the resistor.

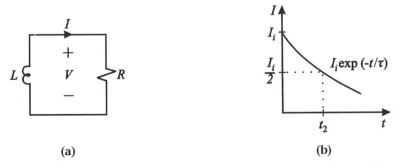

(a)   (b)

Figure 2.20. Inductor carrying current with a resistor (a), and the decaying current (b).

We can analyze the circuit by writing two expressions for the voltage:

$$V = IR = -LI'. \tag{2.54}$$

There is a minus sign because the current arrow points out from the inductor:

$$(L/R)I' + I = 0. \tag{2.55}$$

The time constant is given by

$$\tau = L/R. \tag{2.56}$$

We write the solution as

$$I(t) = I_i \exp(-t/\tau). \tag{2.57}$$

The current decays in an inductor just as voltage decays in a capacitor (Figure 2.20b). However, in an inductor, the current decays quickly if the resistor is large. In a capacitor, it is the other way around – the discharge is fast if the resistor is small. Since an oscilloscope measures voltage rather than current, it is not as convenient to measure this decay in an inductor as it is in a capacitor. Usually we will end up measuring the voltage across a series resistor and divide by the resistance to get the current.

In the lab, you will drive an RL circuit with a function generator. To understand the behavior, we will analyze an RL circuit with a current source I_s (Figure 2.21a). The approach is to solve for a current that decays to zero, the resistor current, and then to use the resistor current to find the inductor current that we are really interested in. Kirchhoff's current law gives us an expression for I_s:

$$I_s = I_l + I_r, \tag{2.58}$$

where I_l is the inductor current and I_r is the resistor current. We can write the voltage V as

$$V = RI_r = LI_l'. \tag{2.59}$$

For the derivative I_l' we can substitute $-I_r'$. This gives us

$$V = RI_r = -LI_r' \tag{2.60}$$

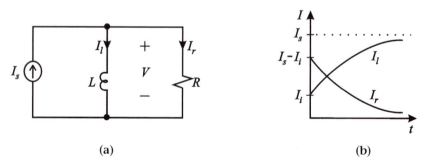

(a)   (b)

Figure 2.21. RL circuit with a current source I_s added (a), and the resistor and inductor currents (b).

or

$$(L/R)I_r' + I_r = 0, \tag{2.61}$$

which gives us a decaying exponential with the same time constant L/R as before. The initial value of I_r is given by $I_s - I_i$, where I_i is the initial inductor current. We find the inductor current I_l as the difference between I_s and I_r (Figure 2.21b). The inductor current builds up exponentially from the initial current I_i to I_s, with the same L/R time constant.

FURTHER READING

The classic book on electronics is the encyclopedic *The Art of Electronics*, by Horowitz and Hill, published by Cambridge University Press. This is a brilliant book, and it is more likely to be on the bookshelf of working engineers than any other book that I know of. I recommend consulting this book continuously as you learn about electronics. For a serious discussion of Thevenin's theorem, see Desoer and Kuh's *Basic Circuit Theory*, published by McGraw-Hill. This is also an excellent reference for matrix solutions of circuits. A good book for information on diodes is *Device Electronics for Integrated Circuits*, by Muller and Kamins, published by Wiley.

PROBLEM 1 – RESISTORS

A. Figure 2.22 shows a Thevenin source with a load resistor R_l. Find the formula for the power in the load. Find the load resistance R_l that gives the maximum power. What is the maximum load power? As a check, it is a good idea to find the formula for the maximum available power for a Norton source with a source conductance G_s and a load conductance G_l. Your result should be equivalent to the Thevenin result.

B. Figure 2.23 shows two resistive circuits that appear often in *attenuators*, which are circuits that reduce the power of a signal. Attenuators can prevent radios from

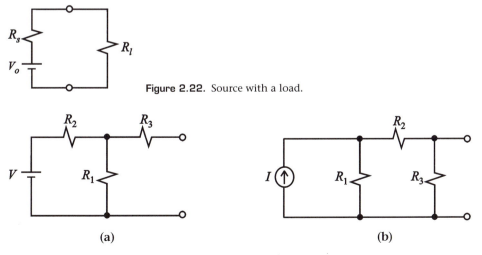

Figure 2.22. Source with a load.

(a)

(b)

Figure 2.23. *T* resistor circuit (a), and π resistor circuit (b).

overloading and sensitive instruments from burning out. Figure 2.23a is called a *T* network, and Figure 2.23b is a π network. They get their names because the outlines resemble these letters. For each circuit, find the parameters needed to make Thevenin and Norton equivalent circuits: V_o, I_s, and R_s. Make sure that you express each quantity in the simplest form.

PROBLEM 2 - SOURCES*

For these measurements, we use a 12-V, 0.8-A-hr battery that is manufactured by Yuasa. You should start with four 510-Ω, quarter-watt resistors. Use a multimeter to measure the open-circuit voltage. Be careful not to attach the leads to the current jack on the multimeter. This jack is only for measuring current. Its resistance is quite low, and several amps will flow, blowing the fuse in the multimeter.

A. Connect the positive lead from the battery to the top row of holes on the breadboard and the negative lead to the bottom row (Figure 2.24). It is a good idea to follow the tradition of using black for low voltage and red for high voltage. This will save you from blowing out circuits later. Now you can add the resistors in parallel by plugging them into any of the holes in the top and bottom rows. You should wait two minutes after adding each resistor before you take a measurement for the battery voltage to stabilize. Plot the voltage you measure on the *y* axis versus the current on the *x* axis as the number of resistors increases from 0 to 4. To calculate the current, assume the nominal resistance value of 510 Ω. Use linear graph paper and choose scales carefully to show what is happening in the plot. Label your axes with units. Draw a smooth curve through the data points.

B. Find an equivalent circuit for the battery with an ideal voltage source V_o and a resistor R_s when the current is the neighborhood of 75 mA. You should notice that this circuit will not be accurate at currents that are much lower or higher than this.

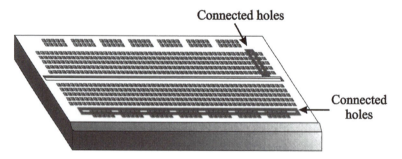

Figure 2.24. Hole pattern for a solderless breadboard. On the top and bottom are four rows of connected holes. In the center are columns of five connected holes. If you are confused about the connections, you can measure the resistance between holes with a multimeter.

* *Note:* Please see Appendix A for a complete list of the supplies and equipment that are used in each problem.

C.   When the NorCal 40A is receiving, it draws 20 mA. What voltage would you expect from the battery? If the battery has an amp-hour rating of 0.8 A-hr, how long would you expect to be able to operate the radio as a receiver?

PROBLEM 3 – CAPACITORS

A combination of a series resistor and parallel capacitor is used in many circuits to give a time delay of about RC. In our transceiver, this delay is used to make sure the receiver is muted while the transmitter is turning on and off – otherwise there would be a loud pop, because the transmitter voltages are much larger than the ordinary signals that are received. In other circuits, delays may be unintentional. A major factor limiting the speed of computers is the resistance and capacitance of the metal patterns that connect different parts of a circuit.

Connect a function generator and an oscilloscope, using test hooks attached directly to the resistor and capacitor leads (Figure 2.25). Make sure that the red leads are across the resistor and that the black leads from the scope and the function generator are connected together. The black leads are connected to the ground through the AC outlets, but this is not a reliable connection. You should use a sync cable from the function generator to trigger the scope. Do not use a breadboard, because it adds capacitance that confuses the measurements.

A.   The function-generator settings should be for a 20-Hz, 1-Vpp (peak-to-peak) square wave. For a function generator with a 50-Ω source resistance, this amplitude, 1 V peak-to-peak, is the voltage that we *would* see if the load were 50 Ω. For an

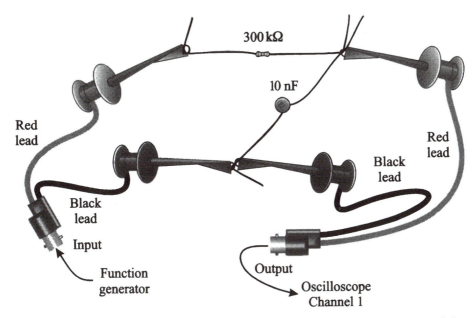

Figure 2.25. RC delay circuit with an input square wave from a function generator and the output to an oscilloscope. The output contains sections of exponential waveforms with a time constant RC.

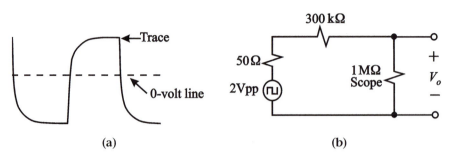

(a)   (b)

Figure 2.26. Oscilloscope waveform at output of the delay network (a), and simplified circuit diagram without capacitors (b).

open-circuit load, the amplitude is twice this, or 2 Vpp. The frequency, 20 Hz, is low enough that it allows the capacitor to charge fully each time the voltage rises and to discharge fully each time the voltage falls. With the scope voltage and time scales properly set, you should see the waveform in Figure 2.26a. It is a square wave with rounded corners. Measure the peak-to-peak output voltage on the oscilloscope.

B. Now calculate what the voltage should be. A good way to start is to consider the circuit without capacitors. Figure 2.26b shows the open-circuit voltage of the function generator (2 Vpp), the function-generator resistance, 50 Ω, the 300-kΩ load resistance, and a 1-MΩ resistance for the oscilloscope. This is a divider circuit. The function-generator resistance is much smaller than the others, and you can ignore it. Make a Thevenin equivalent for this circuit by finding V_o and the look-back resistance R_s. The open-circuit voltage V_o should be close to the value you measured, and you will need R_s later to calculate the delays.

C. The delays come from the time required to charge the capacitor when the voltage rises and to discharge the capacitor when the voltage falls. Expand the time scale on the oscilloscope so that the falling part of the waveform occupies the entire screen (Figure 2.27a). When the voltage drops to zero volts, the capacitor has discharged halfway. The time for it to reach 0 V is the time t_2 that we related to the time constant τ. Measure t_2.

D. Now calculate what t_2 should be, from the Thevenin source resistance R_s and the load capacitance 10 nF (Figure 2.27b).

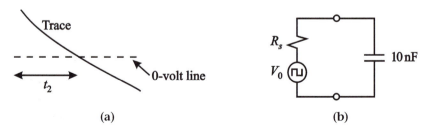

(a)   (b)

Figure 2.27. Output waveform with time scale expanded to show detail of discharging capacitor with the measuring time t_2 (a), and Thevenin circuit with 10-nF load (b).

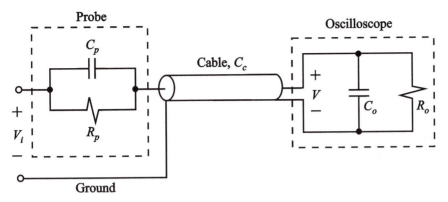

Figure 2.28. Construction of a scope probe.

E. We can try to eliminate the delay by removing the capacitor from the circuit. This reduces the delay considerably, but not as much as you might expect, because there is capacitance in the scope and the cable. Measure t_2 again. To make the measurement accurate, you should expand the time scale so that the delay is several large divisions long. Use this delay measurement to figure out the total scope and cable capacitance C.

F. The oscilloscope capacitance C_o is usually written by the scope input jack. Subtract it from C to find the cable capacitance C_c. We will study cable capacitance when we discuss transmission lines later, but for now you should know that it is proportional to the length. This can be a good reason to make cables short. Divide the cable capacitance by the length of the scope cable to get the capacitance per unit length. This is called the *distributed capacitance.* The cable from the function generator does *not* contribute to this capacitance, and we will see why when we study transmission lines.

G. We can reduce the delay further with a high-impedance probe. Figure 2.28 shows a simplified view of the construction of a high-impedance scope probe. We will calculate the values for R_p, C_p, and C_c. Start by considering only the resistances. What value must the probe resistance R_p have to make the resistance marked on the probes correct, given the resistance R_o marked on the scope? You may want to check the probe resistance R_p with a multimeter to verify your answer. For these values of R_p and R_o, what is the ratio of the input voltage V_i to the output voltage V?

H. Now consider only the capacitances in the circuit. You should find the values that the series probe capacitance C_p and parallel cable capacitance C_c must have to make the capacitance marked on the probe correct. You should use the capacitance C_o marked on the scope and the same ratio of input voltage to output voltage that you calculated for the resistors.

I. Replace the scope cable with a high-impedance probe and measure t_2 again.

J. Now calculate what t_2 should be, using the capacitance marked on the probe. You should take into account that R_s has changed because the probe resistance is larger than the scope resistance.

K. Measure the peak-to-peak output voltage again, and calculate what it should be for comparison.

PROBLEM 4 – DIODE DETECTORS

In Chapter 1, we discussed amplitude modulation, which is used by AM radio stations. We wrote the voltage as

$$V(t) = V_c \cos(2\pi ft) + a(t) \cos(2\pi ft). \tag{2.62}$$

The first term, $V_c \cos(2\pi ft)$, is the carrier, and $a(t)$ is the audio modulating signal. We characterize the modulation by the *modulation depth*, which is written as

$$m = a_p / V_c, \tag{2.63}$$

where a_p is the peak value of $a(t)$. Usually m is expressed as a percentage. Commercial broadcasters monitor the modulation depth to make sure that it does not reach 100% when $a(t)$ is negative, or the receiver output becomes distorted. We will see this in our measurements.

We can use a diode circuit to detect the audio signal (Figure 2.29). The circuit is a half-wave rectifier with a capacitor. Whenever the diode is on, the capacitor charges up to the input voltage. The bleeder resistor adjusts how fast the current leaks out of the capacitor. This current needs to be small enough that the capacitor voltage does not drop much during an RF cycle, but it should allow the voltage to follow the audio signal.

The function-generator settings should be for a 1-MHz, 5-Vpp sine wave with a modulating frequency of 1 kHz and a modulation depth of 70%. You need to connect a cable from the sync output of the function generator to the oscilloscope trigger input and to use external triggering. The modulated waveforms that you see will be difficult to trigger on if you do not do this. Adjust the scope controls for a good display of the modulated waveform on channel 1. Now connect the detector circuit, using the breadboard, with the output on channel 2. The detector output should be a 1-kHz sine wave like the modulating waveform.

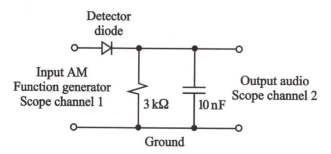

Figure 2.29. An AM detector circuit.

A.   The RC time constant τ needs to be considerably less than the period of the modulating waveform, or the output will not be able to follow the modulation. Calculate τ and compare it with the period of the modulating waveform.

B.   Compare the maximum voltage of the input AM signal with the maximum voltage of the output audio. It is convenient to use the same voltage scale and zero-voltage reference line on the scope for both the input and output. What would you expect the difference of these voltages to be?

C.   The time constant τ should be considerably longer than the period of the carrier, or the voltage will droop between each cycle. To see this effect, reduce the carrier frequency to 100 kHz. Measure the voltage droop. Now calculate what you would expect for the droop.

D.   Return the carrier frequency to 1 MHz. Adjust the modulation depth to 100% and sketch the distorted output waveform. Why does this distortion occur?

PROBLEM 5 – INDUCTORS

Inductors come in small packages that resemble resistors. Inside is a small magnetic rod with fine wire wrapped around it. Inductors use the same color code as resistors, except that the units are μH rather than Ω. We will measure the time it takes current to build up and decay in an inductor. Make the connections shown in Figure 2.30, using function-generator settings for a 1-kHz, 5-Vpp square wave. At the input, use a tee to connect channel 1 of the scope. At the output, use a tee at the scope to connect a 50-Ω load.

A.   At the output, you should see a square wave with rounded corners. The zero-voltage line is the half-way point for current building up or for current decaying. Measure the time it takes to reach the zero-voltage line, t_2. Deduce the peak-to-peak inductor current from the voltage across the 50-Ω load.

B.   Now calculate the peak-to-peak inductor current and the delay t_2 that we should expect.

C.   Sketch the input voltage and interpret it.

In electronic circuits it is common to use transistors as switches. Transistors have three terminals, and by applying a current at one of the terminals, called the base, we can make the other two terminals, the collector and the emitter, act like a switch. There may be times when we want to switch the current in an inductor without producing a

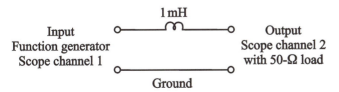

Figure 2.30. Circuit for observing current buildup and decay in inductors.

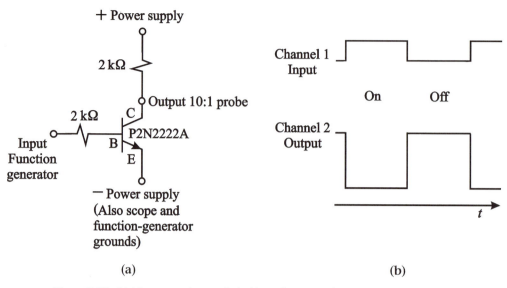

Figure 2.31. Making a transistor switch (a), and input and output scope traces (b).

large voltage. One example is in driving a relay. Relays are mechanical switches that are switched on and off by magnetic forces. The magnetic forces are generated by the current in a drive coil. Relays can control large voltages and currents with little loss, but the drive coils are quite inductive, and they generate large voltages if the drive current changes suddenly. You will make a transistor switch, first with a resistor load, and then with an inductor load, and finally, you will include a diode to prevent large inductive voltages.

To start, you will need to connect the circuit in Figure 2.31a on a breadboard. You should leave out the inductor and diode at first. Use a tee to connect the function generator to both channel 1 of the scope and your circuit at the same time. At the output, use a 10:1 probe on channel 2. The voltages in these measurements get rather large and will go off the oscilloscope screen unless you use a 10:1 probe. You will need to plug in a 12-V power supply.

When you put together circuits with a transistor and a power supply, there are many opportunities to destroy components. If your circuit does not work, and you have checked that the connections are correct, it is possible that a component has been destroyed. You can check resistors by measuring the resistance on a multimeter. You can also check an inductor by checking its resistance. The 1-mH inductors should have a resistance of around 10 Ω. The other inductors you will use have much smaller inductances, and their resistances are smaller also, an ohm or less. Inductors can either fail as open circuits if a wire melts or as short circuits if the insulation melts.

For testing diodes, some multimeters have a diode check setting. Other multimeters can provide a fixed current during a resistance measurement of 1 mA, and take the ratio of voltage to current. For example, for a diode with a forward voltage of 0.6 V, the reading would be 600 Ω. You do have to orient the diode correctly, by connecting the anode to the high-voltage terminal, and the cathode to the low-voltage terminal. The same approach can be used to check transistors. The base–emitter and base–collector connections are diodes, so that a transistor can be checked as if it were two diodes.

The P2N2222A transistor comes in a plastic package with three terminals. ("P" is for plastic.) The base is the input, the collector is the output, and the emitter is the ground connection. Check the data sheet in Appendix D to identify the terminals. The base-to-emitter connection is a diode, so that only positive current flows into the base, and there is a forward voltage of about 0.6 V. When no current flows in the base, the resistance between the collector and emitter is high, and the switch is effectively open. When this happens, there is no current in the 2-kΩ collector resistor, and the output voltage is the same as the supply voltage. However, when enough current flows in the base, the resistance between the collector and emitter drops, and a large current can flow, effectively closing the switch. This causes a large voltage across the resistor, and it means that the output voltage will be small, nearly zero. The purpose of the 2-kΩ base resistor is to keep too much current from flowing in the base.

The function generator should be set for a 100-kHz, 100-mVpp square wave. Increase the input voltage until the switch begins to turn on. The output voltage should be low when the transistor is on. When the input-voltage setting is 1 Vpp, the scope traces should look like Figure 2.31b.

D. Now reduce the input voltage until the transistor does not turn on at all, and add a 1-mH inductor as shown in Figure 2.32a. Increase the function-generator setting gradually to 1 Vpp again. Sketch the output. When the transistor switches off, the current drops rapidly, and this causes a large voltage in the inductor. This voltage can destroy a transistor. Measure the maximum voltage across the transistor. You should notice that when the transistor turns off, the output voltage oscillates. We call this *ringing*. The ringing comes from a resonance between the inductor and circuit capacitance.

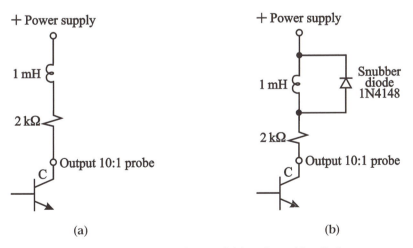

(a)   (b)

Figure 2.32. Adding an inductor to the switch (a), and a snubber diode to suppress ringing (b). These figures show only the collector circuit. The base and emitter connections do not change.

E. We can reduce the voltage by adding a diode across the inductor as in Figure 2.32b. This diode is called a *snubber*. The snubber diode limits the voltage across the inductor to the forward voltage of the diode. Sketch the transistor voltage with the snubber in the circuit.

PROBLEM 6 – DIODE SNUBBERS

A. Put together the circuit for the previous problem again, with the inductor in the circuit, but no snubber diode. Examine the ringing that you see when the transistor switches off. Measure the ringing frequency. You can do this by measuring the time it takes for several cycles of the oscillation to be completed. You should expand the time scale so that a few cycles of the oscillating waveform take up most of the scope display.

B. The ringing results from a resonance between the inductor and circuit capacitance. The circuit capacitance comes from many places, including the inductor itself, the 10 : 1 probe, the transistor, and the breadboard. Calculate the circuit capacitance from the following formula that relates the inductance and capacitance to the resonant frequency f_0:

$$f_0 = \frac{1}{2\pi\sqrt{LC}}. \tag{2.64}$$

Sketch the inductor voltage with the snubber diode in the circuit, using a scale of 2 V per large division and 2 μs per large division. Identify the time when the diode is on and the time when the diode is off. This measurement takes some thought. When you measure the inductor voltage, it is important not to move a scope ground connection. The two oscilloscope grounds are connected by the scope case, and they are connected to the minus lead of the power supply through the wall plugs. If you move a scope ground clip to the inductor, you will put the full power-supply voltage across it. This will destroy the inductor. One way to approach the measurement is to use the plus lead of the power supply as a reference. Attach the scope probe to it and adjust the vertical position control to set the trace on the center line of your screen. This becomes your reference, and you can then move the probe to the other end of the inductor to measure the inductor voltage.

C. You should find that the snubber diode is on all of the time that the transistor is off. Measure the forward voltage of the diode during the middle of the off time. Calculate how long the diode would stay on if the current were allowed to run down completely. Start by setting the diode voltage equal to the inductor voltage so that you can calculate the derivative of the current. You can figure out the initial current by measuring the collector-resistor voltage.

D. Now decrease the square-wave frequency to 30 kHz so that you can see the diode turn off. How long does the diode actually stay on?

3

Phasors

Complex numbers find one of their best applications in analyzing electronic circuits, because cosine signals can be efficiently represented by complex numbers called phasors. With phasors we can analyze circuits with inductors and capacitors almost as easily as resistor circuits, without worrying about calculating derivatives and integrals. In addition, we can use phasors to calculate average power and stored energy.

3.1 Complex Numbers

Complex numbers are often introduced in a mathematics class by writing $\sqrt{-1}$ as i. Electrical engineers use j instead of i, so that i can be reserved for current. It is a good idea to use j in electrical-engineering problems and i in mathematics and physics problems, because the fields follow different sign conventions. Typically

$$j = -i. \tag{3.1}$$

Using j will let people know that you are following the electrical-engineer's sign convention, and i will tell them that you are following the mathematician's or physicist's convention.

In electrical engineering, it may be best to start by thinking of a complex number as a pair of numbers that we call the *real* and *imaginary* parts. In this sense, a complex number is like a two-dimensional vector, and we can draw it like a vector in a plane (Figure 3.1a). We call this the *complex plane*. The horizontal axis is used for the real part, and the vertical axis for the imaginary part. We will use several different notations for writing a complex number, depending on what we want to emphasize. If we let z be a complex number, we can write

$$z = x + jy, \tag{3.2}$$

where x is the real part and y is the imaginary part. The *complex conjugate* z^* is given by

$$z^* = x - jy. \tag{3.3}$$

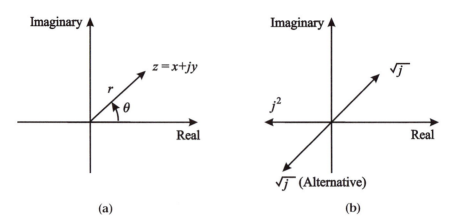

Figure 3.1. Representing a complex number z in a plane by drawing a line from the origin to the point $z = x + jy$ (a). The square and square roots of j (b).

We will also indicate that x and y are the real and imaginary parts of z by writing

$$\text{Re}(z) = x, \tag{3.4}$$
$$\text{Im}(z) = y. \tag{3.5}$$

We add and subtract complex numbers by adding and subtracting the real and imaginary parts separately. This is like vector addition.

We can also represent a complex number in terms of its magnitude and phase. The *magnitude* is the distance from the origin to z in the complex plane. We let the magnitude be given by r and calculate it from the Pythagorean theorem as

$$r = \sqrt{x^2 + y^2}. \tag{3.6}$$

The *phase* is the angle from the real axis, and we write it as θ (the Greek letter *theta*),

$$\theta = \tan^{-1}(y/x). \tag{3.7}$$

We may use either degrees or radians to represent the angle. As a shorthand notation we can write

$$z = r \angle \theta. \tag{3.8}$$

The number j itself can be written as

$$j = 1 \angle 90° \tag{3.9}$$

and -1 is given by

$$-1 = 1 \angle 180°. \tag{3.10}$$

With trigonometry, we can express the real and imaginary parts in terms of the magnitude and phase as

$$x = r \cos \theta, \tag{3.11}$$
$$y = r \sin \theta. \tag{3.12}$$

We can also indicate that r and θ are the magnitude and phase of z by writing

$$|z| = r, \tag{3.13}$$
$$\angle z = \theta. \tag{3.14}$$

So far complex numbers only seem to be a funny form of vector notation, and if this were all there was to it, we would not need complex numbers. A key difference is in how we multiply and divide. The magnitude and phase are convenient for this calculation. If we have two complex numbers s and t, we can write the magnitude and phase of the product as

$$|st| = |s|\,|t|, \tag{3.15}$$
$$\angle(st) = \angle s + \angle t. \tag{3.16}$$

This means that the magnitude of the product of two complex numbers is the product of the magnitudes, and the phase is the sum of the phases. For example, the product $-1 \cdot z$ is given by

$$|-z| = 1 \cdot |z| = |z|, \tag{3.17}$$
$$\angle(-z) = \angle z + 180°. \tag{3.18}$$

Similarly the quotient s/t is given by

$$\left|\frac{s}{t}\right| = \frac{|s|}{|t|}, \tag{3.19}$$
$$\angle(s/t) = \angle s - \angle t. \tag{3.20}$$

In words, the magnitude of the quotient is given by the quotient of the magnitudes, and the phase is the difference of the phases. As a special case, the quotient $1/s$ is given by

$$|1/s| = 1/|s|, \tag{3.21}$$
$$\angle(1/s) = -\angle s. \tag{3.22}$$

We can deduce the formulas for squares and square roots from the product formulas. We write z^2 as

$$|z^2| = |z|^2, \tag{3.23}$$
$$\angle(z^2) = 2\angle z. \tag{3.24}$$

For example, we can write j^2 as

$$|j^2| = 1, \tag{3.25}$$
$$\angle(j^2) = 180°, \tag{3.26}$$

so that $j^2 = -1$ (Figure 3.1b). If we think of taking the square root as the inverse of squaring, we can write

$$|\sqrt{z}| = \sqrt{|z|}, \tag{3.27}$$

$$\angle(\sqrt{z}) = \frac{\angle z}{2}. \tag{3.28}$$

As in the ordinary square root of a positive number, we have a choice of two roots that differ only in sign. This is shown in Figure 3.1b. The other root can be written as

$$|\sqrt{z}| = \sqrt{|z|}, \tag{3.29}$$

$$\angle(\sqrt{z}) = \frac{\angle z}{2} + 180°. \tag{3.30}$$

For example, consider the square root of j:

$$|\sqrt{j}| = 1, \tag{3.31}$$

$$\angle(\sqrt{j}) = 45° \quad \text{or} \quad 225°. \tag{3.32}$$

In rectangular coordinates, we would write

$$\sqrt{j} = 1/\sqrt{2} + j/\sqrt{2} \quad \text{or} \quad -1/\sqrt{2} - j/\sqrt{2}. \tag{3.33}$$

3.2 Exponential Function

The exponential function $\exp(x)$ has a deep connection to the cosine and sine functions through complex numbers. We will start with a fundamental definition of the exponential function. There are two parts. First, the exponential function is its own derivative:

$$\frac{d \exp(x)}{dx} = \exp(x). \tag{3.34}$$

To completely determine the function, we must specify its value at some point, because any multiple of the exponential function also satisfies this equation. We set

$$\exp(0) = 1. \tag{3.35}$$

It is interesting to consider the exponential of an imaginary number $j\theta$. We can write the derivative with the chain rule as

$$\frac{d \exp(j\theta)}{d\theta} = j \exp(j\theta). \tag{3.36}$$

This expression indicates that the derivative has the same magnitude as the exponential, but the angle differs by 90°. If we start at $\theta = 0$, where the exponential is just 1, then the function will move up as θ increases. The interesting thing is that the function always moves at right angles to the arrow that represents it.

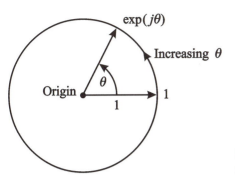

Figure 3.2. Locus of $\exp(j\theta)$ as θ increases from 0.

This causes the function to follow a circular path (Figure 3.2). It is as if the arrow were a rod pinned at one end, so that movement is always at right angles to the rod. More than this, the derivative always has magnitude 1, so that the distance traveled around the circle is equal to θ. Thus θ is the angle in radians as we move around the circle. The path followed by a function as its argument changes is called a *locus*. We would say that the locus of $\exp(j\theta)$ is a unit circle centered at the origin.

From Figure 3.2, we can see by trigonometry that the real and imaginary parts of $\exp(j\theta)$ are $\cos\theta$ and $\sin\theta$. We can write

$$\exp(j\theta) = \cos\theta + j\sin\theta. \tag{3.37}$$

This is Euler's formula, and it is one of the most elegant (and surprising) formulas in all of mathematics. We can use Euler's formula to represent the cosine and sine functions in terms of exponentials:

$$\cos(\theta) = \frac{\exp(j\theta) + \exp(-j\theta)}{2}, \tag{3.38}$$

$$\sin(\theta) = \frac{\exp(j\theta) - \exp(-j\theta)}{2j}. \tag{3.39}$$

If you are not familiar with these formulas, it is a good idea to work out the details by substituting Euler's formula into these expressions.

3.3 Phasors

Let us summarize the circuit relations that we have learned for resistors, capacitors, and inductors:

$$V(t) = RI(t), \tag{3.40}$$

$$V(t) = LI'(t), \tag{3.41}$$

$$I(t) = CV'(t). \tag{3.42}$$

The primes denote derivatives. Often it is not very convenient to work with derivatives. However, our radio signals can often be described by cosine functions, and these have simple derivatives, particularly when we consider the relation between the cosine function and an exponential.

We may write a cosine voltage $V(t)$ as

$$V(t) = A\cos(\omega t + \theta), \tag{3.43}$$

where A is the *peak amplitude* in volts, ω (the Greek lower-case *omega*) is the *frequency* in radians per second, and θ is the *phase* in radians. The frequency in radians per second differs from the frequency in cycles per second, or hertz, by a factor of 2π, and so we can write

$$\omega = 2\pi f. \tag{3.44}$$

We will be careful to write the frequency in radians per second as ω and the frequency in hertz as f, so that we can distinguish them. We can write a current $I(t)$ at the same frequency in a similar form:

$$I(t) = B\cos(\omega t + \phi), \tag{3.45}$$

where ϕ (the Greek letter *phi*) is the phase of the current. If the current phase ϕ is different from the voltage phase, then the current can either be ahead of the voltage or behind it (Figure 3.3). If $\phi > \theta$, then we say the current *leads* the voltage, and if $\phi < \theta$, we say the current *lags* the voltage.

If the voltage $V(t) = A\cos(\omega t + \theta)$ is applied to a capacitor C, we can write the current $I(t)$ as

$$I(t) = CV'(t) = -CA\omega \sin(\omega t + \theta) = CA\omega \cos(\omega t + \theta + \pi/2). \tag{3.46}$$

We would say that the current in a capacitor leads the voltage by $\pi/2$, or 90°. In an inductor, the situation is reversed, and the current lags the voltage. An interesting thing happens if we use Euler's formula to express the cosine as the real part of an exponential and repeat this calculation. We write

$$V(t) = \text{Re}[A \exp(j\omega t + j\theta)]. \tag{3.47}$$

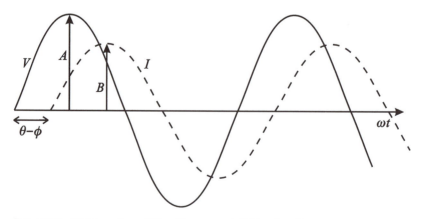

Figure 3.3. Cosine voltage $V(t)$ with a current $I(t)$ lagging it.

We can write the current as

$$I(t) = CV'(t) = \text{Re}\left(C\frac{d}{dt}[A\exp(j\omega t + j\theta)]\right). \tag{3.48}$$

We take the derivative of the exponential by multiplying by $j\omega$. This gives us

$$I(t) = \text{Re}[j\omega CA\exp(j\omega t + j\theta)]. \tag{3.49}$$

This formula is equivalent to Equation 3.46, and you should work out the details to show this. The exponential allowed us to replace a derivative with a multiplication by $j\omega$. We can put this approach on firmer ground if we define complex numbers V and I, given as

$$V = A\exp(j\theta), \tag{3.50}$$
$$I = B\exp(j\phi). \tag{3.51}$$

In terms of the magnitude and phase, we can write

$$|V| = A, \tag{3.52}$$
$$\angle V = \theta, \tag{3.53}$$
$$|I| = B, \tag{3.54}$$
$$\angle I = \phi. \tag{3.55}$$

V and I are called *phasors*. Because they are fixed complex numbers rather than functions of time, t does not appear.

The magnitude of the phasor is equal to the peak amplitude of the original cosine voltage or current, and the phase is the same. To recover the cosine function, we multiply by $\exp(j\omega t)$ and take the real part:

$$V(t) = \text{Re}[V\exp(j\omega t)] = |V|\cos(\omega t + \angle V), \tag{3.56}$$
$$I(t) = \text{Re}[I\exp(j\omega t)] = |I|\cos(\omega t + \angle I). \tag{3.57}$$

Taking the derivative with respect to time is equivalent to multiplying by $j\omega$ for a phasor. For example, for a capacitor we write

$$I = j\omega CV \tag{3.58}$$

as the phasor equivalent of $I(t) = CV'(t)$. We can write a similar relation between current and voltage phasors for an inductor, if we repeat these steps. This gives us

$$V = j\omega LI \tag{3.59}$$

as the equivalent of $V(t) = LI'(t)$. For a resistor we have

$$V = RI \tag{3.60}$$

for phasors, which looks the same as before.

3.4 Impedance

We will be writing voltage and current as phasors most of the time. The ratio of V and I is called the *impedance* and it is written as Z:

$$V = ZI. \tag{3.61}$$

The units of impedance are ohms, like resistance. However, because V and I are complex numbers, the impedance is a complex number with real and imaginary parts. It is traditional to write the real and imaginary parts as

$$Z = R + jX, \tag{3.62}$$

where R is the resistance and X is the *reactance*. We can compare this formula to Equation 3.59, and say that the reactance of an inductor is given by

$$X = \omega L. \tag{3.63}$$

The reactance of an inductor is positive. It is trickier to get the reactance of a capacitor. If we invert Equation 3.58, we get

$$V = \frac{I}{j\omega C}, \tag{3.64}$$

and so we would say that the reactance of a capacitor is given by

$$X = -1/\omega C. \tag{3.65}$$

The minus sign takes the j in the denominator into account. The reactance of a capacitor is negative. Be forewarned: We will often work with the absolute value of the reactance, given by

$$|X| = 1/\omega C. \tag{3.66}$$

People often call this quantity "the reactance," even though it is positive. This is ambiguous but convenient. You have to get the sign from the context.

Impedance is a powerful idea, because it lets us include inductors and capacitors in our analysis without having to take derivatives and integrals. The arithmetic is like that for resistors, except that we use complex numbers, although we have to be careful to remember that impedance is only used for cosine voltages and currents. For example, the impedance Z_s of a series connection of components is the sum of the impedances,

$$Z_s = \sum_i Z_i, \tag{3.67}$$

and the impedance Z_p of a parallel connection is given by the formula

$$\frac{1}{Z_p} = \sum_i \frac{1}{Z_i}. \tag{3.68}$$

You can also find Thevenin and Norton equivalent circuits and voltage and current dividers for impedances in just the same manner that you did for resistances. Even the name impedance suggests the same idea as resistance – a large impedance will *impede* current.

We will often use the inverse of impedance. This is the *admittance*, and the units are siemens (S). We write admittance with a Y:

$$I = YV. \tag{3.69}$$

The real and imaginary parts of the admittance are traditionally written as

$$Y = G + jB, \tag{3.70}$$

where G is the conductance and B is the *susceptance*. We say that the susceptance of a capacitor is ωC and the susceptance of an inductor is $-1/\omega L$. Admittances behave like conductances, so that we write

$$Y_p = \sum_i Y_i \tag{3.71}$$

for components in parallel. Using admittances in parallel circuits is convenient because we can just add the admittances. For components in series we get

$$\frac{1}{Y_s} = \sum_i \frac{1}{Y_i}. \tag{3.72}$$

3.5 RC Filters

We can use phasors to analyze the RC circuits that we build in the lab. These act as low-pass or high-pass filters, selecting either the high or low frequencies in a signal. For example, the circuit in Figure 3.4a allows signals at low frequencies through but blocks higher frequencies. This is a low-pass filter. We find the response of the circuit with phasors and impedances.

We can write the current in terms of the input voltage V_i as

$$I = \frac{V_i}{Z} = \frac{V_i}{R + 1/(j\omega C)}. \tag{3.73}$$

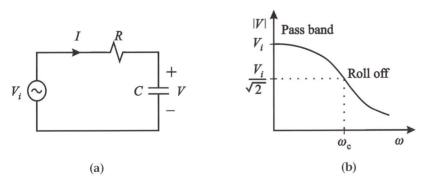

(a)   (b)

Figure 3.4. RC low-pass filter (a), and the response (b).

The output voltage V is given by

$$V = \frac{I}{j\omega C} = \frac{V_i}{1 + j\omega RC} = \frac{V_i}{1 + j\omega \tau}, \tag{3.74}$$

where $\tau = RC$ is the time constant. Figure 3.4b is a plot of $|V|$. In the pass band, where $\omega \tau \ll 1$, the output voltage is close to the input voltage. When $\omega \tau = 1$, the output voltage is given by

$$|V| = \frac{V_i}{|1 + j|} = \frac{V_i}{\sqrt{2}}. \tag{3.75}$$

This means that the output voltage has dropped by a factor of $\sqrt{2}$. Because power is proportional to the square of the voltage, we can think of this as the half-power frequency, or the 3-dB frequency. This means that we can write the cut-off frequency as

$$\omega_c = 1/\tau. \tag{3.76}$$

In the roll-off region above the cut-off frequency, the response drops as the frequency increases. For $\omega \gg \omega_c$, we can write

$$V \approx \frac{V_i}{j\omega \tau}. \tag{3.77}$$

For phasors, multiplying by $j\omega$ is equivalent to differentiating, and dividing by $j\omega$ is equivalent to integrating. This means that in the roll-off region, this circuit acts as an integrator. One application of this filter would be in an audio system to remove the hiss that you often hear. The hiss comes primarily from frequencies that are higher than the frequency range we use for speaking. A filter with a cut-off frequency of about 3 kHz can remove hiss without hurting speech quality.

The circuit in Figure 3.5a acts as a high-pass filter, letting high frequencies through and blocking low frequencies. We can write the response with the potential-divider formula:

$$V = \frac{V_i R}{R + 1/(j\omega C)} = \frac{V_i}{1 + 1/(j\omega RC)} = \frac{V_i}{1 + 1/(j\omega \tau)}. \tag{3.78}$$

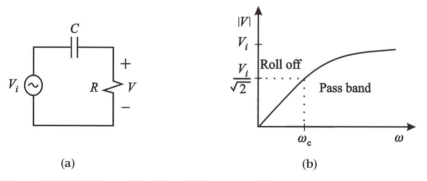

(a)   (b)

Figure 3.5. RC high-pass filter (a), and the response (b).

This is a high-pass response (Figure 3.5b). The cut-off frequency is the same as for the low-pass filter. This time, however, the pass band is above the cut-off frequency, and the stop band is below it. In the stop band, where $\omega \ll \omega_c$, we can write

$$V \approx j\omega\tau V_i. \tag{3.79}$$

The roll off is proportional to frequency. A circuit like this could be used in an audio system to remove hum. Hum is the low-frequency buzzing that is associated with the AC wall supply.

3.6 Series Resonance

Consider a voltage source with an inductor, capacitor, and load resistor (Figure 3.6a). This is a common circuit for band-pass filters that select signals near a particular frequency. We can use the potential-divider formula to write the output voltage as

$$V = \frac{V_i R}{Z}, \tag{3.80}$$

where Z is the circuit impedance, given by

$$Z = R + jX = R + j\omega L + 1/(j\omega C). \tag{3.81}$$

Let us consider the reactance X first, which is the imaginary part:

$$X = \omega L - 1/(\omega C). \tag{3.82}$$

At low frequencies, the capacitive reactance dominates, and the reactance is large and negative. At high frequencies, the inductive reactance dominates, and the reactance is large and positive. The frequency where the reactance is zero is called the *resonant* frequency, and we write it as

$$\omega_0 = \frac{1}{\sqrt{LC}}. \tag{3.83}$$

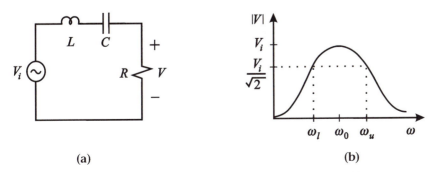

(a) (b)

Figure 3.6. Series resonant circuit with a source (a), and the response (b).

At the resonant frequency, the inductive and capacitive reactance cancel, and Equation 3.80 becomes

$$V = V_i \tag{3.84}$$

so that the output voltage is equal to the input voltage.

Away from the resonant frequency, the reactance increases, and the output voltage drops (Figure 3.6b). When the reactance and resistance are equal, the output voltage is given by

$$|V| = \frac{V_i}{|1 \pm j|} = \frac{V_i}{\sqrt{2}}. \tag{3.85}$$

This means that the upper and lower half-power frequencies ω_u and ω_l are where the reactance and resistance are equal. We can find formulas for these frequencies by setting $X = \pm R$:

$$\omega_u L - 1/(\omega_u C) = +R, \tag{3.86}$$

$$\omega_l L - 1/(\omega_l C) = -R. \tag{3.87}$$

Working with these formulas is messy, but we need to go through the details because the results are important. Let us divide through by the resonant inductive reactance $\omega_0 L$ and substitute $1/(\omega_0^2 L)$ for C. We can write

$$\omega_u/\omega_0 - \omega_0/\omega_u = +R/(\omega_0 L), \tag{3.88}$$

$$\omega_l/\omega_0 - \omega_0/\omega_l = -R/(\omega_0 L). \tag{3.89}$$

The ratio of reactance to resistance in a series circuit is called the *quality factor*, or Q for short:

$$Q = \frac{\omega_0 L}{R} = \frac{1}{\omega_0 C R}. \tag{3.90}$$

We will see later that this corresponds physically to the ratio of the energy stored in reactive elements to the energy lost in the resistor. We can write the quality factor in terms of either the inductive reactance or capacitive reactance, but it is important to realize that in a resonant circuit, it is not the total reactance that we are talking about, but one or the other. The idea of Q is useful because we can relate it to the bandwidth in a simple way. In terms of Q our formula becomes

$$\omega_u/\omega_0 - \omega_0/\omega_u = +1/Q, \tag{3.91}$$

$$\omega_l/\omega_0 - \omega_0/\omega_l = -1/Q. \tag{3.92}$$

If you study these formulas, you can see that ω_u and ω_l must be related by

$$\omega_u/\omega_0 = \omega_0/\omega_l. \tag{3.93}$$

We can rewrite this relation as

$$\sqrt{\omega_l \omega_u} = \omega_0. \tag{3.94}$$

In words, the resonant frequency is the geometric mean of the upper and lower half-power frequencies. Now we substitute Equation 3.93 back in Equation 3.91 to get

$$\omega_u/\omega_0 - \omega_l/\omega_0 = 1/Q. \tag{3.95}$$

I have skipped arithmetic here, but you should fill in the details. It is easier to relate this formula to measurements if we rewrite it in terms of the frequency f by dividing ωs by 2π. We get

$$Q = \frac{\omega_0}{\omega_u - \omega_l} = \frac{\omega_0}{\Delta\omega} = \frac{f_0}{\Delta f}, \tag{3.96}$$

where $\Delta\omega$ is the half-power bandwidth in radians per second and Δf is the half-power bandwidth in hertz. In words, Q is the ratio of the resonant frequency to the bandwidth. If we want a selective filter with a small bandwidth, then we need a large Q. The Q of the resonant circuits that you build with inductors and capacitors is rather low, less than 100. Later on, we will study quartz crystal resonators that have Qs in the range of 50,000 to 100,000; these make extremely selective filters.

Now consider the behavior of the circuit in the stop band, far away from the resonant frequency. At high frequencies, where $\omega L \gg 1/(\omega C)$ and $\omega L \gg R$, the inductive reactance dominates the circuit, and we can write the circuit impedance approximately as

$$Z \approx j\omega L. \tag{3.97}$$

The output voltage becomes

$$V = \left(\frac{R}{j\omega L}\right) V_i = \frac{V_i}{j\omega\tau_l}, \tag{3.98}$$

where $\tau_l = L/R$ is the inductive time constant. This resembles the equation for the roll off in a low-pass filter (Equation 3.77).

At low frequencies, where $1/\omega C \gg \omega L$ and $1/\omega C \gg R$, the capacitive reactance dominates, and we can write the circuit impedance approximately as

$$Z \approx 1/j\omega C. \tag{3.99}$$

The output voltage becomes

$$V = j\omega RC V_i = j\omega\tau_c V_i, \tag{3.100}$$

where $\tau_c = RC$ is the capacitive time constant. This resembles the equation for the roll off in a high-pass filter (Equation 3.79). We can use Equations 3.98 and 3.100 to predict the rejection ratio of filters at different frequencies in the stop band.

3.7 Parallel Resonance

We have learned that if we want a small bandwidth, we need a large Q. For a large Q in a series resonant circuit, the reactance must be large compared to the resistance. This makes it convenient to use series circuits when the resistances are low, like the 50-Ω input of an antenna and a receiver. However, if the resistance is large, it becomes more difficult to make a high-Q series resonant circuit. For example, the input resistance of the mixers in the NorCal 40A is 1,500 Ω, and a high-Q filter would require extremely large reactances. A parallel resonant circuit may be a good choice in this case. This is because a high-Q parallel circuit requires that the reactance be small compared to the resistance. Let us consider a current source with an inductor, capacitor, and load resistor in parallel (Figure 3.7a). Our analysis will be like that of the series resonance, except that we use admittance instead of impedance. We write the output voltage as

$$V = I/Y \tag{3.101}$$

and the load admittance Y as

$$Y = G + jB = G + j\omega C + 1/(j\omega L). \tag{3.102}$$

We start with the susceptance B, which is the imaginary part of the admittance:

$$B = \omega C - 1/(\omega L). \tag{3.103}$$

At low frequencies, the inductive susceptance dominates, and the susceptance is large and negative. At high frequencies, the capacitive susceptance dominates, and the susceptance is large and positive. The susceptance is zero at the resonant frequency ω_0 given by

$$\omega_0 = \frac{1}{\sqrt{LC}}. \tag{3.104}$$

This is the same formula we found for series resonant circuits. At the resonant frequency, the inductive and capacitive susceptance cancel, and we have the

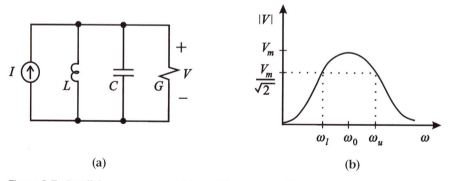

(a)   (b)

Figure 3.7. Parallel resonant circuit (a), and the response (b).

maximum output voltage. We can write it as

$$V_m = I/G. \tag{3.105}$$

Away from the resonant frequency, the susceptance increases, and the voltage falls (Figure 3.7b). The arithmetic for finding the upper and lower half-power frequencies is similar to that for the series circuit. The half-power frequencies occur when the susceptance equals the conductance. We can write the upper and lower half-power frequencies as

$$\omega_u/\omega_0 - \omega_0/\omega_u = +G/(\omega_0 C), \tag{3.106}$$

$$\omega_l/\omega_0 - \omega_0/\omega_l = -G/(\omega_0 C). \tag{3.107}$$

We define the Q for a parallel circuit as the ratio of susceptance to conductance:

$$Q_p = \frac{\omega_0 C}{G} = \frac{1}{\omega_0 L G}. \tag{3.108}$$

We write this as Q_p, with p standing for parallel. We can rewrite these as

$$Q_p = \frac{R}{\omega_0 L} = \omega_0 R C, \tag{3.109}$$

where $R = 1/G$. This is the inverse of the series formula (Equation 3.90). This means that for high Q in a parallel resonant circuit, we need small reactances. This is different from series resonance. The bandwidth formula, however, is the same as before:

$$Q_p = \frac{f_0}{\Delta f}. \tag{3.110}$$

3.8 Phasor Power

We can write the instantaneous power as

$$P(t) = V(t)I(t). \tag{3.111}$$

Here power is a function of time. This expression includes power that is dissipated as heat in resistors or radiated from antennas and power that goes into inductors and capacitors. In Equation 2.9, we wrote the average power for a resistor with a cosine current as

$$P_a = I_p^2 R/2, \tag{3.112}$$

where I_p is the peak amplitude. We can also write the power in terms of phasors. We define the *complex power P* as

$$P = VI^*/2, \tag{3.113}$$

where * denotes the complex conjugate. We substitute in terms of the circuit impedance $V = ZI$ and get

$$P = ZII^*/2. \tag{3.114}$$

We can write this in terms of the magnitude $|I|$ as

$$P = Z|I|^2/2. \tag{3.115}$$

We can interpret this expression if we write Z in terms of the resistance R and the reactance X:

$$P = R|I|^2/2 + jX|I|^2/2. \tag{3.116}$$

The first term on the right side is real. It is equal to the average power (Equation 3.112), so that we can write

$$P_a = \text{Re}(P) = \text{Re}(VI^*/2). \tag{3.117}$$

The second term on the right side of Equation 3.116 is imaginary. This is the *reactive power*, and it is related to the energy stored in the inductors and capacitors. We can illustrate this for a series combination of an inductor and capacitor. We write the reactive power P_r as

$$P_r = \text{Im}(P) = \frac{\omega L|I|^2}{2} - \frac{|I|^2}{2\omega C} = \omega\left(\frac{L|I|^2}{2} - \frac{C|V_c|^2}{2}\right), \tag{3.118}$$

where V_c is the capacitor voltage. We can rewrite this in terms of the stored energy as

$$P_r = \omega(E_l - E_c), \tag{3.119}$$

where E_l is the peak energy stored in the inductor and E_c is the peak energy stored in the capacitor. This calculation is for a series RLC circuit, but the result also holds for more complicated circuits. The reactive power is proportional to the difference between the peak magnetic energy and the peak electric energy. At resonance, the reactive power is zero, and the peak electric energy equals the peak magnetic energy.

Equation 3.119 allows us to develop a more general formula for Q that includes the series and parallel circuits as special cases. We can rewrite the series Q as

$$Q = \omega\frac{L}{R} = \omega\frac{L|I|^2/2}{R|I|^2/2}, \tag{3.120}$$

or

$$Q = \omega\frac{E_l}{P_a}. \tag{3.121}$$

At resonance the peak inductor energy E_l is equal to the peak capacitor energy E_c, and this energy oscillates back and forth between the inductor and capacitor. When the stored energy in the inductor is at its peak, the stored energy in the capacitor is zero, and this means that E_l is actually the total energy stored in the circuit. We drop the subscript and get

$$Q = \omega\frac{E}{P_a}, \tag{3.122}$$

where E is the total stored energy. This says that Q is proportional to the ratio of the stored energy to the average power. To raise Q, we should increase the stored energy or decrease the loss. You should verify that this general formula is equivalent to the Q_p we defined for parallel resonant circuits. We will also apply the formula to resonant transmission lines in the next chapter.

FURTHER READING

Complex numbers are a fascinating part of mathematics, and students who would like to learn more should read Paul Nahin's *An Imaginary Tale: The Story of $\sqrt{-1}$*, published by Princeton University Press, on the history and application of complex numbers. Nahin has developed geometric interpretations that provide powerful insights into the solution of many physics and engineering problems. The classic textbook on complex numbers is *Theory of Functions of a Complex Variable*, by A. I. Markushevich, published by Chelsea Publishing Company.

PROBLEM 7 – PARALLEL-TO-SERIES CONVERSION

A. It is often useful in discussing circuits to be able to convert a parallel combination of reactance and resistance to an equivalent series combination. Starting with the parallel circuit in Figure 3.8a, find expressions for the components in a series circuit (Figure 3.8b) that give the same impedance. One way to approach this problem is to define a Q for each circuit that is the ratio of the reactance to the resistance. We let

$$Q_s = X_s/R_s \tag{3.123}$$

and

$$Q_p = R_p/X_p. \tag{3.124}$$

First show that if the two circuits are to have the same impedance, the two Qs must be the same. This means that in the rest of the problem, you can drop the subscripts, and just write Q.

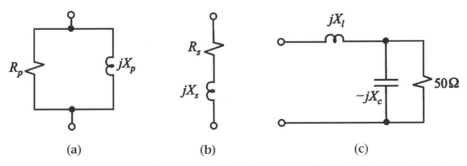

Figure 3.8. Parallel circuit with a resistance R_p and a reactance X_p (a), equivalent series circuit with R_s and X_s (b), and matching circuit with a parallel capacitor and a series inductor (c).

B. Find an approximate formula for X_s when Q is large. Find an approximate formula for R_s when Q is small.

There is one thing you should think about. The Q we define here is not quite the same as the one we use for resonant circuits. This Q involves the total reactance, whereas the resonant circuit Q uses only one of the two reactances. In practice, people use the same letter Q for both situations, and you have to figure out which is intended by the context.

Many transmitters have a low output impedance so that the output power varies inversely with the load resistance. For example, if an amplifier has an output of 1 W with a 50-Ω load, we would hope for 10 W with a 5-Ω load.

C. We will use the network in Figure 3.8c to transform a 50-Ω antenna to 5 Ω. We need our parallel–series conversion formulas. The first step is to find the capacitor reactance X_C. When the capacitor and resistor are converted to a series circuit, the resistance should be 5 Ω. Next choose the inductor reactance X_l to cancel the capacitive reactance. What capacitance (in nF) and inductance (in nH) are required at a frequency of 7 MHz?

PROBLEM 8 – SERIES RESONANCE

In this problem, we solder an inductor and capacitor on the NorCal 40A circuit board and make measurements. It is convenient to mount the board in an electronics vise for soldering. The components mount on the side with the white lettering, and the solder is applied to the other side. Insert the parts that you plan to solder. They should be close to the board, but you may want to leave a millimeter of space so that you can hook up scope probes conveniently. You may need to bend the wires a bit so that the parts do not fall out. Before you solder, check that the parts are in the right holes. They can be unsoldered if you make a mistake, but this is difficult if the part has more than two leads.

Before you start, put some water on a sponge. Turn on a soldering iron, and when it is warm, apply solder to the tip of the iron to tin it. Wipe the tip on the sponge to remove the excess solder. This wiping leaves a shiny surface on the tip that heats up parts much better than a tip without solder. Apply the tip and solder at the same time to the hole and the wire. Be alert when soldering parts with plastic packages, or the plastic will melt. Do not use more solder than you need to flow through the hole and coat the wire, or you run the risk that there will be short circuits to other holes. Clip off the wire ends close to the board after you finish so that they will not touch other wires. Inspect the hole and the wire. The solder should flow completely through the hole and coat the wire. If the wire is not hot enough, the solder will not coat the wire well. This is called a cold solder joint. Cold solder joints often cause open circuits.

If you do make a mistake and put the parts in the wrong holes, be careful when you take them out so that you do not damage the parts or the board. I like to remove solder with wick before I remove the part. Solder wick is a copper braid that absorbs molten solder. Melt the solder with the iron and coax the solder into the wick. Cut off pieces of the wick that get solder on them and throw the pieces away. When you have taken off as much solder as you can, apply the soldering iron at the joints to melt the remaining solder and loosen the part with pliers. You may have to do this repeatedly with each lead

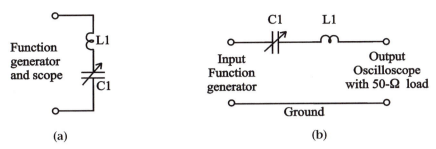

Figure 3.9. Connections for observing a series resonance (a), and band-pass filter connection (b).

until the part comes out. Finally, use the solder wick to remove the solder from the hole before you insert another part.

Solder contains lead, which is dangerous if you eat it; therefore you should wash your hands with soap and water after soldering. It is a good idea to remember to turn off the soldering iron when you have finished working. Leaving the irons on is hard on the tips.

Install C1 and L1 on the NorCal 40A board. C1 is a variable capacitor with a nominal range from 8 to 50 pF, and L1 is an inductor with an inductance of 15 μH. The variable capacitor has meshing metal vanes separated by ceramic insulators. It is best adjusted with a plastic screwdriver. Metal screwdrivers have extra capacitance that shifts the resonant frequency. We will tune the capacitor for resonance at 7 MHz. The inductor and capacitor form part of the RF Filter. Look inside the front cover of this book to see where this circuit goes in the receiver. Appendix D also has data on the components that you use.

A. Make the connections shown in Figure 3.9a, with a tee connection to the scope. Set the function generator for a 7-MHz sine wave with an amplitude of 1 Vpp. When a circuit is resonant, the capacitive reactance cancels the inductive reactance, leaving us with the source resistance of the function generator (50 Ω) and the resistance of the capacitor and the inductor. If we tune the capacitor through resonance, we will see a dip in the input voltage when the circuit is resonant. Adjust the capacitor for minimum input voltage. This sets the resonant frequency to 7 MHz. What is this voltage? Use the voltage to calculate the total resistance of the capacitor and inductor.

B. Now make the connections for the band-pass filter shown in Figure 3.9b, with a 50-Ω scope load. Adjust the capacitor for maximum output voltage. What is this voltage? Calculate the voltage that you expect.

C. Find the half-power bandwidth by measuring the frequencies f_u and f_l where the output voltage has dropped by a factor of $\sqrt{2}$. One way to do this is to use a larger amplitude setting of 1.41 Vpp, and look for the two frequencies that give the same output voltage as 1 Vpp at 7 MHz.

D. Now we will calculate the half-power bandwidth Δf that we expect. Start by finding the resonant inductive reactance. Calculate the Q from the inductive reactance and the total circuit resistance, and then calculate the half-power bandwidth.

E. Return the amplitude setting to 1 Vpp. Measure the output voltage at 1-MHz intervals from 1 MHz to 15 MHz and make a plot. The response changes dramatically between 6 MHz and 8 MHz, and you will need some additional data points to keep them from getting too far apart.

AM radio transmitters in the frequency range from 0.5 to 1.5 MHz are a big problem for receivers because they are usually close and powerful. For example, some broadcasters use 50 kW, and they may only be a few miles away. The NorCal 40A is for 2-W stations that might be a thousand miles away. We will find the AM *voltage rejection factor* R_{am}, given by

$$R_{am} = V_{rf}/V_{am},\qquad\qquad(3.125)$$

where V_{rf} is the output at 7 MHz and V_{am} is the output at 1 MHz.

F. Measuring V_{am} is tricky, because the output is small at 1 MHz. One way to approach the problem is to use an amplitude setting of 10 Vpp at 1 MHz. This increases the output voltage by a factor of 10, making it easier to measure. You will need to divide your output voltage by a factor of 10 to take this into account before you compare it with the 7-MHz voltage. It is not a good idea to use an amplitude of 10 Vpp at 7 MHz, because the voltages on the inductor and capacitor get large enough to change their response. What is R_{am}?

G. Use the low-frequency approximation (Equation 3.100) to calculate the value of R_{am} that we would expect.

PROBLEM 9 – PARALLEL RESONANCE

In the NorCal 40A, the transmitter signal is produced by mixing the VFO at 2.1 MHz with the Transmit Oscillator at 4.9 MHz. The transmitter frequency is the sum of these two frequencies, 7.0 MHz. The Transmit Mixer also produces other frequencies that are removed by the Transmit Filter. This filter uses a parallel resonance. A parallel resonance is a good choice for a band-pass filter if the source and load resistances are large, because we can easily make capacitors and inductors with much smaller reactances to give high Q. This filter is made up of C37, C38, C39, and L6. You should study the endpaper to see how this circuit works in the transmitter.

Start by soldering C37 (5-pF disk) and C38 (100-pF disk) on the board. Do not include the variable capacitor C39 yet – we will make some measurements first. L6 is the first inductor that you make yourself by winding wire on a toroidal core. *Toroidal* means donut-shaped. This shape is good for radio inductors because it keeps the magnetic field inside the magnetic material. Compared with the smaller rod inductors we have worked with so far, the toroidal inductors have a better Q and can operate at higher power. L6 uses a T37–2 core. "T" indicates toroidal core, 37 is the outside diameter in hundredths of an inch, and 2 refers to the particular mix of material. Material #2 is an iron powder mix that is useful from 1 to 30 MHz. #2 cores are traditionally painted red to distinguish them from other mixes.

The L6 coil has 28 turns of #28 wire. Cut a 40-cm length and wrap it around the core, being careful with the count (Figure 3.10a). It is easy to be low by one turn. For example, the figure shows a core with 6 turns, not 5. After you finish winding, spread the turns evenly around the core, leaving a gap between the first and last turn so that the

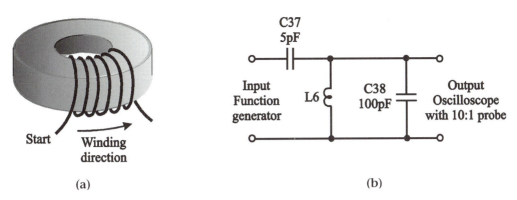

Figure 3.10. (a) Winding a toroid. Six turns are shown. (b) Initial band-pass filter connections.

wire ends line up with the holes in the board. Strip the wire ends using a cigarette lighter to burn the enamel. Sand the ends until the enamel is completely gone. If the enamel is not completely removed, the solder may not stick to the wire. If you are lucky, you get an open circuit. More likely is an intermittent contact that depends on temperature, pressure, and the phase of the moon.

Solder L6 onto the board. Connect the function generator and 10:1 probe as shown in Figure 3.10b. The coil wire is thin, and it is a bad idea to attach probes to it. It will take a little practice to follow the traces on the circuit board so that you can tell where to connect the probes. Most of the traces are on the solder side of the board, but there are also a few on the component side. Moreover, most of the component side is a single connected ground plane. If you see solder pads that do not appear to lead anywhere, they are likely to be ground connections.

A. The function generator should be set for a sine wave, with an amplitude setting of 1 Vpp. Find the resonant frequency f_0 that gives the largest output voltage. From f_0 and the total capacitance (C37, C38, and the probe capacitance), calculate the inductance of the coil that you wound.

B. We will discuss inductance calculations in Chapter 6, but for now you need to know that the inductance is proportional to the square of the number of turns. We write

$$L = A_l N^2,$$ (3.126)

where A_l is an inductance constant and N is the number of turns. Core manufacturers provide the inductance constant in their data sheets. For the T37–2 core, A_l is 4.0 nH/turn^2. Calculate the inductance that you expect for L6.

C. Now solder the variable capacitor C39 into the circuit. Set the frequency to 7 MHz and adjust the capacitor carefully for maximum output. Record the output voltage. Measure the half-power bandwidth Δf, and calculate the Q.

D. Calculate the inductor reactance X at 7 MHz. Use this reactance and the Q you measured to find the effective parallel resistance R. This resistance is not a separate component but is associated with the inductor, the capacitors, the function generator, and the scope probe.

E. We can also calculate the output voltage that we expect. One way to start is to find a Norton equivalent circuit for the series combination of the function generator and the 5-pF capacitor. The output voltage can be calculated from the Norton current and the effective parallel resistance R.

F. In addition to the sum frequency at 7 MHz, the mixer produces a strong difference-frequency signal at 2.8 MHz. We do not want to transmit the difference frequency, because it might interfere with other services. Measure the response of the filter at the difference frequency. Express the difference-frequency voltage rejection factor R_- as

$$R_- = V_{rf}/V_-, \tag{3.127}$$

where V_{rf} is the 7-MHz voltage and V_- is the difference voltage. At 2.8 MHz, you should turn up the function generator to 10 Vpp to make the output signal as large as you can, and you should take this into account in calculating the voltage ratio. The output signal will be quite small, and the trace will become fuzzy because of scope noise. You need to be careful to measure at the same place in the noise at the top and bottom of the sine wave.

G. Although dB are units for comparing power levels, we can also write dB expressions in terms of voltage or current if we take into account the fact that the power is proportional to the square of the voltage or current. We write

$$10 \log(P_1/P_2) = 20 \log(V_1/V_2) = 20 \log(I_1/I_2) \text{ dB.} \tag{3.128}$$

For example, if V_1 is twice V_2, then we would say that the first signal is 6 dB bigger than the second. For these voltage and current formulas to make sense, the resistance associated with each power must be the same, because the power depends on the resistance. This is appropriate for the rejection factor of a filter. Now express the rejection factor as a dB difference, using the formula

$$R_- = 20 \log(V_{rf}/V_-) \text{ dB.} \tag{3.129}$$

H. Calculate what the difference-frequency rejection should be. You will need to consider how the circuit quantities vary with frequency.

I. What would the Q of the filter be if the 5-pF input capacitor (C37) were bypassed and the 50-Ω function generator were connected directly to C38?

Transmission Lines

Cables allow us to transmit electrical signals from one circuit to another. For example, we might attach coaxial cable between a function generator and an oscilloscope (Figure 4.1a) and plastic-coated twin lead between an antenna and a television (Figure 4.1b). Usually, when we analyze the circuit, we assume that the voltage at one end of the cable is the same as the voltage at the other end and that the current at the beginning is the same as the current at the end. This is appropriate if the frequency is low. However, at high frequencies the cable itself begins to have an effect. A fundamental limitation is the speed of light. If the voltage at one end of the cable changes appreciably in less time than it takes light to propagate to the other end, we should expect the voltage to be different at the two ends. Another way of saying this is that we would expect the voltages at the ends to be different when the length of the cable becomes an appreciable fraction of a wavelength.

4.1 Distributed Capacitance and Inductance

However, even when the cable is considerably shorter than a wavelength, it can have a large effect. We found in Problem 3 that a cable has capacitance. This capacitance is associated with the charges that the voltages on the line induce. We can take the capacitance into account in a circuit by adding a capacitance between the wires (Figure 4.1c). Some of the current will return through this capacitance. This means that the current at the end of the cable will not be the same as the current at the beginning. This is apparently a violation of Kirchhoff's current law. In addition, the cable has inductance. The inductance comes from the magnetic field that the currents make. We can include this effect by adding a series inductor (Figure 4.1c). There will be a voltage drop across the inductance, so that the voltage at the end of the cable will not be the same as at the beginning. This is an apparent violation of Kirchhoff's voltage law. Now we have an equivalent circuit for our cable with a series inductance and a parallel capacitance. There is another effect, resistance in the wires, that we will take into account later.

It is not obvious whether the inductance or capacitance is more important. It depends on the load impedance. If the impedance is high, the current is relatively small, and the inductance has little effect. However, the capacitive current will

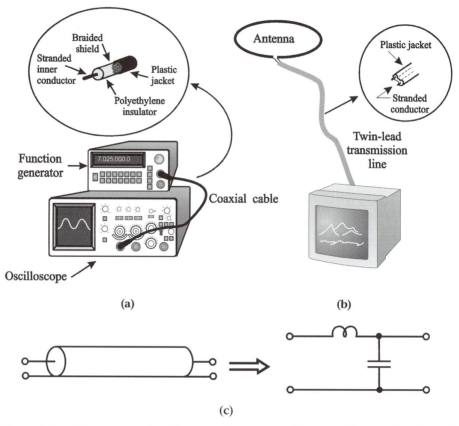

Figure 4.1. (a) Connecting a function generator to an oscilloscope with coaxial cable, and coaxial-cable construction (inset). (b) Connecting an antenna to a television with twin lead, and twin-lead construction (inset). (c) An equivalent circuit for the cable that includes a series inductance and a shunt capacitance.

be relatively important. For example, let us assume that we are making a coaxial-cable connection to an oscilloscope with an input resistance of 1 MΩ and a parallel capacitance of 20 pF. This is a relatively high impedance, and it is usually more important to consider the effect of the cable capacitance than the inductance. The capacitance of a typical coaxial cable is 100 pF/m. A one-meter cable increases the capacitance of the oscilloscope connection from 20 pF to 120 pF, and we would notice delays that are much larger than we would expect without the cable. However, if the load impedance is small, the load current will be large, and the inductance will be more important.

Our circuit model is really a simplification. We cannot really say that the inductor should go before the capacitor, or the other way around, because the capacitance and inductance are spread out along the cable. This capacitance and inductance are called *distributed* elements to distinguish them from ordinary *lumped* capacitors and inductors. There is an elegant approach to calculate the effect of distributed elements, called *transmission-line theory*. We will derive the transmission-line theory by analyzing a network of small inductors and capacitors.

4.2 Telegraphist's Equations

Our transmission line will have two parallel conductors with uniform cross section. We assume that they are long enough that we need not worry about the ends. We do not assume anything about the shape of the conductors – they could be two adjacent wires, or they could be coaxial. They should not touch each other, because then there would be just one conductor. We divide the line into small sections of length l (Figure 4.2a). Each of these sections has an inductor L_l and a capacitor C_l associated with it. We can draw a network that represents our transmission line and define voltages and currents (Figure 4.2b). We can write the inductor voltage as

$$V_{n+1} - V_n = -L_l \frac{dI_{n+1}}{dt} \tag{4.1}$$

and the capacitor current as

$$I_{n+1} - I_n = -C_l \frac{dV_n}{dt}. \tag{4.2}$$

When we draw a model for a transmission line with small inductors and capacitors, we are implicitly assuming that the inductance and capacitance are proportional to the length. This can be shown by electromagnetic theory. If the inductance and capacitance are proportional to the length, then we can let L and C be equal to the proportionality constants and write

$$L = L_l/l, \tag{4.3}$$
$$C = C_l/l. \tag{4.4}$$

Here L and C are called the distributed inductance and capacitance. These are the fundamental quantities that characterize a transmission line. They are determined by the shape of the conductors and the nature of the insulators. The units of distributed inductance are henries per meter, and the units of distributed capacitance

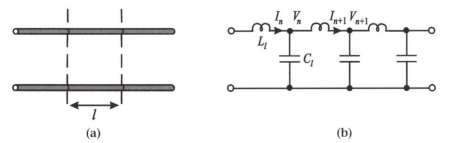

(a)   **(b)**

Figure 4.2. Dividing a transmission line into sections of length l (a). Representing the transmission line as a network of inductors and capacitors (b). This kind of network is called a *ladder network*.

are farads per meter. We can rewrite our equations in terms of L and C as

$$\frac{V_{n+1} - V_n}{l} = -L \frac{dI_{n+1}}{dt}, \tag{4.5}$$

$$\frac{I_{n+1} - I_n}{l} = -C \frac{dV_n}{dt}. \tag{4.6}$$

Notice that if we take the limit as l approaches 0, the quotients on the left become derivatives with respect to distance. To be more precise, these are *partial derivatives*, since the current and voltage are also functions of time. We will let our distance variable be z. In the limit, the equations become

$$\frac{\partial V}{\partial z} = -L \frac{\partial I}{\partial t}, \tag{4.7}$$

$$\frac{\partial I}{\partial z} = -C \frac{\partial V}{\partial t}. \tag{4.8}$$

Here we use the partial derivative sign, ∂ (which is a funny d), to show that we are taking a derivative with respect to a particular variable. These formulas are known as the *telegraphist's equations* or *transmission-line equations*. They were developed by Oliver Heaviside for telegraph cables more than one hundred years ago. They are extremely important in science and engineering. Similar equations describe radio waves, light, sound, and heat. Consequently, once you understand how to work with the equations, you can solve a wide variety of problems.

The telegraphist's equations predict the propagation of waves. We can derive a wave equation by differentiating the first formula with respect to z and the second with respect to t:

$$\frac{\partial^2 V}{\partial z^2} = -L \frac{\partial^2 I}{\partial t \partial z}, \tag{4.9}$$

$$\frac{\partial^2 I}{\partial t \partial z} = -C \frac{\partial^2 V}{\partial t^2}. \tag{4.10}$$

We can eliminate I between these two formulas to get the voltage wave equation,

$$\frac{\partial^2 V}{\partial z^2} = LC \frac{\partial^2 V}{\partial t^2}. \tag{4.11}$$

4.3 Waves

We can write a voltage wave in the form $V(z-vt)$, where V is a voltage function and v is the velocity. We will consistently use an upper-case V for voltage and a lower-case v for velocity to keep them distinct. We will assume that V is a pulse function centered around $z=0$ at $t=0$ (Figure 4.3a). Some time t_0 later, we sketch the function again (Figure 4.3b). We get the same pulse, displaced to the right by an amount $z=vt_0$. The wave moves in the $+z$ direction, and we call this a *forward* wave. We can write a voltage wave that propagates in the $-z$ direction in the form $V(z+vt)$.

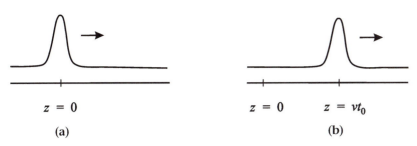

Figure 4.3. A forward wave of the form $V(z - vt)$. At $t = 0$, the wave peaks at $z = 0$ (a). At time $t = t_0$ later, the wave has moved a distance $z = vt_0$ to the right (b).

This is a *reverse* wave. We can have both a forward and a reverse wave on a transmission line at the same time. The reverse wave is often the reflection from a load.

Now we can show that these wave functions satisfy Equation 4.11. We substitute a forward wave $V(z - vt)$ into the equation, and use the chain rule to write the partial derivatives in terms of the second derivative of V, which we write as V'':

$$\frac{\partial^2 V}{\partial z^2} = V'' = LC\frac{\partial^2 V}{\partial t^2} = LC\,v^2V''. \tag{4.12}$$

This gives us

$$v = 1/\sqrt{LC}. \tag{4.13}$$

This formula allows us to predict the velocity if we know L and C. For coaxial cable, the velocity is typically $2/3$ the speed of light, or 2×10^8 m/s. The twin lead that is commonly used for connecting FM and TV antennas has a velocity of $4/5$ the speed of light, or 2.4×10^8 m/s.

Now we can use our transmission-line equations to relate the current to the voltage. The wave equation for current is the same as that for the voltage; thus the solutions are also waves with the same velocity. We can use the chain rule again to rewrite Equation 4.7 as

$$V' = vLI'. \tag{4.14}$$

Notice that both the voltage and the current appear only as derivatives. When we integrate this equation we will have arbitrary constants, which correspond to constant voltages and currents on the line. We will neglect these, because we already know how transmission lines work at DC. We integrate this formula, setting the integration constants to zero, and substitute for v from Equation 4.13 to find the ratio of the voltage to the current:

$$V/I = \sqrt{L/C}. \tag{4.15}$$

This ratio of voltage and current in a forward wave is called the *characteristic impedance*, and it is written as Z_0:

$$Z_0 = \sqrt{L/C}. \tag{4.16}$$

Coaxial cables usually have a characteristic impedance of 50 Ω or 75 Ω, whereas twin lead is typically 300 Ω. If we know Z_0 and v for a transmission line, we can work backwards and calculate L and C. Using Equation 4.13 and Equation 4.16 we can write

$$L = Z_0/v, \tag{4.17}$$
$$C = 1/(Z_0 v). \tag{4.18}$$

We can repeat this analysis for a reverse wave of the form $V(z + vt)$. Equation 4.12 does not change, and so the velocity is the same for a reverse wave as it is for a forward wave. This makes sense, because we have not assumed anything about the line that would make the wave go faster in one direction than the other. We can find the ratio of the voltage and current by substituting into Equation 4.7, and we find

$$V' = -vLI'. \tag{4.19}$$

We integrate to find that

$$V/I = -\sqrt{L/C}. \tag{4.20}$$

This tells us that the ratio of voltage and current in a reverse wave changes sign. We can write formulas that relate the voltage and current as

$$V_+/I_+ = +Z_0, \tag{4.21}$$
$$V_-/I_- = -Z_0, \tag{4.22}$$

where the $+$ subscript is for a forward wave and the $-$ subscript is for a reverse wave. Figure 4.4 shows how these voltages and currents look.

We can understand why the ratio changes sign if we consider the power. Power is positive if it flows to the right and negative if it flows to the left. For a forward wave the power is

$$P_+(t) = V_+(t)I_+(t) = V_+^2(t)/Z_0. \tag{4.23}$$

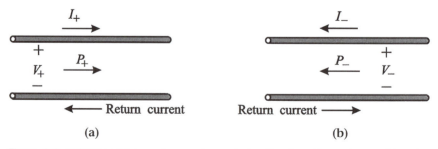

(a)   (b)

Figure 4.4. Voltages and currents on a transmission line for a forward wave (a) and a reverse wave (b). Both V_+ and V_- are taken to be positive. The current I_+ is positive and it flows to the right in the top conductor. In addition there is a return current in the bottom conductor. The current I_- is negative and it flows to the left in the top conductor.

This is a positive number, indicating that power flows to the right, in the direction of propagation. Notice that the power does not depend on the sign of the voltage. For a reverse wave, the voltage–current ratio changes sign, and we have

$$P_-(t) = V_-(t)I_-(t) = -V_-^2(t)/Z_0. \tag{4.24}$$

This is a negative number, and so power flows to the left, again in the direction of propagation. The sign change reverses the direction of power flow.

4.4 Phasors for Waves

We found that signals that vary in time as cosines can be described in a simple way by phasors, and this allows many circuits to be solved by algebra alone. Waves generated by cosine signals can also be represented by phasors. Let us consider a forward wave of the form

$$V(z - vt) = A\cos(\omega t - \beta z), \tag{4.25}$$

where β (the Greek letter *beta*) is called the *phase constant*, because it determines the phase. The units of β are radians per meter. We can write the corresponding expression for a reverse wave by changing the $-$ signs to $+$ signs. You should work through the details to show that the cosine expression actually has the correct form to be a forward wave. If we compare the right and left sides of the equation, we find that we can write the following expression for v:

$$v = \omega/\beta. \tag{4.26}$$

In addition, the wave is periodic in z, and its wavelength, written as λ, given by

$$\lambda = 2\pi/\beta. \tag{4.27}$$

To convert to phasors, we start by writing the wave as the real part of a complex exponential,

$$V = A\cos(\omega t - \beta z) = \text{Re}[A \exp +j(\omega t - \beta z)]. \tag{4.28}$$

We can rewrite the equation as

$$V = \text{Re}[A\exp(-j\beta z)\exp(j\omega t)]. \tag{4.29}$$

In phasor notation, we consider the complex factor of $\exp(j\omega t)$, given by

$$V = A\exp(-j\beta z). \tag{4.30}$$

It is interesting to plot the locus for wave phasors in the complex plane. For example, consider a forward wave given by

$$V_+ = \exp(-j\beta z). \tag{4.31}$$

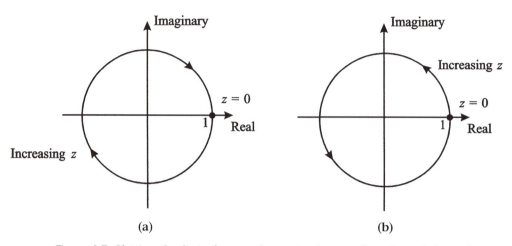

Figure 4.5. Plotting the loci of wave phasors in the complex plane. A forward wave, $V_+ = \exp(-j\beta z)$ (a), and a reverse wave, $V_- = \exp(+j\beta z)$ (b). The phasor for the forward wave rotates clockwise as z increases, and the phasor for the reverse wave rotates counterclockwise.

This path is shown in Figure 4.5a. The phase lags as z increases, and the phasor traces out a clockwise circle. For comparison, in Figure 4.5b we also plot the locus for a reverse wave,

$$V_- = \exp(+j\beta z). \tag{4.32}$$

We see a progressive phase lead as z increases and a counterclockwise circle. Notice that for both waves, the magnitude is constant but the phase varies along the line.

Now we can develop power formulas for phasor waves. We write the complex power P as

$$P = VI^*/2, \tag{4.33}$$

where V and I are phasors. For a forward wave, we have

$$P_+ = \frac{V_+ I_+^*}{2} = \frac{V_+ V_+^*}{2Z_0} = \frac{|V_+|^2}{2Z_0}, \tag{4.34}$$

assuming that Z_0 is real. The power is real and positive. For a reverse wave, the sign of the impedance changes and we get

$$P_- = \frac{V_- I_-^*}{2} = -\frac{V_- V_-^*}{2Z_0} = -\frac{|V_-|^2}{2Z_0}, \tag{4.35}$$

and the average power is negative.

4.5 General Lines

We have seen that phasors allow us to define impedances and admittances for circuit elements. This makes it natural to consider a transmission line with a

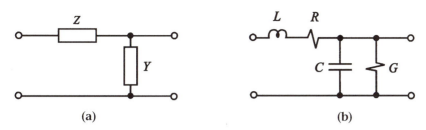

Figure 4.6. General transmission lines. (a) A transmission line with distributed series impedance Z and parallel admittance Y. It is traditional to use a thin rectangle as the symbol for impedances and admittances, because they could be combinations of capacitors, resistors, and inductors. We can consider an LC transmission line as a special case with $Z = j\omega L$ and $Y = j\omega C$. (b) The distributed circuit elements for a transmission line with series resistance R and parallel conductance G.

distributed series impedance and a distributed parallel admittance. We will analyze a transmission line having a distributed impedance Z, with units of ohms per meter, and an admittance Y, with units of siemens per meter (Figure 4.6a). We can follow the same limiting procedure that we used for LC transmission lines to derive more general telegraphist's equations:

$$\frac{dV}{dz} = -ZI, \tag{4.36}$$

$$\frac{dI}{dz} = -YV. \tag{4.37}$$

Now consider a forward wave with a voltage V and current I that vary as $\exp(-jkz)$. We use k here rather than β because we want to allow for the possibility that k will be complex. k is called the *propagation constant*. It is traditional to characterize the real and imaginary parts of k by writing

$$jk = \alpha + j\beta, \tag{4.38}$$

where α is the Greek letter *alpha*. The forward wave phasor is then of the form

$$\exp(-jkz) = \exp(-\alpha z - j\beta z). \tag{4.39}$$

We can see that α determines the loss of the wave as it propagates, and for this reason, it is called the *attenuation constant*. It should be positive, or else the wave will grow instead of decay. The units of α are given their own name, *nepers*/m. The word neper is pronounced "neeper," and it is derived from a Latin version of the name Napier. John Napier was the Scottish mathematician who invented logarithms. Since we often quote losses in decibels, we need to figure out how to convert between nepers and dB. An attenuation of 1 neper corresponds to a voltage reduction by a factor of e. This means we can relate dB and nepers by the formula

$$\alpha_{\mathrm{dB/m}} = \alpha_{\mathrm{nepers/m}} \cdot 20 \log_{10}(e) = 8.686 \cdot \alpha_{\mathrm{nepers/m}}. \tag{4.40}$$

Now we return to the general telegraphist's equations, assuming forward waves of the form $\exp(-jkz)$. The equations become

$$jkV = ZI, \tag{4.41}$$
$$jkI = YV. \tag{4.42}$$

In addition, if we let the ratio of V to I be Z_0, we get

$$jkZ_0 = Z, \tag{4.43}$$
$$jk/Z_0 = Y. \tag{4.44}$$

We can write the solutions as

$$jk = \sqrt{ZY}, \tag{4.45}$$
$$Z_0 = \sqrt{Z/Y}. \tag{4.46}$$

In general, all these quantities are complex, and there are two complex roots differing only in sign. It can be difficult to choose the correct sign. Ordinarily, we should choose the sign of jk so that α is positive, to keep the wave from growing as it propagates. In addition, Z_0 should have a positive real part to keep the average power positive.

As an example, let us consider loss in transmission lines. Loss is associated with either the metal or the insulator. We can model the metal loss as a distributed series resistance R, with units of Ω/m (Figure 4.6b). In practice, R is not a constant but usually increases as the square root of the frequency because of an electromagnetic phenomenon called the skin effect. We can write the distributed impedance Z as

$$Z = j\omega L + R. \tag{4.47}$$

We can take the insulator loss into account by a distributed parallel conductance G with units of S/m. The conductance also varies with frequency. In practical transmission lines, G is often small enough that it can be neglected. With conductance, the distributed admittance Y becomes

$$Y = j\omega C + G. \tag{4.48}$$

When we substitute these into our formulas for jk and Z_0, we get

$$jk = \sqrt{(j\omega L + R)(j\omega C + G)}, \tag{4.49}$$
$$Z_0 = \sqrt{(j\omega L + R)/(j\omega C + G)}. \tag{4.50}$$

In both formulas, the correct root is the one with a positive real part.

4.6 Dispersion

The velocity v and the attenuation constant α may vary with frequency. This frequency variation is called *dispersion*, and it is a problem. For example, if v depends on frequency, then different frequency components travel at different velocities and one part of a message interferes with another. If α increases with frequency, then we lose the high-frequency information in our signal. There is in principle a simple solution to this problem, however, that was discovered by Oliver Heaviside. If the transmission-line parameters can be adjusted to satisfy

$$R/L = G/C \tag{4.51}$$

then the attenuation and velocity become constants. Equation 4.49 can be written as

$$jk = j\omega\sqrt{LC}\sqrt{\left(1 + \frac{R}{j\omega L}\right)\left(1 + \frac{G}{j\omega C}\right)}. \tag{4.52}$$

The terms in parentheses are the same, and so we can rewrite this as

$$jk = j\omega\sqrt{LC}\left(1 + \frac{R}{j\omega L}\right), \tag{4.53}$$

or

$$v = \omega/\beta = 1/\sqrt{LC} \tag{4.54}$$

and

$$\alpha = \sqrt{RG}. \tag{4.55}$$

The velocity is the same as that of a lossless line and is independent of frequency. There is loss, but it is independent of frequency, and an amplifier can compensate for it. The impedance is also independent of frequency. Equation 4.50 can be written as

$$Z_0 = \sqrt{L/C}\sqrt{\left(1 + \frac{R}{j\omega L}\right)\bigg/\left(1 + \frac{G}{j\omega C}\right)}. \tag{4.56}$$

Again the terms in parentheses are the same, and so we have

$$Z_0 = \sqrt{L/C}, \tag{4.57}$$

as it is in a line with no loss.

The telephone company uses an approach like this in phone lines. Typically R is considerably larger than ωL and this causes v and α to depend strongly on frequency. In practice, Heaviside's zero-dispersion condition is hard to satisfy, because G is usually close to zero. However, we can come close to zero dispersion by making ωL much larger than R. The phone company does this by adding inductor coils to the lines, usually 88-mH inductors at intervals of one mile. To see how this works, consider a line where $j\omega L \gg R$ and $G = 0$. This is a *large-reactance* approximation. We start with the exact formula and derive approximate expressions for Z_0 and jk:

$$Z_0 = \sqrt{(j\omega L + R)/(j\omega C)} \approx \sqrt{L/C} \tag{4.58}$$

and

$$jk = \sqrt{(j\omega L + R)j\omega C} \approx j\omega\sqrt{LC} + (R/2)\sqrt{C/L}, \tag{4.59}$$

where we have used the first-order Taylor-series formula

$$\sqrt{1+z} \approx 1 + z/2, \tag{4.60}$$

which holds when $|z| \ll 1$. From Equation 4.59, we can write α and v as

$$\alpha = R/(2Z_0), \tag{4.61}$$

$$v = \omega/\beta = 1/\sqrt{LC}. \tag{4.62}$$

These are independent of frequency. As an example, in 50-Ω coaxial cable at 5 MHz, the series resistance might be 0.5 Ω/m and the inductance 250 nH/m. The reactance ωL is 7.9 Ω/m, and thus the high-reactance approximation is justified. The loss is given by

$$\alpha = R/(2Z_0) = 0.005 \text{ nepers/m}. \tag{4.63}$$

Now consider a *high-resistance line*, where $R \gg \omega L$. We write

$$jk = \sqrt{(j\omega L + R)j\omega C} \approx \sqrt{j\omega RC}. \tag{4.64}$$

The square root of an imaginary number has an angle of 45°. This means that α and β are equal. We can write

$$\alpha = \sqrt{\omega RC/2}, \tag{4.65}$$

$$v = \sqrt{2\omega/(RC)}. \tag{4.66}$$

Because both α and v vary as $\sqrt{\omega}$, the line is highly dispersive. As an historical example, we can analyze the first transatlantic telegraph cable, laid in 1865. This cable was 3,600 km long and weighed 5,000 tons. The insulator was a vegetable gum called gutta-percha. For this cable $L = 460$ nH/m, $C = 75$ pF/m, and $R = 7$ mΩ/m. At a frequency of 2.4 kHz, $\omega L = R$, and so the high-resistance assumption is well satisfied for frequencies below 100 Hz. At 12 Hz, we can write α and v as

$$\alpha = \sqrt{\omega RC/2} = 4.4 \times 10^{-3} \text{ nepers/km}, \tag{4.67}$$

$$v = \sqrt{2\omega/(RC)} = 17,000 \text{ km/s}. \tag{4.68}$$

The loss for the entire line is $\alpha l = 140$ dB and the delay is $l/v = 210$ ms. For comparison, at 3 Hz, the loss in dB and the delay change by a factor of 2, to 70 dB and 420 ms. Thus the 12-Hz component attenuates 70 dB more than the 3-Hz component. In addition, the 12-Hz component arrives 210 ms ahead of the 3-Hz component. In order to improve these characteristics, the signalling speed had to be drastically reduced, to about one word per minute, which was twenty times slower than hoped for. You might be interested to know that the renowned physicist, Lord Kelvin, did this analysis, but the project chief ignored it. His name was

Dr. Whitehouse (a medical doctor), and he said, "In electricity, there is seldom any need of any mathematical or other abstractions, ... and the formulas may for all practical purposes be dispensed with." The end was tragic. The cable operators thought they could improve the signalling rate by increasing the voltage, and they drove the line with 2-kV pulses. The insulation was not good enough to take this large voltage, and within two weeks, a short developed, somewhere along the 3,600-km line.

4.7 Reflections

Until now, we have not worried about what happens at the end of a transmission line. However, most of the time we are really more interested in what is going on at the ends than we are in the middle, because our sources and loads are usually at the ends. First we consider a transmission line with a load (Figure 4.7). Let the line impedance be real and be given by Z_0 and the load impedance be Z. Assume a forward wave V_+ is incident on the load. The effect of the load will be to make a reflected wave V_-. We will see that the amplitude of the reflected wave is determined by how different Z is from Z_0.

We call the ratio of V_- to V_+ the *reflection coefficient*. It is given by

$$\rho = V_-/V_+, \tag{4.69}$$

where ρ is the Greek letter *rho*. Notice that ρ is a *voltage* reflection coefficient. Sometimes we need a *current reflection coefficient*, ρ_i, defined in a dual way as

$$\rho_i = I_-/I_+. \tag{4.70}$$

The current reflection coefficient has the same magnitude as the voltage reflection coefficient because the voltages and currents in the waves are proportional. However, since the current in the reverse wave changes sign, we can write

$$\rho_i = -\rho. \tag{4.71}$$

The load voltage V is also proportional to the incident voltage, and this ratio is called the *transmission coefficient*. We write this as τ:

$$\tau = V/V_+. \tag{4.72}$$

Now we can find a simple formula that relates ρ and τ. The load voltage V is the

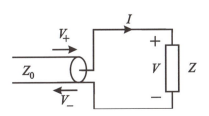

Figure 4.7. Reflection and transmission at a load.

sum of the incident wave V_+ and the reflected wave V_-. We can write

$$V = V_+ + V_-. \tag{4.73}$$

If we divide by V_+, we get

$$\tau = 1 + \rho. \tag{4.74}$$

This is an important expression, because it means that ρ and τ are not independent – we can calculate one if we know the other.

Next we can find formulas that relate ρ to Z. We can write an expression for the load current I as the sum of the current in the incident wave I_+ and the current in the reflected wave I_-:

$$I = I_+ + I_-. \tag{4.75}$$

Now divide this formula into Equation 4.73 to get

$$\frac{V}{I} = \frac{V_+ + V_-}{I_+ + I_-} = \frac{V_+}{I_+} \frac{1 + V_-/V_+}{1 + I_-/I_+}. \tag{4.76}$$

We can substitute for all these ratios and rewrite the formula as

$$\frac{Z}{Z_0} = \frac{1 + \rho}{1 - \rho}. \tag{4.77}$$

This formula lets us calculate Z if we know ρ. Microwave instruments measure reflection coefficient rather than impedance, and we can use this formula to see what Z really is. We can also solve for ρ as

$$\rho = \frac{Z - Z_0}{Z + Z_0}. \tag{4.78}$$

This is one of a family called *bilinear transforms*, and because of this it turns out that the loci of constant resistance, reactance, conductance, and susceptance are circles or straight lines. It is typical to plot ρ in the complex plane (Figure 4.8).

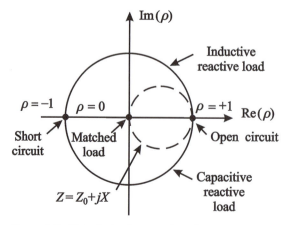

Figure 4.8. Plot of ρ in the complex plane.

These plots are usually called Smith charts, after Philip Smith, the engineer at Bell Labs who thought of this approach.

Let us consider some special cases and see what the reflection coefficients are. When $Z = Z_0$, the load is said to be *matched*. The reflection coefficient is 0; hence there is no reflection. We use this idea in the lab when we work with oscilloscopes and fast signals, where we may see ringing on waveforms from repeated reflections. If we put a matching resistor in parallel with the oscilloscope input we can stop the ringing. When Z is real, then ρ is also real. This means that if the load is a resistor R, then ρ lies along the real axis. If $R > Z_0$, then ρ is positive, and the reflected wave has the same phase as the incident wave. If $R < Z_0$, then ρ is negative, and the reflected wave is $180°$ out of phase with the incident wave. At the extremes, a short circuit has a reflection coefficient of -1, and an open circuit has a reflection coefficient of $+1$. Now consider a reactive load, with $Z = jX$. We can write

$$\rho = \frac{jX - Z_0}{jX + Z_0}. \tag{4.79}$$

The absolute values of the real and imaginary parts of the numerator and denominator are the same, and so the magnitudes of the numerator and denominator are the same. This means that the reflection coefficient lies on the unit circle. The inductive reactances are along the top half. Let us start at $X = 0$ and consider the locus as X increases. When $X = 0$, the load is a short, and $\rho = -1$. As X increases we move clockwise along the top half of the unit circle. When $X = Z_0$, we are at the top, and as X approaches ∞, we reach $\rho = 1$. The capacitive reactances are along the bottom half. One other interesting case to consider is an impedance of the form $Z = Z_0 + jX$. This locus appears in Figure 4.8 as a dashed circle that passes through $\rho = 0$ and $\rho = +1$.

4.8 Available Power

We can find the transmission coefficient τ by combining Equation 4.74 and Equation 4.78 to get

$$\tau = 1 + \rho = \frac{2Z}{Z + Z_0}. \tag{4.80}$$

Notice that τ can be larger than one. For an open-circuited load, $\tau = 2$, so that the transmitted voltage is twice as large as the incident voltage. We can use this fact to find the Thevenin equivalent circuit for a transmission line. The open-circuit voltage V_0 is given by

$$V_0 = 2V_+. \tag{4.81}$$

The look-back resistance R_s is just the characteristic impedance of the cable. We get

$$R_s = Z_0. \tag{4.82}$$

This gives us the Thevenin circuit shown in Figure 4.9.

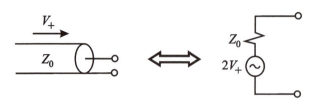

Figure 4.9. Thevenin equivalent circuit for a transmission line with a characteristic impedance Z_0 and an incident wave V_+.

There is more to this result than might appear at first. The Thevenin circuit produces the same voltages and currents in a load as the transmission line. However, another way to see this is to note that the transmission line produces the same results as the Thevenin circuit. We can turn things around, and think of the transmission line as an equivalent source. From this point of view, it is easy to calculate the maximum power from a Thevenin source. The power in the incident wave is given by Equation 4.34 as

$$P_+ = \frac{V_+^2}{2Z_0},$$ (4.83)

where we have taken V_+ to be real. This is the power that is delivered to a matched load, where there is no reflection. It is the maximum power that can be delivered to any load. In terms of the Thevenin parameters, we can rewrite this as

$$P_+ = \frac{V_o^2}{8R_s}.$$ (4.84)

We call $V_o^2/(8R_s)$ the *available power* from a Thevenin source. This is the AC version of the DC formula we derived in Problem 1. It is a good idea to learn it, because we will use it repeatedly. The formula is for peak voltages, but in the lab, we use peak-to-peak voltages, which are twice as large. In addition, function generators read half the open-circuit voltage. Usually these factors of two cancel, and this makes it easy to apply the formula.

4.9 Resonance

We found in the last chapter that when we combine inductors and capacitors, we make resonant circuits. Because a cable has both inductance and capacitance, it can also resonate. An open-circuited transmission line turns out to be much more interesting than an ordinary open circuit. It shows the effects of delays and reflections and can even be used as a filter. To start, we connect a function generator to a transmission line with the same impedance (Figure 4.10a). We assume that the line is long enough that we do not have to worry about reflections from the far end. We represent the generator by a Thevenin equivalent circuit, with open-circuit voltage V_o and impedance Z_0. The forward voltage V_+ is given by

$$V_+ = V_0/2.$$ (4.85)

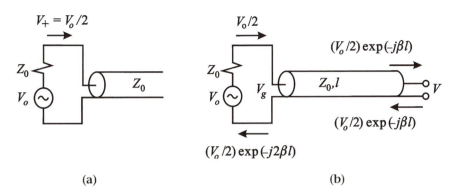

Figure 4.10. Connecting a sine-wave generator to a long transmission line (a) and to an open-circuited line (b).

Now cut the transmission line at some point, leaving the end open-circuited (Figure 4.10b). Starting with the forward voltage at the generator, we can calculate the other voltages on the line by multiplying by phase factors and the reflection coefficient. The forward voltage at the end is given by $(V_0/2)\exp(-j\beta l)$, where l is the length of the line. The reflection coefficient of an open circuit is $+1$; thus the reflected reverse wave is also $(V_0/2)\exp(-j\beta l)$. The total voltage V is

$$V = V_+ + V_- = V_0 \exp(-j\beta l). \tag{4.86}$$

This makes sense. It is just the Thevenin voltage with a phase lag due to the transmission line. At low frequencies β approaches zero, and V approaches the Thevenin voltage V_0.

The generator voltage is more surprising. The reflected wave propagates back to the generator where it is absorbed without further reflection. There is an additional phase lag due to the line, so that the reverse wave at the generator is given by

$$V_- = (V_0/2)\exp(-j2\beta l). \tag{4.87}$$

The total voltage at the generator V_g is given by

$$V_g = V_+ + V_- = V_0/2 + (V_0/2)\exp(-j2\beta l). \tag{4.88}$$

We can write this in terms of a cosine by pulling out a factor of $\exp(-j\beta l)$:

$$V_g = V_0 \exp(-j\beta l)\cos(\beta l). \tag{4.89}$$

Notice that the phase of V_g is the same as the phase of V (Equation 4.86). In fact, the phase is the same everywhere along the line. We call this a *standing wave*. However, the magnitude of V_g depends on the length of the line. You should notice that V_g is zero when $l = \lambda/4$. In a measurement, we see a resonance at the frequency where the line is a quarter of a wavelength long. This seems mysterious, because the generator output voltage is zero at the same time the voltage at the other end is V_0. However, we do have a current at the generator and we can think

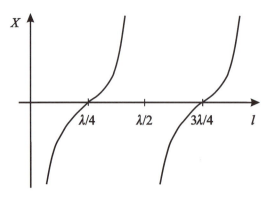

Figure 4.11. Reactance of a section of open-circuited transmission line.

of it as pumping the resonance. We can write the current as

$$I_g = V_+/Z_0 - V_-/Z_0 = jI_s \exp(-j\beta l)\sin(\beta l), \tag{4.90}$$

where $I_s = V_0/Z_0$ is the short-circuit current for the generator. When $l = \lambda/4$, $I_g = I_s$. As far as the generator is concerned, a quarter-wave section of open-circuited line looks like a short. The current is the short-circuit current, and the voltage is zero.

In an open-circuited line, the generator current and voltage are $90°$ out of phase, and this means that the impedance is reactive. Energy is stored in the waves propagating back and forth. We can find the reactance X by taking the ratio of the generator voltage and current:

$$X = \frac{V_g}{jI_g} = -\frac{Z_0}{\tan(\beta l)}. \tag{4.91}$$

This formula is plotted in Figure 4.11. The curve shows that the line can be used as either an inductor or a capacitor, depending on the length and the frequency.

The figure also shows that an open-circuited line can be used as a resonant circuit. When the reactance is zero, we effectively have a series resonance. When the reactance becomes very large, we have a parallel resonance. When you study the series resonance in the lab, you will see that the input voltage does not really go to zero as the theory predicts, because of loss in the line.

Transmission-line resonators are usually not very practical for filters at frequencies in the MHz range, because the lines turn out to be inconveniently long. For example, if we wanted a series resonance at 5 MHz, we would need a 10-meter cable. However, at the frequencies for microwave radars, in the GHz range, the required length might be a few millimeters, and transmission-line elements are very easy to use. There is a simple transmission line called microstrip, which is just a printed-circuit board with a ground plane on the back (Figure 4.12). To make your circuits on this board, you do not even need capacitors or inductors; you can just etch the copper on the top in the shape that you want. For example, let us suppose that we want to make a filter that can stop signals at a particular frequency. This

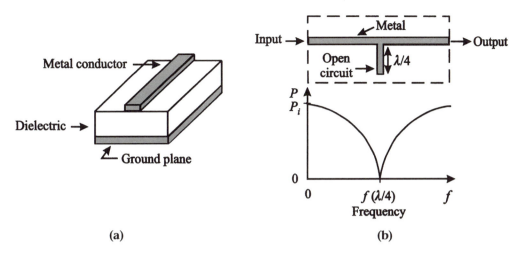

Figure 4.12. Microstrip transmission line (a), and notch filter (b).

is called a *notch* filter. We can add a parallel open-circuited section of transmission line that is a quarter of a wavelength long at the notch frequency. At this frequency, the reactance of the open-circuited line is zero, so that it is effectively a short circuit across the line.

4.10 Quality Factor

We characterize a resonance by a quality factor Q, given in the last chapter as

$$Q = \omega \frac{E}{P_a},$$ (4.92)

where E is the stored energy and P_a is the average power lost. In a transmission line, stored energy is in the form of power propagating down the line. We can write

$$E = P_+(l/v),$$ (4.93)

where P_+ is the power in the forward wave and l/v is the delay time for the cable. Next we calculate the dissipated power P_a. As the forward wave travels along the transmission line, the voltage decays by a factor of $\exp(-\alpha l)$. Because the power is proportional to the square of the voltage, it decays as $\exp(-2\alpha l)$. The lost power can thus be written

$$P_a = P_+ - P_+ \exp(-2\alpha l)) \approx 2\alpha l P_+.$$ (4.94)

Here we are using the first-order Taylor series approximation

$$\exp(x) \approx 1 + x,$$ (4.95)

which is valid when $|x| \ll 1$. When we substitute for E and P_a into Equation 4.92, we get

$$Q = \omega \frac{E}{P_a} = \frac{\beta}{2\alpha}.$$

(4.96)

We have calculated Q in terms of the forward wave only. However, the result is the same for a reverse wave. Thus the formula can be applied to a resonator as a whole. A typical Q for a transmission-line resonator is between 10 and 100.

4.11 Lines with Loads

We have found that the impedance of an open-circuited line depends strongly on the length of the line and the frequency. We can also find formulas for lines with loads. Let us consider a section of length l connected to a load with a re-flection coefficient $\rho(0)$ (Figure 4.13). We will calculate the reflection coefficient at the other end of the line. To start, let the forward wave at the input be V_+. We can write the forward wave at the load as $V_+ \exp(-j\beta l)$. To find the reverse wave at the load, we multiply the forward wave by the reflection coefficient, $\rho(0)$, to give $\rho(0) V_+ \exp(-j\beta l)$. We can write the reverse wave at the input V_- as

$$V_- = \rho(0) V_+ \exp(-j2\beta l).$$

(4.97)

The reflection coefficient at the generator $\rho(l)$ is given by

$$\rho(l) = V_- / V_+ = \exp(-j2\beta l)\rho(0).$$

(4.98)

The magnitude of the reflection coefficient does not change, only the phase. No-tice that the reflection coefficient at the input lags the reflection coefficient at the load. There are actually two phase lags. One comes from the propagation of the forward wave from the generator to the load, and the other from the propaga-tion of the reflected wave from the load back to the generator. When we plot the reflection coefficient in the complex plane (Figure 4.14), the locus is a clockwise circle.

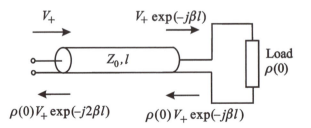

Figure 4.13. Reflection coefficient calculation for a lossless line with a load.

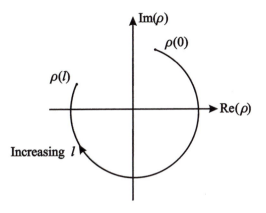

Figure **4.14.** Locus of the reflection coefficient as the length increases.

If the transmission line is a half wavelength long, the reflection coefficient is
the same as the load reflection coefficient,

$$\rho(\lambda/2) = \rho(0). \tag{4.99}$$

This means that at this frequency, the transmission line has no effect except for
a propagation delay. This idea is used to make protective covers for radars on
airplanes. The cover is called a *radome* or *half-wave window*. The problem with
mounting a radar antenna on the front of an airplane is that it will blow away
unless it is protected. If the covering is made a half wavelength thick, the waves
go right through it.

The other interesting length to consider is a quarter wavelength. When the line
is a quarter wavelength long, the reflection coefficient changes sign. We can write

$$\rho(\lambda/4) = -\rho(0). \tag{4.100}$$

Changing the sign of the reflection coefficient transforms the impedance. We can
write the impedance at the generator end, $Z(\lambda/4)$, with Equation 4.77 as

$$\frac{Z(\lambda/4)}{Z_0} = \frac{1 + \rho(\lambda/4)}{1 - \rho(\lambda/4)} = \frac{1 - \rho(0)}{1 + \rho(0)} = \frac{Z_0}{Z(0)}, \tag{4.101}$$

where $Z(0)$ is the load impedance. We can rewrite this expression as

$$\frac{Z(\lambda/4)}{Z_0} = \frac{Z_0}{Z(0)}. \tag{4.102}$$

One way to understand this formula is to define a *normalized impedance*, which is
the impedance scaled to Z_0. We will use lower-case letters for normalized impedance and write

$$z = Z/Z_0. \tag{4.103}$$

The *normalized admittance* is given by

$$y = 1/z = YZ_0. \tag{4.104}$$

In terms of normalized impedances, Equation 4.102 becomes

$$z(\lambda/4) = 1/z(0).$$ (4.105)

This means that we can think of the quarter-wave transmission line as an impedance inverter. We will study other impedance inverters in the next chapter. We will see that we can make excellent band-pass filters by combining impedance inverters and resonators. Another application of quarter-wave sections is in eliminating reflections. We know that if the load resistance is different from the characteristic impedance of a cable, there will be a reflection. We can use a quarter-wave section to transform the impedance of a load to match the cable. To see this, rewrite Equation 4.102 as

$$Z_0 = \sqrt{Z(\lambda/4)Z(0)}.$$ (4.106)

In words, the characteristic impedance of the transmission line is the geometric mean of the load impedance $Z(0)$ and the transformed impedance $Z(\lambda/4)$. This means that if we choose Z_0 to be the geometric mean of the load resistance R_l and the source resistance R_s, all of the available power from the source will be delivered to the load. We write the matching condition as

$$Z_0 = \sqrt{R_s R_l}.$$ (4.107)

This idea is also used in optics. Lenses are coated with matching layers that are a quarter-wavelength thick to eliminate reflections from the surface of the lens. Typically several layers are used so that the reflections can be reduced for the full range of wavelengths we can see. These are called *antireflection*, or AR, coatings.

FURTHER READING

The classic textbook is *Fields and Waves in Communication Electronics* by Simon Ramo, John Whinnery, and Theodore Van Duzer, published by Wiley. It is comprehensive, covering this material and much more. There is an excellent discussion of distributed inductance and capacitance and the skin effect. Paul Nahin has also written a terrific biography, *Oliver Heaviside: Sage in Solitude*, published by the IEEE Press, that touches directly on many of these topics. Heaviside was an English engineer who helped develop transmission-line theory, Laplace transforms, and the notation that we use for vector calculus and for Maxwell's equations. Nahin also tells the sad story of the failure of the first transatlantic cable.

PROBLEM 10 – COAXIAL CABLE

Coaxial cable has many advantages for transmitting electrical signals. It can be used from DC to very high frequencies (cables are available that operate as high as 100 GHz). A common laboratory cable is RG58/U, costing about a dollar per meter. The shield is a weave of fine tinned-copper wires around an insulating polyethylene tube. This cable

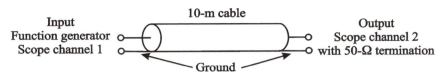

Figure 4.15. Measuring the velocity v.

typically has twist-lock BNC connectors. (A short note is needed on abbreviations. BNC stands for "Bayonet Neill Concelman," after the Bell Laboratories engineers Paul Neill and Carl Concelman, who developed the connector. RG/U is "radio-guide/universal," and different varieties come with different identifying numbers.) RG58 coax and BNC connectors are commonly used up to a frequency of 1 GHz. In this problem, you will measure the velocity and characteristic impedance for the cable and use these to calculate the distributed inductance and capacitance.

A. First measure the velocity on a 10-m cable with the connections in Figure 4.15. Set the function generator for 5-V pulses with a width of 50 ns that repeat at a frequency of 20 kHz. To measure the delay accurately, we need a fast time scale on the scope. A convenient scale is 10 ns per division. You should be able to see an incident pulse on channel 1 and a delayed pulse on channel 2. Measure the delay, and calculate the velocity v. Express v as a fraction of the speed of light c, where $c = 3.00 \times 10^8$ m/s.

B. Disconnect the 10-meter cable and plug an antenna cable into the channel-1 tee. Now your pulses are sent up to the antenna. At the antenna, the pulses are reflected and come back down the cable. Use the delay to deduce the length of the antenna cable, assuming that the velocity is the same as before.

C. Next we find the characteristic impedance Z_0 with the circuit shown in Figure 4.16. The voltage is measured by a tee connection to channel 1 of the oscilloscope. The current is measured through a 1:1 transformer. We will study transformers in Chapter 6, but for now you should know that the 1-Ω resistor effectively appears in series for our signals. This means that the voltage on channel 2 is numerically equal to the current in amperes. The transformer is needed to avoid a short to the scope ground. Measure the voltage and current in the middle of the pulse and calculate Z_0.

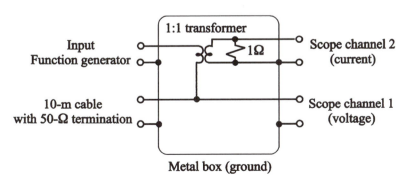

Figure 4.16. Circuit for measuring the characteristic impedance Z_0.

D. Now remove the 50-Ω load from the end of the cable so that the cable is open-circuited. Sketch and interpret the voltage and current waveforms.

E. Use your measurements of v and Z_0 to calculate L and C.

PROBLEM 11 - WAVES

A. We saw that for either a forward wave or a reverse wave alone, the magnitude remains the same at different positions, but the phase changes. We call this a *traveling wave*. However, if both a forward wave and a reverse wave are present at the same time, we need to add the two phasors, and the locus changes dramatically. Sketch the locus as z varies when both a forward voltage wave $\exp(-j\beta z)$ and a reverse wave $\exp(+j\beta z)$ are present. How does the locus change if the reverse wave becomes $-\exp(+j\beta z)$? Sketch the new locus. How does the locus change if the reverse wave becomes $\rho \exp(+j\beta z)$, where $|\rho| \leq 1$? The *standing wave ratio* (SWR) is defined as the ratio of the maximum magnitude to the minimum magnitude. Find a formula for the SWR in terms of $|\rho|$. The SWR is often used to characterize connectors, filters, and antennas.

B. We analyzed a transmission line by breaking it up into short sections and letting the length of the sections go to zero. We found a formula for the characteristic impedance, which is the ratio of the voltage and current in a forward wave. This would be the impedance we would measure at the input of a transmission line that is sufficiently long that we do not have to consider the effect of the far end. It is interesting to consider the impedance of a ladder network of discrete cascaded components (Figure 4.17). We let the impedance of the series element be Z and the admittance of the parallel element be Y. We will assume that the number of elements is large enough that we do not need to consider the effect of the far end. Find the input impedance Z_0 of the discrete line in terms of Z and Y. One way to approach this problem is to consider that adding another Z and Y section at the beginning should not change the input impedance. You can use this fact to find Z_0.

C. Suppose we want to transmit voice signals over 100 km of cable with $L = 250$ nH/m and $C = 100$ pF/m. The distributed resistance at voice frequencies is 50 mΩ/m. The distributed conductance may be neglected. Using the high-resistance approximation, calculate the total loss in dB and the delay in ms at 500 Hz, 1 kHz, and 2 kHz.

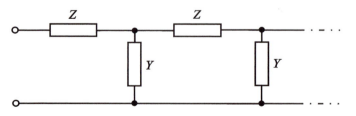

Figure 4.17. A discrete transmission line with lumped series impedance Z and parallel admittance Y.

D. Now add 100-mH series inductor coils at 1-km intervals. You may assume that the added inductance is effectively distributed uniformly along the line, and you may neglect the resistance of the coils. Using Equations 4.49 and 4.50, calculate the total loss in dB and the delay in ms at 500 Hz, 1 kHz, and 2 kHz. For comparison, calculate the total loss and delay using the high-reactance approximation.

PROBLEM 12 - RESONANCE

A. We will consider an open-circuited section of transmission line connected to a generator (Figure 4.18). Let the attenuation constant of the line be α and the phase constant be β. Derive an expression for the ratio $|V_g/V|$ at the first series resonant frequency. Find a first-order approximation for the ratio, assuming that α is small.

B. Now we find α. Make the connections shown in Figure 4.19, with the end of the cable connected to channel 2 of the oscilloscope. Do not use a 50-Ω load. Use an amplitude setting of 1 Vpp. Adjust the frequency to find the first series resonance where $|V_g|$ is a minimum. Use the ratio $|V_g/V|$ to calculate α.

C. Next we use the resonant frequency to find the velocity. Because the scope capacitance shifts the resonance, you should disconnect the cable from channel 2 for this part. Readjust the frequency for resonance. Use the frequency and the length to calculate the cable velocity v. How large was the frequency shift caused by the scope capacitance? Calculate the frequency shift that you would expect, using the scope and cable capacitance.

D. Next we consider the bandwidth. For a series resonance, we defined the half-power frequencies f_l and f_u where R and X are equal and the load voltage changed by a factor of $\sqrt{2}$. Here the load is effectively the distributed resistance of the cable and we do not have access to the resistance by itself. However, at resonance, the cable resistance is only a few ohms, and the function generator current is very close to the short-circuit current $I_s = V_o/Z_0$. The voltage $|V_g|$ will be a minimum at the resonant

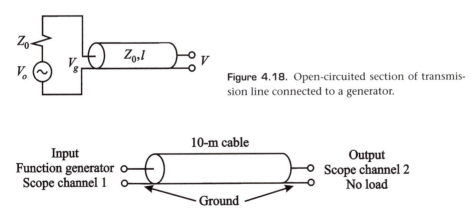

Figure 4.18. Open-circuited section of transmission line connected to a generator.

Figure 4.19. Measuring the attenuation constant α.

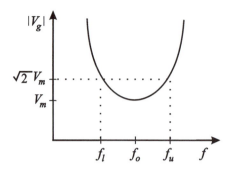

Figure 4.20. Variation of $|V_g|$ near the resonant frequency. V_m is the voltage minimum at resonance.

frequency and will increase as we move away from it. At f_u and f_l, where the input resistance and reactance are equal, $|V_g|$ will rise by a factor of $\sqrt{2}$. One way to make the measurement is to first measure $|V_g|$ at resonance and then reduce the amplitude setting by $\sqrt{2}$. Now measure the frequencies f_l and f_u that give the same value of $|V_g|$ that you measured before. What Q does this bandwidth indicate?

E. Now calculate the Q that you expect from the energy formula

$$Q = \frac{\beta}{2\alpha}. \tag{4.108}$$

Filters

So far the filters we have made have had only two elements: a capacitor and a resistor or inductor. We can improve the response of our filters by adding more elements. This allows us to make the pass band flatter and the roll-off steeper. Multielement filters behave somewhat like transmission lines, and we need to have the right input and output resistance to avoid problems with reflections. Analyzing these filters by hand is quite difficult, but the calculations are easy on a computer. For this we will use a computer program called *Puff*, which is included with this book. Instructions for running the program are given in Appendix C.

5.1 Ladder Filters

We will consider ladder networks with alternating series and shunt elements like the discrete transmission line we studied in Problem 11. If the series elements are inductors and the shunt elements are capacitors, then the circuit acts as a low-pass filter (Figure 5.1a, b). At low frequencies, the impedance of the inductors and the admittance of the capacitors are small, and the input signal passes through to the output with little loss. In contrast, at high frequencies the inductors begin to act as voltage dividers and the capacitors as current dividers. This reduces the power transmitted to the load. We can also make high-pass filters with series capacitors and shunt inductors (Figure 5.1c, d).

Many different filters have been developed, giving a wide choice of amplitude, phase, pass-band, and stop-band characteristics. We will focus on amplitude and consider two different types – the Butterworth filter, which gives a flat pass band, and the Chebyshev filter, which gives an excellent roll-off. Mathematically, we write the loss factor L for the Butterworth low-pass filter as

$$L = P_i/P = 1 + (f/f_c)^{2n}, \tag{5.1}$$

where the input power P_i is the available power from the source, P is the output power delivered to the load, and f_c is the 3-dB cut-off frequency. This is shown in Figure 5.2a. The loss characteristic is quite flat in the pass band. In fact, the first $2n - 1$ frequency derivatives are zero at $f = 0$. For this reason, people call these *maximally flat* filters. When we are well into the stop band, the loss factor can be

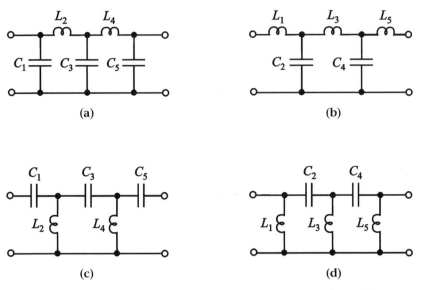

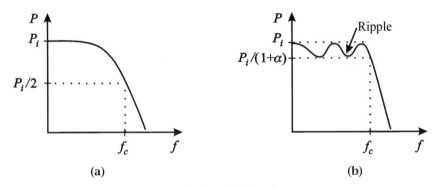

Figure 5.1. Low-pass (a, b) and high-pass (c, d) ladder filters. Often inductors are larger and more expensive than capacitors, and so filters that use fewer inductors, like (a) and (c), are more common. The number of elements is called the *order* of the filter. These are all fifth-order filters.

Figure 5.2. Butterworth (a) and Chebyshev (b) filter characteristics.

written approximately as

$$L \approx (f/f_c)^{2n}. \tag{5.2}$$

This means that the loss increases by $6n$ dB each time the frequency doubles. We say that the loss increases by 6 dB per octave per element.

It turns out that we can get a faster roll-off if we allow a ripple in the pass band. The Chebyshev filter takes advantage of this (Figure 5.2b). Its loss is given by

$$L = 1 + \alpha C_n^2(f/f_c), \tag{5.3}$$

where α is the ripple size and $C_n(x)$ is the Chebyshev polynomial of order n. Chebyshev polynomials have the interesting property that they oscillate between -1

Order	Polynomial
0	1
1	x
2	$2x^2 - 1$
3	$4x^3 - 3x$
4	$8x^4 - 8x^2 + 1$
5	$16x^5 - 20x^3 + 5x$
6	$32x^6 - 48x^4 + 18x^2 - 1$
7	$64x^7 - 112x^5 + 56x^3 - 7x$

(a)   (b)

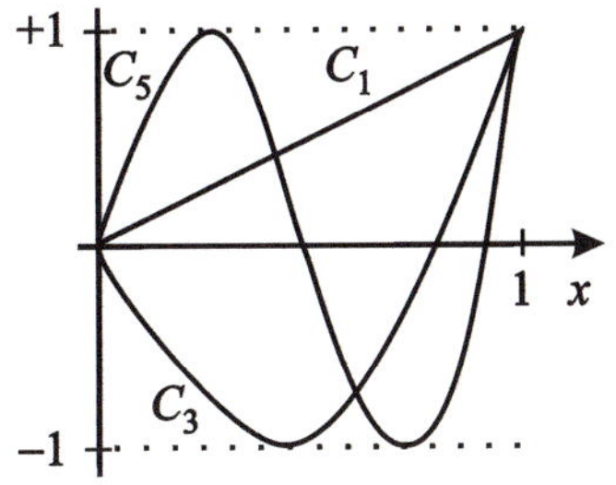

Figure 5.3. The Chebyshev polynomials (a), and plots of C_1, C_3, and C_5 (b).

and $+1$ as x varies from -1 and $+1$. The first two polynomials are given by

$$C_0 = 1, \tag{5.4}$$
$$C_1 = x. \tag{5.5}$$

We can calculate the rest from the following formula:

$$C_i(x) = 2xC_{i-1}(x) - C_{i-2}(x). \tag{5.6}$$

You need to calculate the polynomials in order to use this formula. Figure 5.3a gives the Chebyshev polynomials through order 7.

We will use only the odd-order polynomials, because the even-order polynomials are for filters with different source and load resistances. The odd-order polynomials are 0 when x is zero, and then oscillate between $+1$ and -1, finally ending up at $+1$ when $x = 1$ (Figure 5.3b). This is really rather surprising, considering how large the coefficients of the polynomials are. For example, the cubic coefficient in C_5 is 20. However, the different terms offset each other as long as $x \leq 1$. Now consider what this means for the loss. Each time the Chebyshev polynomial hits either $+1$ or -1, the loss factor becomes $1 + \alpha$. This means that in the pass band, the loss factor swings between 1 and $1 + \alpha$ repeatedly. For this reason we call this an *equal-ripple* filter. This happens once for C_1, twice for C_3, and three times for C_5. This is shown in Figure 5.2b for a fifth-order filter. The last time the loss factor is $1 + \alpha$ is when $f = f_c$, at the pass-band edge. As the frequency increases above f_c, the loss factor increases sharply, and the filter rolls off rapidly. The roll-off advantage for the Chebyshev filters comes from the large leading coefficient, given by 2^{n-1}. By comparison, this coefficient for Butterworth filters is 1. In dB terms, the loss factor of the Chebyshev filter in the stop band is $6(n - 1)$ dB larger than that of a Butterworth filter with the same pass-band loss. For example, for fifth-order filters this is 24 dB, a considerable improvement.

5.2 Filter Tables

A good way to start a filter design is to consult a table that lists component values for different filters. With these values, you can simulate the filter response on a computer, and from there you can make adjustments to account for the available components. Manufacturers make only a limited range of capacitor values. Inductors also must have an integral number of turns. In practice, however, the turns can be squeezed to increase the inductance somewhat, or spread to reduce it. Also, the components themselves have loss, and this effect can be included in the computer simulations. Finally, it is a good idea to see the effect of component variation. Chebyshev filters are particularly sensitive to this. Deriving the formulas for the tables is quite difficult, and so I will just give the results. These formulas are for filters with the same source and load resistance, and we will use this resistance for normalizing. The normalized susceptances and reactances for a Butterworth filter at f_c are given by

$$a_i = 2\sin\left(\frac{(2i-1)\pi}{2n}\right),$$ (5.7)

where i is an index for the components and n is the order of the filter. These values are given in Table 5.1a through seventh order.

Calculating the values for a Chebyshev filter is quite involved. We usually specify a maximum ripple loss in dB in the pass band. In practice, these specifications vary over a wide range from 0.01 dB to 1 dB. We relate this loss L_r to α by

$$1 + \alpha = 10^{L_r/10}.$$ (5.8)

We calculate an auxiliary quantity β as

$$\beta = \sinh\left(\frac{\tanh^{-1}(1/\sqrt{1+\alpha})}{n}\right),$$ (5.9)

where sinh is the hyperbolic sine function and $\tanh^{-1}$ is the inverse of the hyperbolic tangent. We calculate the Chebyshev components in order, starting with

$$c_1 = a_1/\beta,$$ (5.10)

where a_1 is given by Equation 5.7. Then we proceed sequentially to c_2 and the rest, using the formula

$$c_i = \frac{a_i a_{i-1}}{c_{i-1}(\beta^2 + \sin^2[(i-1)\pi/n])}.$$ (5.11)

The Chebyshev components for 0.2-dB ripple are given in Table 5.1b.

Table 5.1. Component Values for Ladder Filters.

Order	a_1	a_2	a_3	a_4	a_5	a_6	a_7
1	2						
2	$\sqrt{2}$	$\sqrt{2}$					
3	1	2	1				
4	0.765	1.848	1.848	0.765			
5	0.618	1.618	2	1.618	0.618		
6	0.518	$\sqrt{2}$	1.932	1.932	$\sqrt{2}$	0.518	
7	0.445	1.247	1.802	2	1.802	1.247	0.445

(a)

Order	c_1	c_2	c_3	c_4	c_5	c_6	c_7
1	0.434						
3	1.228	1.153	1.228				
5	1.339	1.337	2.166	1.337	1.339		
7	1.372	1.378	2.275	1.500	2.275	1.378	1.372

(b)

Note: The values for Butterworth filters from Equation 5.7 are given in (a), and for Chebyshev filters with a ripple of 0.2 dB from Equation 5.11 in (b). These are the normalized susceptances of the shunt elements at f_c, and the normalized reactances of the series elements. People call these *immittance* values to indicate that the numbers can be used for either susceptance or reactance. Any of the filters shown in Figure 5.1 can be designed from these tables. For a low-pass filter, the series elements are inductors, and the shunt elements are capacitors. For a high-pass filter, the series elements are capacitors and the shunt elements are inductors, and the reactances and susceptances are negative. Because the values are symmetric, we can count from either end. Start with either a series element or a shunt element, and alternate throughout the filter. One warning: By tradition, f_c for a Butterworth filter means the 3-dB frequency, but for Chebyshev filters, f_c is defined by the frequency that gives the maximum ripple, 0.2 dB in this case.

5.3 Examples

Using the tables is surprisingly complicated, and it is a good idea to simulate filters on the computer to make sure that you actually get the characteristics you want. To show how the tables work, we will go through several examples in detail. Let us assume that we need a filter for a 50-Ω antenna cable with a 3-dB cut-off frequency of 10 MHz and a loss of at least 20 dB at 20 MHz. A fourth-order Butterworth filter with a cut-off frequency of 10 MHz should have a loss at 20 MHz of

$$L(20\,\text{MHz}) = 6n = 24\,\text{dB}, \tag{5.12}$$

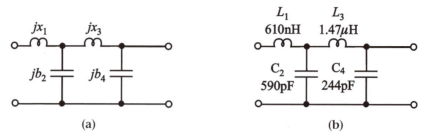

Figure 5.4. Designing a 50-Ω low-pass Butterworth filter with $f_c = 10$ MHz. Defining the normalized reactances for the series elements and the normalized susceptances for the shunt elements (a), and the component values (b).

which is sufficient. We start with the filter structure in Figure 5.4a. There are two series inductors and two shunt capacitors. From Table 5.1a, we can write the normalized reactance of the first inductor as

$$x_1 = a_1 = 0.765. \tag{5.13}$$

We can find the actual reactance X_1 at 10 MHz by multiplying by the characteristic impedance of the cable, $Z_0 = 50\,\Omega$. This gives us

$$X_1 = x_1 \cdot Z_0 = 38\,\Omega, \tag{5.14}$$

and the inductance L_1 is given by

$$L_1 = X_1/\omega_c = 610\,\text{nH}. \tag{5.15}$$

Now we proceed to the other values. From the table, the normalized susceptance of the first shunt capacitor is given by

$$b_2 = a_2 = 1.848. \tag{5.16}$$

We can find the actual susceptance B_2 at 10 MHz by dividing by Z_0. This gives us

$$B_2 = b_1/Z_0 = 37\,\text{mS}, \tag{5.17}$$

and the capacitance C_2 is given by

$$C_2 = B_2/\omega_c = 590\,\text{pF}. \tag{5.18}$$

The inductance L_3 is given by

$$L_3 = a_3 Z_0/\omega_c = 1.47\,\mu\text{H}, \tag{5.19}$$

and finally

$$C_4 = \frac{a_4}{Z_0 \omega_c} = 244\,\text{pF}. \tag{5.20}$$

Figure 5.4b shows the complete filter.

Now consider a high-pass filter with a 3-dB cut-off frequency of 10 MHz and a loss at 5 MHz of at least 20 dB. Figure 5.5a shows the structure, with series

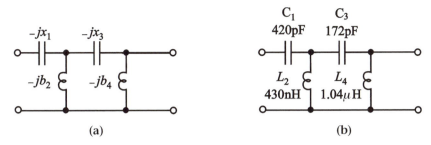

Figure 5.5. Designing a high-pass Butterworth filter. Defining the normalized reactances for the series elements and the normalized susceptances for the shunt elements (a), and the final component values (b).

capacitors and shunt inductors. We use the same table values as before. This time the reactances and susceptances are negative. We write

$$C_1 = \frac{1}{a_1 Z_0 \omega_c} = 420 \, \text{pF}, \tag{5.21}$$

$$L_2 = \frac{Z_0}{a_2 \omega_c} = 430 \, \text{nH}, \tag{5.22}$$

$$C_3 = \frac{1}{a_3 Z_0 \omega_c} = 172 \, \text{pF}, \tag{5.23}$$

$$L_4 = \frac{Z_0}{a_4 \omega_c} = 1.04 \, \mu\text{H}. \tag{5.24}$$

These values are shown in Figure 5.5b.

Figure 5.6 shows a simulation of the loss of these two filters with the computer program *Puff*. The response of these two filters is complementary, and they cross at the 3-dB level at 10 MHz. One interesting fact is that the filters show *reciprocity*. This means that it does not matter which end we use for the input. This is not obvious, because the filters are not symmetric end to end, but it is easily checked on the computer by swapping the input and output. We will see another example of reciprocity when we study antennas in Chapter 15.

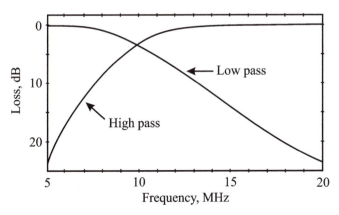

Figure 5.6. *Puff* simulation for the low-pass Butterworth filter in Figure 5.4b and the high-pass Butterworth filter in Figure 5.5b.

5.4 Band-Pass Filters

The ladder structure can also be used for band-pass and band-stop filters. For band-pass filters, the series elements are series resonant circuits, and the parallel elements are parallel resonant circuits (Figure 5.7a). Each of the elements is resonant at the center frequency f_0, so that the signal passes through unaffected. For band-stop filters, it is the other way around (Figure 5.7b), and the resonant circuits are arranged to block the signal at f_0. We can make Butterworth and Chebyshev filters with the same tables as before. For the band-pass filter, we find the series inductors and shunt capacitors as we did in the low-pass filter, except we use the filter bandwidth $\Delta\omega$ instead of ω_c in the reactance and susceptance calculations. For the band-stop filter, the values for the series capacitances and shunt inductors are calculated like the ones in the high-pass filters, but with $\Delta\omega$ instead of ω_c.

As an example, let us design a second-order band-pass Butterworth filter for 7 MHz, by adding a parallel resonant element to the series resonant circuit that we tested in Problem 8. The series resonant circuit had a 15-μH inductor and a variable capacitor that was adjusted for resonance at 7 MHz. This means that we can write

$$L_1 = 15\,\mu\text{H} \tag{5.25}$$

and

$$C_1 = \frac{1}{\omega_0^2 L_1} = 34.5\,\text{pF}. \tag{5.26}$$

For the first element, Table 5.1a gives us the value

$$a_1 = \sqrt{2}. \tag{5.27}$$

This is the normalized reactance of L_1 evaluated at $\Delta\omega$, and therefore we write

$$\Delta\omega L_1 = a_1 Z_0 \tag{5.28}$$

so that

$$\Delta\omega = a_1 Z_0 / L_1 = 4.71 \times 10^6\,\text{radians/s}. \tag{5.29}$$

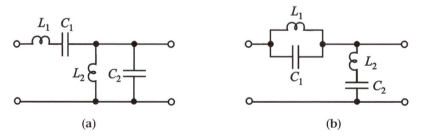

(a)   (b)

Figure 5.7. Band-pass filter (a), and band-stop filter (b). These are second-order filters.

In hertz, this is

$$\Delta f = \Delta\omega/(2\pi) = 750\,\text{kHz}. \tag{5.30}$$

Thus the 3-dB bandwidth of the filter is 750 kHz. Now we find C_2. We write

$$C_2 = \frac{a_2}{Z_0\Delta\omega} \tag{5.31}$$

and substitute for $\Delta\omega$ from Equation 5.29 to get

$$C_2 = L_1/Z_0^2 = 6.0\,\text{nF}, \tag{5.32}$$

where we have used the fact that $a_1 = a_2$. The inductance L_2 is given by

$$L_2 = \frac{1}{\omega_0^2 C_2} = 86\,\text{nH}. \tag{5.33}$$

These component values are shown in Figure 5.8a.

We plot the response of the filter for two different loss scales. Figure 5.9 shows the response from 0 to 3 dB. For comparison, I have plotted the loss for the

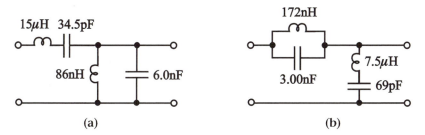

Figure 5.8. Component values for second-order, 7-MHz Butterworth band-pass (a) and band-stop (b) filters. The 3-dB bandwidth for each filter is $f_c = 750$ kHz.

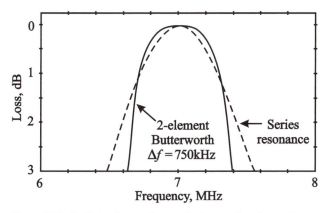

Figure 5.9. *Puff* simulation for the 2-element band-pass Butterworth filter in Figure 5.8a showing the range from 0 to 3 dB. The 3-dB bandwidth is 750 kHz, as predicted by Equation 5.30. Also shown is a plot for the series resonant element alone for comparison.

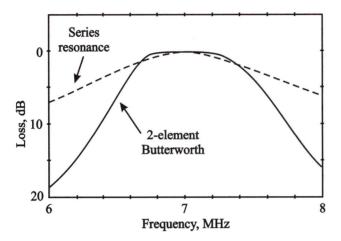

Figure 5.10. The same *Puff* simulation as in Figure 5.9, but with the range extended to 20 dB to show how the 2-element Butterworth filter rolls off much faster than the single resonant circuit.

series resonant circuit alone. This plot shows that near the center frequency, the response of the Butterworth filter is flatter than the series resonant circuit. Figure 5.10 is the same simulation with the loss scale extended to 20 dB. This shows that the Butterworth filter rolls off much more quickly than the series resonant circuit.

As a final example, we design a 2-element band-stop Butterworth filter with a 3-dB bandwidth of $\Delta f = 750$ kHz centered on 7 MHz. We use the circuit shown in Figure 5.7b. We write

$$C_1 = \frac{1}{a_1 Z_0 \Delta\omega} = 3.00 \,\text{nF}, \tag{5.34}$$

$$L_1 = \frac{1}{\omega_0^2 C_1} = 172 \,\text{nH}, \tag{5.35}$$

$$L_2 = \frac{Z_0}{a_2 \Delta\omega} = 7.5 \,\mu\text{H}, \tag{5.36}$$

$$C_2 = \frac{1}{\omega_0^2 L_2} = 69 \,\text{pF}. \tag{5.37}$$

These component values are shown in Figure 5.8b. The response of this filter is shown in Figure 5.11, together with the band-pass filter plot for comparison.

One thing to notice about the 2-element Butterworth filters is that the immittance values a_1 and a_2 are equal. This means that the normalized reactance of the series components is the same as the normalized susceptance of the shunt components. This gives us a way to recognize this filter in a circuit. The RF Filter in the NorCal 40A is a second-order Butterworth band-pass filter. We will study it in Problem 16. In addition, we can think of the IF Filter as a cascade of a pair of these band-pass filters. The IF Filter is a critical circuit in the receiver.

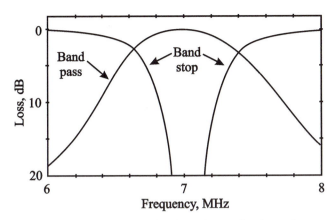

Figure 5.11. *Puff* simulation for the 2-element band-stop Butterworth filter in Figure 5.8b with the band-pass filter in Figure 5.8a shown for comparison.

It uses high-Q quartz-crystal resonators to achieve an extremely narrow bandwidth.

5.5 Crystals

Crystal quartz is an important material in electronics. Quartz crystals allow watches to keep precise time, and they control the master oscillators in microprocessor systems. In radios, they set oscillation frequencies and act as extremely narrowband filters. This crucial role may seem surprising, because quartz is an insulator. However, quartz has several interesting properties. It is *piezoelectric*, which means that when we apply a voltage across it, it moves. The piezoelectric effect allows us to couple electrical signals to mechanical vibrations. This works both ways. Voltages cause motion, and motion causes voltages. For example, gas stoves and water heaters often use piezoelectric starters. In a piezoelectric starter, a force deforms the crystal, causing a large voltage across the contacts to make a spark.

Quartz mechanical resonators have very high Qs in the range of 50,000 to 100,000. The main loss is not within the quartz itself, but to the air and the supports for the crystal. These Qs are much higher than those of LC and transmission-line resonators, which are limited by metal resistance and are usually 100 or less. The resonant frequencies are in the range from 1 kHz to 100 MHz. Chemically, quartz is silicon dioxide. The raw material for quartz is sand, and this means that quartz crystals can be manufactured inexpensively. In addition, quartz crystals can be cut precisely so that the resonant frequencies change only slightly with temperature. The orientation in our crystals is called an AT cut, and this gives a temperature coefficient as low as 1 part in a million per degree Celsius. In a clock,

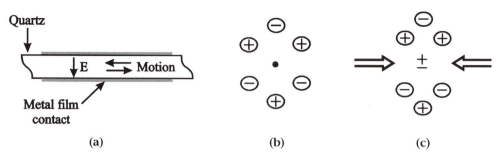

(a)   (b)   (c)

Figure 5.12. (a) Quartz crystal with metal contacts. The thickness needed for AT-cut quartz in the lowest shear mode is approximately 1.67 mm/f, where f is the frequency in MHz. (b) Structure of a material with a piezoelectric effect. The center of balance of the positive and negative charge is the same. (c) The charge movement when a force is applied. The charge movement causes the center of balance for the positive and negative charges to separate. The + indicates the center of positive charge, and the − the center of negative charge.

this corresponds to an error of one second per day for a 10°C change in temperature. This stability is also important for transceivers, because the frequency should not shift when the temperature changes. Figure 5.12a shows the structure. A thin wafer of quartz has evaporated metal film contacts on each side. A voltage between the metal contacts creates a vertical electric field E between the plates, which causes a horizontal *shear* movement in the crystal.

The details of the piezoelectric effect are complicated, but we can say qualitatively why it happens. Inside a solid, different atoms carry different charges. For example, in quartz the oxygen atoms have a net negative charge, and the silicon atoms have a net positive charge. In Figure 5.12b, we show the charged atoms in a triangular arrangement. The center of balance of the positive charges and the center of balance of the negative charges are in the same position, indicated by the dot. When a force is applied, the charge centers separate (Figure 5.12c). This causes a voltage across the material.

In circuit terms, a crystal has both a series and a parallel resonance. Figure 5.13a shows the schematic symbol for a crystal, and Figure 5.13b shows an equivalent circuit. It includes a series RLC circuit. However, this does not represent an electrical effect, but a mechanical one. L and C are called *motional inductance* and

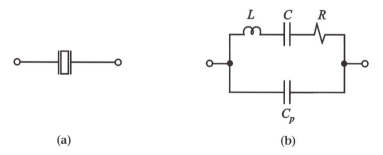

(a)   (b)

Figure 5.13. Schematic symbol for a crystal (a), and the equivalent circuit (b).

motional capacitance to make this clear. Here L accounts for the crystal density, C the stiffness, and R the loss during mechanical vibration. The parallel capacitance C_p, in contrast, is purely electrical. It arises from the capacitance between the two metal contacts, and it is usually a few picofarads. C_p gives the crystal a parallel resonance a few kilohertz above the series resonance. This affects filters that have wide pass bands that extend clear to the parallel resonance. You can ignore it in the filter that you make, which has a pass band of only a few hundred hertz.

5.6 Impedance Inverters

The band-pass filter that we designed requires both series and parallel resonant circuits. This is a problem if we try to make a narrow-band filter, because we need high-Q circuits. We can make excellent high-Q series resonant circuits with quartz crystals, but we do not have equivalent high-Q parallel resonant circuits. However, there are circuits that act as impedance inverters that effectively turn a series resonance into a parallel resonance. This allows us to make a band-pass filter with impedance inverters and series resonant quartz crystals. We have already studied one impedance-inverter circuit. We saw in the last chapter that a quarter-wave transmission line acts as an inverter. However, at our frequencies, a quarter-wave cable would be quite long. Fortunately, we can also make an impedance inverter with inductors and capacitors. This circuit is shown in Figure 5.14.

To see how this circuit works, write the input impedance Z_i as

$$Z_i = jX + \frac{1}{1/(-jX) + 1/(jX + Z_l)}. \tag{5.38}$$

After some arithmetic, this simplifies to

$$Z_i = \frac{X^2}{Z_l}. \tag{5.39}$$

If we define impedances normalized to the inverter reactance X, we can rewrite this as

$$z_i = 1/z_l. \tag{5.40}$$

The normalized input impedance is the inverse of the normalized load impedance. We could also say that the normalized input admittance is equal to the normalized

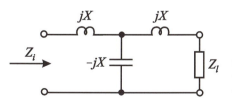

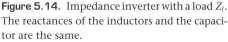

Figure 5.14. Impedance inverter with a load Z_l. The reactances of the inductors and the capacitor are the same.

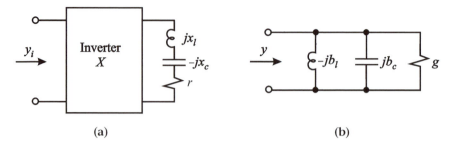

Figure 5.15. Inverter in front of a series resonant circuit (a), and the equivalent parallel resonant circuit (b). All quantities are normalized to the inverter reactance X.

load impedance:

$$y_i = z_l. \tag{5.41}$$

Notice that this kind of formula only makes sense for normalized quantities, because impedance and admittance ordinarily have different units. Also, for the inverter to work, we must be near the frequency where the reactances of the inductors and capacitors are equal. This is a reasonable assumption in crystal filters because the bandwidths are so narrow.

Now consider what happens when we put an inverter in front of a series resonant circuit (Figure 5.15a). We write the input admittance y_i with Equation 5.41 as

$$y_i = jx_l - jx_c + r. \tag{5.42}$$

Now let us compare this with the admittance of the parallel resonant circuit shown in Figure 5.15b. We can write this as

$$y = jb_c - jb_l + g. \tag{5.43}$$

If we compare these formulas, the two circuits are equivalent if

$$b_c = x_l, \tag{5.44}$$

$$b_l = x_c, \tag{5.45}$$

$$g = r. \tag{5.46}$$

This means that the combination of the inverter and the series resonant circuit behaves as a parallel resonant circuit.

Now we can understand the IF Filter in the NorCal 40A. The circuit is shown in Figure 5.16a. It is called a Cohn filter, after the American engineer, Seymour Cohn, who invented it. In the figure, all five capacitors are identical. The shunt capacitors act as impedance inverters. To make a proper impedance inverter we need series inductors to go with the shunt capacitors. We could include the inductors, but we can get an equivalent effect by adding capacitors at each end of the filter. To see how this works, consider that if we add a series combination of an

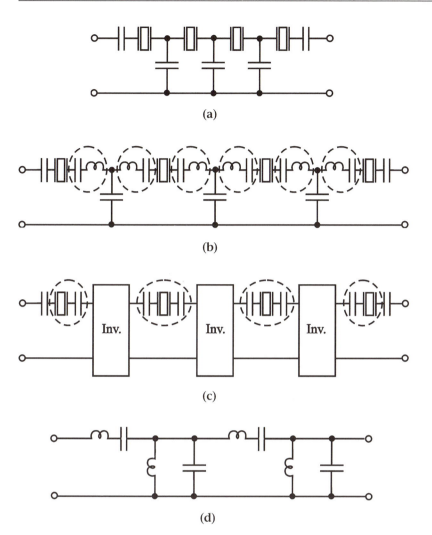

(a)

(b)

(c)

(d)

Figure 5.16. Fourth-order Cohn filter used as the IF Filter in the NorCal 40A (a). Adding series LC pairs on each side of the inverter capacitors (b). Redrawing the circuit to show the inverters (c). The equivalent band-pass filter circuit (d).

inductor and a capacitor with the same reactance, there will be no change in the circuit behavior. In Figure 5.16b, we add these series LC pairs on each side of the inverter capacitors. I have enclosed them in dotted lines to indicate that no real components are added and that the behavior of the circuit does not change. Now we associate the inductors with the shunt capacitors to form inverters, and redraw the circuit in Figure 5.16c. This effectively leaves the two inside crystals with two extra series capacitors, and the two outside crystals with one extra capacitor. This is indicated with the dotted lines. Series capacitors increase the resonant frequency, and to make sure that all the crystals have the same resonant frequency, we add a capacitor at each end. Finally we go through the circuit, removing the inverters and swapping series and parallel circuits, one by one. A crystal inverted once becomes a parallel resonant circuit, a crystal inverted twice is a series circuit, and three times, a parallel circuit. We get the equivalent band-pass filter shown in Figure 5.16d.

FURTHER READING

The *ARRL Handbook*, published annually by the American Radio Relay League, has extensive filter tables. This book also has an excellent discussion of quartz crystals. Wes Hayward's book *Radio Frequency Design*, also published by the American Radio Relay League, gives the formulas for the Butterworth and Chebyshev filter components. Useful information on the hyperbolic functions is given in *Tables of Integrals and Other Mathematical Data*, by Herbert Dwight, published by McMillan. For more depth on the mathematics of filter and inverter design, see *Foundations for Microwave Engineering*, by Robert Collin, published by McGraw-Hill.

PROBLEM 13 – HARMONIC FILTER

The Power Amplifier in the NorCal 40A produces a 7-MHz carrier with 2 watts of power. In addition, the amplifier produces a small amount of power at the harmonic frequencies 14, 21, and 28 MHz. Signals at the wrong frequencies are called *spurious emissions*, or spurs for short. Spurs are bad, because they interfere with other radio services. The FCC sets limits on spurs. For HF transmitters with an output of less than 5 W, each spur must be at least 30 dB below the carrier.

The NorCal 40A has a low-pass ladder filter to reduce the harmonics, consisting of the toroidal inductors L7 and L8 and the disk capacitors C45, C46, and C47 (Figure 5.17). The inductors use the same T37-2 cores as the Transmit Filter. These are the cores with red paint. However, they have only 18 turns, and this lets us use thicker wire (#26 instead of #28) to accommodate the large transmitter currents. Start with a 30-cm piece of wire for each core. Solder in the filter components, leaving the C45 leads partly exposed so that you can attach test hooks. Also solder on the BNC Antenna jack J1. The two small pins are the electrical connections, and the two large pins are the mechanical connections. Solder all four pins to the board.

Attach the function generator across C45 with test hooks, making sure the ground clip is connected to the ground lead of the capacitor. Connect the oscilloscope with a

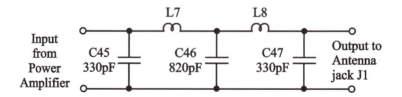

Figure 5.17. NorCal 40A Harmonic Filter.

coaxial cable to the Antenna jack J1. You should use a parallel 50-Ω termination on the scope.

A. Set the function-generator amplitude to 10 Vpp. We do not have a direct measurement of the incident voltage V_{+1}, but it is reasonable to use the amplitude setting on the function generator, 10 Vpp. This makes it convenient to calculate the loss L in dB by the formula

$$L = 20 \log(10/V) \, \text{dB}, \tag{5.47}$$

where V is the peak-to-peak output voltage. Measure the output voltage at 7 MHz and 14 MHz, and express L in dB at these frequencies.

B. From the manufacturer's inductance constant, $A_l = 4.0$ nH/turn^2, calculate the inductance of L7 and L8.

C. Now use *Puff* to simulate the filter response from 0 to 28 MHz (the fourth harmonic). Instructions for installing and running the program are given in Appendix C. The design frequency fd should be 7 MHz. In the F2 Plot Window, set up an s_{21} plot to see the loss. 101 points is sufficient. You should choose the y axis carefully so that the curve does not drop off the bottom. Find the loss in dB at 7 and 14 MHz. Make a screen dump.

In addition to reducing the harmonics, the filter sets the load impedance for the Power Amplifier. The output power of amplifiers often varies inversely with the impedance, so that halving the impedance can double the output power. In addition, having a small inductive component often improves the efficiency, by helping the amplifier approach a Class-E operating condition, where little power is lost in switching the transistor on and off.

D. *Puff* allows you to measure the input impedance of the filter conveniently. Plot the reflection coefficient s_{11}, and move the cursor to the s_{11} line in the F2 Plot Window. Then type =, and the impedance will appear in the Message Box. Find the input impedance of the filter.

E. Assume that we would like to double the output power. You should adjust the components in the filter so that the impedance is cut in half. There are many components that you could change, but to make the problem specific, try varying only L7 and C46. For the capacitor, you should stick to values in the standard 5% series, where the first two digits of the capacitance come from this list: 10, 11, 12, 13, 15, 16, 18, 20, 22, 24, 27, 30, 33, 36, 39, 43, 47, 51, 56, 62, 68, 75, 82, 91. Otherwise you would not be able to buy the capacitors. For the inductor, use only values that you can get by adding or subtracting turns from your cores. What values of L7 and C46 give an impedance closest to half the original impedance?

F. We can improve the harmonic rejection by allowing more ripple. Using the filter table, design a 5th-order, 0.2-dB ripple Chebyshev filter with $f_c = 8$ MHz. Specify the closest 5% capacitor values and the closest number of turns that you can get with T37-2 cores. Simulate your design with *Puff* and make a plot of $|s_{21}|$. What is the loss in dB at 14 MHz?

PROBLEM 14 – IF FILTER

The IF Filter in the NorCal 40A is a 4-element Cohn filter (Figure 5.18). Study the endpaper to see how this filter is connected in the receiver. The filter uses crystals for microprocessor clocks. These are quite inexpensive, costing only about a dollar, but unfortunately, as they come from the dealer, the resonant frequencies are not nearly close enough together to make a good filter. Wilderness Radio sorts crystals for the NorCal 40A so that they match within 20 Hz. You need six matched crystals in all, four for the IF Filter now, and two for mixer oscillators later.

A. First we measure the resonant frequency of one of the crystals with the setup in Figure 5.19. The function generator should be set to a 4,913,500-Hz sine wave with an amplitude setting of 0.5 Vpp. You should set up the function generator so that you can change the frequency in intervals of 1 Hz. Because the crystals have a series

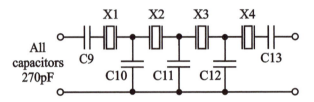

Figure 5.18. The IF Filter in the NorCal 40A.

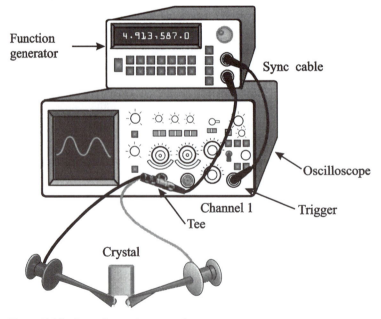

Figure 5.19. Setup for testing crystals.

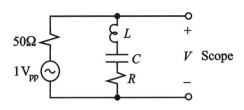

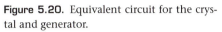

Figure 5.20. Equivalent circuit for the crystal and generator.

resonance, we can recognize the resonant frequency by a dip in the oscilloscope voltage as we vary the frequency. Find the frequency to the nearest hertz that gives the minimum voltage on the scope.

B. Next we will find the components of an equivalent circuit for a crystal, starting with the resistance. Use the equivalent circuit shown in Figure 5.20. Record the output voltage V at resonance and use it to calculate the crystal resistance R.

C. When we shift the frequency off resonance, the scope voltage will increase. Calculate the scope voltage V_x that we would expect when the crystal reactance is equal to R. Notice that this is not simply $\sqrt{2}$ times the minimum voltage, because the crystal resistance is comparable to the resistance of the function generator. Now measure the upper and lower frequencies f_u and f_l that give a scope voltage equal to V_x. Calculate the Q of the crystal from the bandwidth $\Delta f = f_u - f_l$ and the resonant frequency f_0. You need to be careful about the Q here. The crystal Q only includes the resistance of the crystal. It is different from the circuit Q, which also includes the resistance of the generator and is lower because of it. Often people call the crystal Q the *unloaded Q*, and the circuit Q the *loaded Q*.

D. Now calculate the equivalent inductance and capacitance of the crystal. One thing that you need to be careful about here is that we do not have a precise measurement of either L or C individually, but we know their product extremely precisely through the resonant frequency. For one of the components you should use only the number of significant digits that makes sense from your scope measurement, but for the other you will need to use six significant digits, so that the product will give the correct resonant frequency. Check with a calculator that the product of your L and C values gives the resonant frequency correctly to six digits. Otherwise the filter pass band will shift clear off the screen in the *Puff* simulation.

E. Make a model of the Cohn filter with *Puff*, using the equivalent circuit model for the crystal that you have developed and 270-pF capacitors. You should use a range of 2.5 kHz for frequency and 0 to 60 dB for $|s_{21}|$. The design impedance zd should be 200 Ω. Make a plot of $|s_{21}|$.

F. Investigate the effect of changing the port impedance zd to 50 Ω. Make a plot of $|s_{21}|$ and describe the behavior qualitatively.

G. Return the port impedance to 200 Ω, and investigate the effect of changing the capacitors to 200 pF. Make a screen dump and describe the behavior qualitatively.

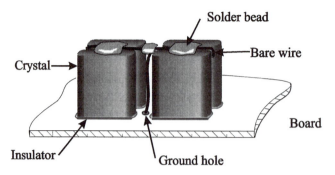

Figure 5.21. Crystal metal cases and the ground connections.

H. In your simulation, return the capacitance to 270 pF. What is the minimum loss in dB in the pass band?

I. One important job of the IF Filter is to reject interference at the upper-sideband frequency, 1,240 Hz above the signal frequency. We hear the upper-sideband frequency as a tone of the same pitch as the signal, and so our ears cannot distinguish the interference from the signal. This is called a *spurious response*. The upper-sideband frequency is a difficult spur to reject, because it is so close to the signal. In the *Puff* simulation, what is the upper-sideband rejection?

Now build the filter. Solder in the 270-pF disk capacitors (C9 through C13). Slide a plastic crystal spacer onto the leads of each of the four crystals, all the way up against the metal case. Now install the filter crystals (X1 through X4) close to the board. The metal cases of the crystals are not connected to the leads, or to any other part of the circuit yet. We say that the cases are *floating*. It is a bad idea to leave large pieces of metal in a circuit floating, because signals can couple capacitively through the metal pieces between different parts of the circuit and end up where you do not want them. To avoid this coupling, we connect each can to ground. There is a small ground hole in the board between the crystals to make this easy. Use bare #22 wire to connect the crystal cans. Figure 5.21 shows how you can do this. Connect the cans with a wire running along the top. It may help to gently bend the cans toward each other until the space between them is small. You should use large solder beads, and make sure that the top of the cans get quite hot so that the solder beads stick well to the cans. If the cases are not hot enough, the wire and solder will pop off the cans. Then solder a wire to the ground hole, hook the other end to the top wire, and solder them together.

The filter is designed for a 200-Ω generator and load. We will add resistors to give the function generator and scope this resistance (Figure 5.22). For the load, solder a 200-Ω resistor from the left L4 hole (connecting to C13 at the filter output) to the left C14 hole, which is a ground connection. Connect the scope across the 200-Ω resistor. The scope connection should be as short as possible, or else the capacitance of the cable will affect the shape of the filter response. The best thing to do is to use a BNC barrel adapter to connect directly to channel 1, and to let the board dangle off the front of the scope. For the function-generator connection, solder one end of a 150-Ω resistor to

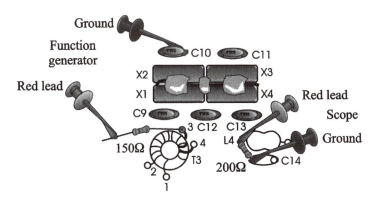

Figure 5.22. Resistor connections to the crystal filter.

the number-3 hole of T3. Attach the function-generator red lead to the other end of the 150-Ω resistor (Figure 5.22). The ground lead can be attached to C10 on the ground side.

J. With an amplitude setting of 0.5 Vpp, measure the minimum loss in dB of the filter to compare with the *Puff* simulation.

K. Next we make a plot of the loss in dB versus frequency. Because we will need to measure very small signals, it is a good idea to switch in a 10-MHz low-pass filter on the oscilloscope if one is available. Much of the noise that blurs the scope trace is at frequencies greater than 10 MHz, and so this will make the trace sharper at low voltage levels. It does, however, reduce the reading somewhat even at 4.9 MHz; thus our plot will be a relative plot. Increase the function-generator amplitude setting to 2.0 V to get a bigger signal. Even though this increases power, it is safe because the power is no longer going into a single crystal, but rather divides between the four crystals and the resistors. Measure the output voltage V over a 2,500-Hz bandwidth centered on the pass band. You should plot the loss L relative to the maximum voltage V_m in dB, by the formula

$$L = 20 \log(V_m/V)\,\text{dB}. \tag{5.48}$$

Use a 60-dB scale, with 0 dB at the top. Use judgment in choosing the frequency intervals. Often 50 Hz is a good spacing in the pass band, and 100 Hz is a good spacing in the stop band. You may need to increase the bandwidth beyond 2,500 Hz if the pass band is not centered in your plot. What is the upper-sideband rejection that you measure? When you have finished the plot, remove and discard the two resistors and remove the solder from the holes with solder wick.

After the signal passes through the IF Filter, it goes to the Product Detector, which converts the signal to a 620-Hz audio signal. The product detector is based on an integrated circuit, or IC, made by Philips, the SA602AN. We will have much more to say about the SA602AN later, because it is the most important IC in the transceiver. We use three of them: the Product Detector and the RF Mixer in the receiver and the Transmit Mixer in the transmitter. The SA602AN has a large input impedance, listed in the data sheets as

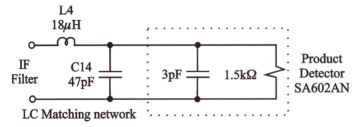

Figure 5.23. LC matching network for connecting the IF Filter to the SA602AN.

1.5 kΩ shunted by 3 pF. This is a bad load impedance for the crystal filter, which should see about 200 Ω. The NorCal 40A has an LC circuit (L4 and C14) that transforms the input of the SA602AN to near 200 Ω (Figure 5.23).

L. Calculate the resistance R and the reactance X that the matching network and the SA602AN present to the IF Filter. Notice that the result is not precisely 200 Ω. Our choice of components is limited to the values that a manufacturer makes. If you could specify any value for L4 and C14, what values would you use to transform the input impedance of the SA602AN to 200 Ω? Solder L4 and C14 into the circuit.

6

Transformers

Previously we have made inductors by wrapping a coil on a toroidal core. We make a *transformer* by adding another coil. Transformers are valuable in radio circuits because they can do several different jobs at once. A transformer blocks DC voltages but transmits AC. It can step up a voltage or current and change the impedance levels of loads to eliminate reflections. Finally, a coil on a transformer has an inductance that can be used in a resonant circuit as a filter.

6.1 Inductance Formulas

To start, let us see how an inductor works. Consider a toroidal core with a single loop of wire (Figure 6.1). We let the current in the wire be I and the voltage across the ends be V. In a magnetic material like iron or ferrite, the current produces a large magnetic flux Ψ (the Greek capital letter *psi*) around the toroid. The units of flux are volt·seconds (Vs). We will assume that the flux is proportional to the current. This is reasonable as long as the current is not too large and the toroid is not permanently magnetized. We write

$$\Psi = A_l I, \tag{6.1}$$

where A_l is called the *inductance constant*. A_l is the inductance of a single turn, but we can also use it to calculate the inductance of larger coils. In addition, Faraday's law says that the voltage V is the time derivative of the flux,

$$V = \frac{d\Psi}{dt}. \tag{6.2}$$

This formula takes getting used to, because it says that in a loop of wire, where we would ordinarily expect no voltage between the ends because the wire is continuous, there will be a voltage if the magnetic field through the loop changes with time. We can rewrite this formula in terms of phasors:

$$V = j\omega\Psi. \tag{6.3}$$

We can combine the two formulas to give the relation

$$V = j\omega A_l I. \tag{6.4}$$

This is the relation between voltage and current for an inductor.

Figure 6.1. A toroidal core with one turn.

Consider what happens if we add more loops. Each additional loop will act like the first in producing flux. We can write the total flux as

$$\Psi = NA_l I. \tag{6.5}$$

Moreover, the flux produces a voltage in each loop. Because the loops are connected in series, the total voltage is proportional to the number of loops, and we can write

$$V = Nj\omega\Psi. \tag{6.6}$$

We combine these two formulas and get

$$V = j\omega N^2 A_l I. \tag{6.7}$$

The inductance is given by

$$L = N^2 A_l. \tag{6.8}$$

This is the formula that we have been using. The inductance is proportional to the square of the number of turns. The inductance constant A_l is given in manufacturer's data sheets for different core sizes and materials. Larger cores have larger inductance constants. However, it is important to realize that A_l can change drastically with frequency. It is also somewhat dependent on the number of turns and how they are spread. Data for the cores in the NorCal 40A are given in Appendix D. There are two iron-powder cores and two nickel–zinc ferrite cores. The iron-powder cores have lower inductance constants that are suitable for inductors. The nickel–zinc ferrites have higher inductance constants that are better for transformers.

6.2 Transformers

Now consider a toroid with two coils (Figure 6.2a). This is called a *transformer*. The coils are called *primary* and *secondary* to distinguish them. It is not always clear which is which, although usually you can think of the primary coil as the input coil and the secondary as the output. We denote the primary coil by the subscript p and the secondary by the subscript s. A voltage will be induced in each coil by the changing flux. We can calculate the voltages using Faraday's law again. We write

$$V_p = N_p j\omega\Psi, \tag{6.9}$$
$$V_s = N_s j\omega\Psi, \tag{6.10}$$

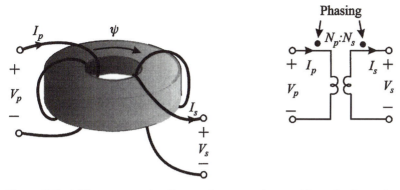

Figure 6.2. Adding a second coil to make a transformer (a), and schematic symbol (b).

where N_p is the number of turns in the primary and N_s the number of turns in the secondary. We can write V_s in terms of V_p if we take the ratio of these two formulas:

$$V_s = \frac{N_s}{N_p} \cdot V_p. \tag{6.11}$$

This says that the voltage ratio is the same as the turns ratio. The schematic symbol for a transformer is a pair of facing coils (Figure 6.2b). There is a choice in how the coils are wound – clockwise or counterclockwise. This determines the sign of the secondary voltage. If the sign is important, people add black phasing dots at the positive voltage terminals.

Calculating currents is more complicated than calculating voltages. We have to be careful about the directions. The usual convention is that the primary current I_p is positive if it is *entering* the transformer, and the secondary current I_s is positive if it is *leaving*. The total flux is the sum of the fluxes produced by each coil. We can write this as

$$\Psi = N_p A_l I_p - N_s A_l I_s. \tag{6.12}$$

There is a minus sign because the directions of I_p and I_s are different. We can rewrite this relation as

$$I_p = \frac{\Psi}{N_p A_l} + \frac{N_s}{N_p} I_s. \tag{6.13}$$

We rewrite the flux term with Equation 6.9 to give

$$I_p = \frac{V_p}{j\omega L_p} + \frac{N_s}{N_p} \cdot I_s, \tag{6.14}$$

where L_p is the inductance of the primary coil. The term $V_p/(j\omega L_p)$ is the current that the primary coil would draw if there were no secondary. It is called the *magnetizing current*. The term $(N_s/N_p)I_s$ is the *transformer current*, and it is controlled by the turns ratio. We now consider the limit where the magnetizing current is small.

6.3 Ideal Transformers

Assume that the primary inductance L_p is quite large so that we can neglect the magnetizing current. We can rewrite the voltage and current equations (Equations 6.11 and 6.14) as

$$V_s = \frac{N_s}{N_p} \cdot V_p \tag{6.15}$$

and

$$I_s = \frac{N_p}{N_s} \cdot I_p. \tag{6.16}$$

These are the voltage and current relations for an *ideal transformer*. These equations mean that we can step up the voltage or current, depending on the turns ratio. If $N_s > N_p$, the transformer will step up the voltage. If $N_s < N_p$, the transformer will step up the current. You might think about why we cannot step up both the voltage and current at the same time. If we take the product of the two equations, we get

$$V_s I_s = V_p I_p, \tag{6.17}$$

which says that the power that comes out of the secondary coil is the same as the power we put in the primary coil. If both voltage and current increased, the output power would be bigger than the input power, which is not possible, because the transformer has no additional source of power. In the NorCal 40A, we use the transformer T1 to step up the current from the Driver Amplifier to the Power Amplifier. The two coils have 14 turns and 4 turns, and so the current increases by a factor of 3.5.

Since the voltage steps up when the current goes down, and vice versa, a transformer changes the impedance of a load. To see this, consider Figure 6.3, which shows a transformer together with a load. If we take the ratio of Equation 6.15 and Equation 6.16, we get

$$\frac{V_s}{I_s} = \left(\frac{N_s}{N_p}\right)^2 \cdot \frac{V_p}{I_p}. \tag{6.18}$$

We can rewrite this in terms of the load impedance $Z_s = V_s/I_s$ and the primary impedance $Z_p = V_p/I_p$ as

$$Z_p = \left(\frac{N_p}{N_s}\right)^2 Z_s. \tag{6.19}$$

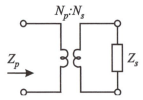

Figure 6.3. Ideal transformer with a load.

In words, the impedance changes by the square of the turns ratio. Transformers are commonly used in matching for maximum power transfer. For example, in the NorCal 40A, the RF Mixer, which has a source resistance of 3 kΩ, is the source for the IF Filter, which has an impedance of 200 Ω. This means that there will be a large mismatch loss without a transformer. We use the transformer T3 with a primary of 23 turns and a secondary of 6 turns. This gives a primary impedance Z_p given by

$$Z_p = \left(\frac{23}{6}\right)^2 \cdot 200\,\Omega = 2.9\,\text{k}\Omega, \tag{6.20}$$

which is quite close to 3 kΩ.

6.4 Magnetizing Current

Now we consider the effect of the magnetizing current. Rewriting the voltage and current equations (Equations 6.11 and 6.14) with the magnetizing current included, we get

$$V_p = \frac{N_p}{N_s} \cdot V_s, \tag{6.21}$$

$$I_p = \frac{V_p}{j\omega L_p} + \frac{N_s}{N_p} \cdot I_s. \tag{6.22}$$

Figure 6.4a shows an equivalent circuit, which includes an inductor L_p in parallel with an ideal $N_p{:}N_s$ transformer. It is important to think through why this equivalent circuit works. In Equation 6.22, the primary current I_p is made up of two parts, the magnetizing current and the transformer current. The magnetizing current is the current for an inductor L_p with voltage V_p. We can calculate L_p from the inductance constant A_l and the number of turns N_p by the usual formula:

$$L_p = N_p^2 A_l. \tag{6.23}$$

The transformer current is that for an ideal $N_p{:}N_s$ transformer with a primary voltage V_p. These relations, where the voltages of the two components are the same, but the current is the sum of the currents for each component, are those

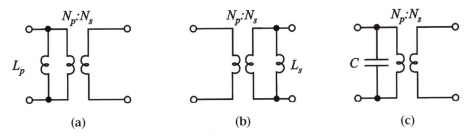

Figure 6.4. The equivalent circuit for a transformer, including the magnetizing current (a). Alternative equivalent circuit for a transformer with a shunt inductor L_s on the secondary side (b). A tuned transformer (c).

for components in parallel. I have put the magnetizing current on the primary side, but we could also have started from the secondary side, and ended up with a magnetizing current on the secondary side. The easy way to do this is to pretend that we are using the secondary as the input, and swap the subscripts. In the equivalent circuit (Figure 6.4a), we use the secondary inductance L_s, given by

$$L_s = N_s^2 A_l,$$

(6.24)

and put it on the secondary side.

Including the magnetizing current in the equivalent circuit raises several issues. First, at low frequencies, the shunt inductor will short out the ideal transformer. This means that the transformer acts as a high-pass filter and that it will isolate DC circuits. Second, we can make a band-pass filter by adding a capacitor to resonate the shunt inductance. A transformer with a tuning capacitor is called a *tuned transformer*. For example, the NorCal 40A uses a shunt capacitor (C6) with the matching transformer T3 to resonate out the transformer inductance. A tuned transformer acts as a band-pass filter for removing interference at the same time that it changes impedance levels for matching. In other situations, we may not need the band-pass filter, but it may be difficult to get a large enough inductance to achieve the full current or voltage step up. The tuning capacitor allows us to resonate the inductance at the frequency we are interested in, so that the ideal transformer equations apply.

FURTHER READING

A good source of practical information on inductors and transformers is *Secrets of RF Circuit Design*, by Joseph Carr, published by McGraw-Hill. Carr also discusses the iron and ferrite materials used for the cores.

PROBLEM 15 – DRIVER TRANSFORMER

In the NorCal 40A, the transformer T1 couples the Driver Amplifier to the final Power Amplifier. The Power Amplifier requires a substantial drive current, and we use the transformer to provide a current step-up. In addition, the transformer isolates the DC voltages and currents in the Driver Amplifier from the Power Amplifier. This makes it possible to connect the output collector of the Driver Amplifier to the supply and the input base of the Power Amplifier to ground at the same time.

The core we use for this transformer is an FT37–43. It has an orange dot to identify the core material as #43 ferrite. The inductance constant for the #43 mix varies greatly with frequency, dropping by almost a factor of three as the frequency increases from 100 kHz to 7 MHz.

Carefully construct the transformer T1 as shown in Figure 6.5. Start by cutting 10-cm and 25-cm lengths of #26 wire for the coils. It is better to wrap the long coil first and to check the turns before you start the short coil. Count carefully as you wrap the second coil, because it is not easy to check the count when there are two coils. Transformers cause more problems in construction than any other component. Make sure that each

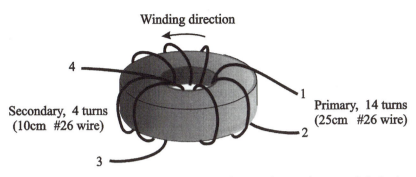

Figure 6.5. Wiring for T1. Not all turns are shown. The numbers match holes in the printed-circuit board.

lead goes to the correct hole and that the wires are properly stripped before you solder them in. One way to check that the insulation is off the wires is to coat the wires with a thin layer of solder before you insert the transformer into the circuit. If the insulation is still there, the solder will not coat the wire smoothly but will form a bead. The transformer should lie flat after it is soldered in.

Install R14 (100 Ω). Leave the resistor a few millimeters above the board surface so that you can attach leads from the oscilloscope. Make oscilloscope connections to R14 with a 50-Ω termination. Solder one end of a 1 k-Ω resistor to the large middle hole in the S1 outline and one end of a 200-Ω resistor into the Q6 hole that is closest to the transformer (Figure 6.6). The other ends of these resistors are the input leads for the function generator. The function generator should be set for a 5-Vpp, 7-MHz sine wave. Figure 6.7 shows the circuit.

A. Measure the output voltage V.

B. Now calculate V from the circuit diagram.

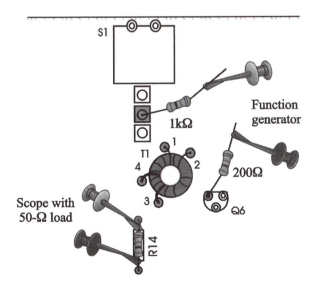

Figure 6.6. Connections for measurements on T1.

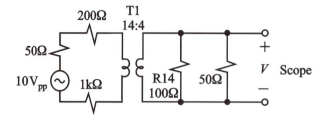

Figure 6.7. Circuit for measurements on T1.

C. Measure the 3-dB low-frequency cut-off f_c.

D. Use your measurement of f_c to deduce the inductance constant A_l, taking all the resistances into account. Note that it may be considerably greater than the 7-MHz value of 160 nH/turn^2 given in Appendix D. After you finish, remove the 1-kΩ and 200-Ω resistors and clean the holes with solder wick.

PROBLEM 16 – TUNED TRANSFORMERS

The NorCal 40A has two tuned transformers, T2 and T3, to match impedances at the input and output of the RF Mixer. Study the endpaper to see how these transformers fit into the circuit. Both transformers use the ferrite core FT37–61, with $A_l = 66$ nH/turn^2. These cores are not painted. T2 combines with the series resonant circuit we studied in Problem 8 to make a 2-element Butterworth band-pass filter at 7 MHz. This is the RF Filter. In addition, the transformer steps up the 50-Ω cable impedance to 1.5 kΩ to match the input of the RF Mixer. T3 is at the output of the RF Mixer. It steps down the 3-kΩ output resistance of the RF Mixer to match the 200-Ω input impedance of the IF Filter at 4.9 MHz.

For T2, start with a 35-cm section of #26 wire and wrap 20 turns (Figure 6.8). For the primary, it is convenient to use a single loop of bare #22 wire. Install T2, the variable capacitor C2, and C4 (5 pF). We also need to connect a temporary 1.5-kΩ resistor to act as a load in place of the RF Mixer U1. The resistor should be soldered to the #1 and #3 holes in U1 (Figure 6.9). These holes are numbered starting at the round solder pad in the lower corner, and proceeding counterclockwise. Attach a 10:1 scope probe across the resistor. Note that the #3 hole is the ground. For an input connection, solder a short piece

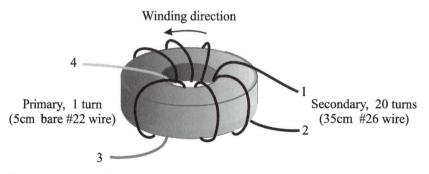

Figure 6.8. Wiring for T2. Not all turns are shown. The numbers match holes in the printed-circuit board.

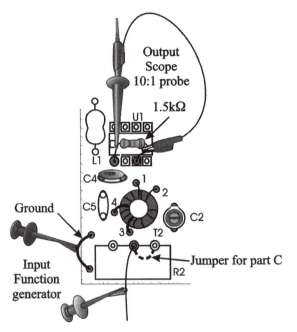

Figure 6.9. Connections for measurements on T2.

of bare #22 wire to the center hole of R2 to attach a lead from the function generator. For a ground connection, solder a loop of bare #22 wire to the two small holes on the edge of the board next to the R2 outline.

A. Set the function generator for a 0.5-Vpp, 7-MHz sine wave. Adjust C2 for maximum output. Find the ratio of the power absorbed by the load P to the available power P_+, and express the loss in dB. Measure the 3-dB bandwidth.

Now we will make a model for the transformer to use in *Puff* (Figure 6.10). The model includes a shunt inductor and an ideal 20:1 transformer. In *Puff*, the ideal transformer would be listed in the Parts Window as x 20, where x indicates a transformer and 20 is the turns ratio expressed as a decimal number. It is important to draw the transformer in the correct direction. It is in the shape of a trapezoid. For a turns ratio greater than one, the wide side is the high-voltage side.

Make sure you include C4, C2, and the capacitance of the 10:1 probe C_p. The output is tricky, because *Puff*'s ports have an impedance equal to zd, which is 50 Ω in this

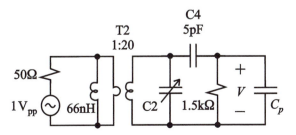

Figure 6.10. Circuit model for T2.

problem. However, if you include an ideal transformer part x 5.5, where 5.5 is the square root of the impedance ratio 1500/50, you can connect it between the filter output and Port 2 to give an impedance of 1.5 kΩ.

B. In the computer model, adjust C2 to give the maximum value of $|s_{21}|$ at 7 MHz. Compute the 3-dB bandwidth from the computer model. Make a screen dump of $|s_{21}|$.

C. Return to your circuit board. Join the tuned transformer to the series resonant circuit, L1 and C1. You can do this by connecting your input wire as a jumper between the center and right holes in R2 (Figure 6.9). The function generator should be connected through the Antenna jack J1. Adjust C1 and C2 to give a maximum output voltage V at 7 MHz. What is the combined loss of the Harmonic Filter and the RF Filter in dB? You should make a note of this loss for the future. We will need it to analyze the receiver performance. If the loss is greater than 7 dB, something is likely to be wrong. You might try tuning C1 and C2 again carefully. If this does not work, you might check the solder joints, and make sure that the coil leads are in the correct holes.

D. A major purpose of the RF Filter is to remove the VFO image. Without the RF Filter, this frequency would be received just as well as the desired signal at 7 MHz. The image frequency f_{vi} is given by Equation 1.38 as

$$f_{vi} = f_{if} - f_{vfo} = 4.9\,\text{MHz} - 2.1\,\text{MHz} = 2.8\,\text{MHz}. \tag{6.25}$$

Find the image rejection ratio R_i in dB, using the formula

$$R_i = 20\log(V_{rf}/V_{vi})\,\text{dB}. \tag{6.26}$$

The image response will be small, and you should increase the function generator setting to 10 Vpp for this measurement. In addition, you will need to switch in the oscilloscope's low-pass filter. You might notice that at the image frequency, the output may not be a pure sine wave because of harmonic content. Your filter rejects the image at 2.8 MHz much better than the harmonics at 5.6 MHz and 8.4 MHz. These components are usually present at a low level in the output of a function generator, but they are made more prominent by the filter.

Remove the temporary 1.5-kΩ load resistor and the jumper in R2 and clean the holes with solder wick. Turn off the low-pass filter on the scope so that it will not throw off later measurements.

E. Extend your *Puff* model to include the series resonant circuit L1 and C1. What does the model predict for R_i?

The NorCal 40A has one more transformer, T3, that connects the RF Mixer to the IF Filter. Like T2, this core is an FT37–61. The primary has 23 turns of #28 wire and the secondary has 6 turns of #26 wire (Figure 6.11). You should start by cutting a 40-cm section of #28 wire and a 15-cm section of #26 wire for the coils. Construct and install T3.

F. At the IF frequency, 4.9 MHz, what capacitance would be needed to tune the transformer on the primary side? on the secondary side?

Install the 47-pF tuning capacitor C6. Do not be concerned if C6 is different from what you calculated. The designer said he chose 47 pF because the radio sounds best with that value!

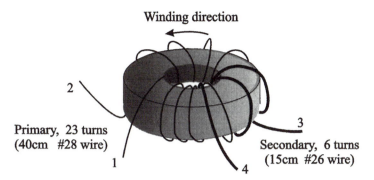

Figure 6.11. Wiring for transformer T3. Not all turns are shown. The numbers match holes in the printed-circuit board.

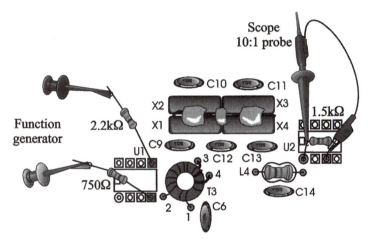

Figure 6.12. Input and output connections for measuring the loss of the complete IF-Filter network.

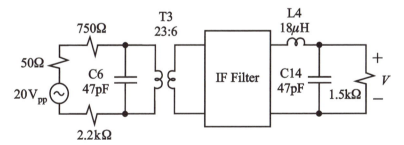

Figure 6.13. Circuit for measuring the loss of the complete IF-Filter network.

For the input connections, solder a 750-Ω resistor in the #4 hole in U1 and a 2.2-kΩ resistor in the #5 hole (Figure 6.12). Connect the function generator to the resistors. The combination of these resistors and the 50-Ω generator resistance gives us 3 kΩ to match the RF Mixer (Figure 6.13). The output connections will be like those for the RF Filter. Solder a 1.5-kΩ resistor between the #1 and #3 holes in U2. Attach the scope across the resistor with a 10:1 probe.

G. Use a 10-Vpp setting on the function generator, and adjust the frequency for maximum scope voltage. Calculate the loss as

$$L = 10 \log(P_+/P) \, dB, \tag{6.27}$$

where P_+ is the power available from the 3-kΩ source and P is the power delivered to the 1.5-kΩ load. Save this number for analyzing the receiver performance later. If the loss is greater than 10 dB, something is likely to be wrong. You might check the solder joints, and make sure that the coil leads are in the correct holes. Remove the resistors and clean the holes with solder wick when you are finished.

Acoustics

Sounds are pressure waves. When an object moves suddenly in air, the air is compressed. This compression causes a change in pressure. The pressure change results in a push on the surrounding air, which causes it to move, disturbing air that is farther away. The result is a disturbance that propagates away from the object. These pressure changes are ordinarily quite small – the changes in a normal conversation are only about a millionth of atmospheric pressure.

7.1 Equations of Sound

We can derive a formula for the speed of sound in terms of the basic properties of air. The equations turn out to be similar to the transmission-line equations, with pressure playing the role of voltage and average velocity the role of current. We start by considering the effect of a pressure disturbance on a section of air of length l (Figure 7.1a). We will write the pressure as $P(z, t)$, where z is the distance and t is the time. If the length l is small, then we can write the difference in pressure between the left and right section approximately as $\frac{\partial P}{\partial z}l$. The pressure difference across the section of air will cause the section to accelerate, because there is more force on one side than the other. We can find the acceleration from Newton's second law,

$$F = ma, \tag{7.1}$$

where F is the force in newtons, m is the mass in kilograms, and a is the acceleration in m/s^2.

In considering acceleration due to pressure, it is convenient to think in terms of the mass per unit area, which we can write as ρl, where ρ is the volume density. We get

$$\frac{\partial P}{\partial z}l = -\rho l \frac{\partial U}{\partial t}, \tag{7.2}$$

where U is the average velocity. We cancel l and write

$$\frac{\partial P}{\partial z} = -\rho \frac{\partial U}{\partial t}. \tag{7.3}$$

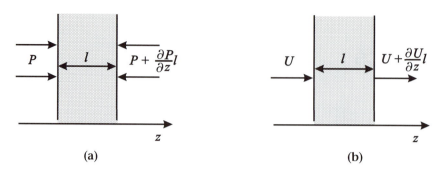

Figure 7.1. Effect of varying the pressure (a) and the velocity (b) across a section of air of length l.

This corresponds to the first transmission-line equation

$$\frac{\partial V}{\partial z} = -L\frac{\partial I}{\partial t}. \tag{7.4}$$

Thus density plays the same role as distributed inductance, and we can write the following correspondences:

$$V \Longleftrightarrow P, \tag{7.5}$$

$$I \Longleftrightarrow U, \tag{7.6}$$

$$L \Longleftrightarrow \rho. \tag{7.7}$$

Deriving the second equation is more difficult. We will consider the effect of a velocity variation across a section of air of length l (Figure 7.1b). If the velocity at each side is different, the length will change with time. For small l, we can write

$$\frac{dl}{dt} = l\frac{\partial U}{\partial z}. \tag{7.8}$$

We can relate the length l to P. For a gas, the pressure P and volume V are related by a power law,

$$PV^\gamma = \text{constant}, \tag{7.9}$$

where γ is an experimental constant, equal to 1.403 for air at room temperature and atmospheric pressure. This is an *adiabatic* relation, which means that the heat does not have time to flow from a compressed region to an expanded region. We can write the relation between l and P as

$$l = \alpha P^{-1/\gamma}, \tag{7.10}$$

where α is a constant. We are assuming here that the gas only compresses and expands in the z direction, and not in the x and y directions. We can write the derivative $\frac{dl}{dP}$ as

$$\frac{dl}{dP} = -\frac{\alpha P^{-1/\gamma - 1}}{\gamma} = -\frac{l}{\gamma P}. \tag{7.11}$$

Now we can find $\frac{dl}{dt}$ by the chain rule as

$$\frac{dl}{dt} = \frac{dl}{dP} \cdot \frac{dP}{dt} = -\frac{l}{\gamma P} \cdot \frac{dP}{dt}. \tag{7.12}$$

If we substitute for $\frac{dl}{dt}$ in Equation 7.8, we get

$$l\frac{\partial U}{\partial z} = -\frac{l}{\gamma P} \cdot \frac{\partial P}{\partial t}. \tag{7.13}$$

If we divide by l, we get

$$\frac{\partial U}{\partial z} = -\frac{1}{\gamma P} \cdot \frac{\partial P}{\partial t}. \tag{7.14}$$

This formula is difficult to interpret. It is analogous to the second transmission-line equation

$$\frac{\partial I}{\partial z} = -C\frac{\partial V}{\partial t}. \tag{7.15}$$

One difficulty with Equation 7.14 is that pressure appears twice, once in the denominator and once in a derivative. This means that the equation is nonlinear. However, at ordinary sound levels, the pressure changes are small enough that we can consider the P in the denominator to be constant. We can then see that the quantity that corresponds to distributed capacitance is

$$C \Longleftrightarrow \frac{1}{\gamma P}. \tag{7.16}$$

We also need to understand that U is the average velocity of the gas molecules at some position. The gas molecules have additional random motion that may be much larger than the average velocity. U is different from the velocity of sound, which is a constant. The average velocity U actually oscillates at the frequency of the sound.

By analogy with transmission-line equations, we write the sound velocity v as

$$v = \sqrt{\gamma P / \rho}. \tag{7.17}$$

The units of pressure are pascals (Pa), where a pressure of 1 Pa exerts a force of 1 newton per square meter. Atmospheric pressure at room temperature and sea level is 101 kPa. The density ρ has units of kg/m^3. At room temperature and sea level, the density of air is $\rho = 1.20$ kg/m^3. If we substitute for γ, P, and ρ in the formula, we find that v is 344 m/s. This is a million times slower than the speed of radio waves, which propagate at the speed of light. We relate frequency and wavelength in the same way as before, except that sound wavelengths are much shorter. Our NorCal 40A filters are tuned for an output audio frequency of 620 Hz. On a piano keyboard, this is near D$\sharp$ (622 Hz) in the second octave above middle C. At 620 Hz, the wavelength λ is given by

$$\lambda = v/f = 55 \text{ cm}. \tag{7.18}$$

There are other analogies between sound and transmission-line waves. We define the characteristic impedance Z_0, given by

$$Z_0 = \sqrt{\rho \gamma P} = v\rho. \tag{7.19}$$

Sound obeys the same reflection formulas as transmission-line waves. The product of the pressure P and velocity U is power density, with units of W/m². We write the power density as S:

$$S = PU. \tag{7.20}$$

The intensity of sound is given by the *sound pressure level*, which is written as L_p, according to the formula

$$L_p = 20 \log(P/P_0) \, \text{dB}, \tag{7.21}$$

where P is the pressure amplitude and P_0 is a reference pressure. These are not for the background atmospheric pressure, just the sound pressure itself. The reference pressure P_0 is 20 μPa, which is close to the lowest sound level that we can hear. We can also write P_0 in terms of the power density S as

$$S = P_0^2/Z_0 = 970 \, \text{fW/m}^2. \tag{7.22}$$

This means that a sound pressure level $L_p = 0$ dB gives a power density quite close to 1 pW/m².

7.2 Hearing

Our ears are remarkable for the range of sound pressure levels they can accommodate. The ratio between the minimum pressure level that we can hear and the pressure level that causes pain is 1 million, or 120 dB. Table 7.1 shows the pressure levels for various sounds. Our perception of loudness is not linear. A pressure level that is 10 dB greater than another is perceived as about twice as loud. We say that our ears have a *logarithmic* response, which means that differences in loudness depend on the ratio of pressures. This makes a dB scale convenient for discussing hearing. We are not particularly sensitive to differences in loudness. It is difficult to perceive changes in L_p that are less than 0.5 dB. This corresponds to a power change of 10%, which is easy to measure with instruments.

We can hear sounds with frequencies over a wide range from 20 Hz to 15 kHz. We perceive a frequency change as a pitch change, and we can distinguish pitch differences of about 3 Hz. Our perception of loudness depends both on the sound pressure level and frequency. This is shown in Figure 7.2. Our ears are most sensitive in the frequency range from 500 to 6,000 Hz. As the frequency drops below 500 Hz, our ears become considerably less sensitive. Hearing acuity decreases with increasing age, particularly for frequencies above 1 kHz. The drop is about 1 dB per year, starting around age 40.

The solid lines in Figure 7.2 show contours of constant loudness. The loudness level is expressed in *phons* (pronounced to rhyme with "johns"), which are marked

Table 7.1. Pressure Levels and Power Densities for Typical Sounds.

Sound	Loudness	L_p	Power Density
		0 dB	1 pW/m^2
rustling leaves	barely audible	10 dB	10 pW/m^2
broadcast studio		20 dB	100 pW/m^2
bedroom at night	quiet	30 dB	1 nW/m^2
living room		40 dB	10 nW/m^2
classroom	moderate	50 dB	100 nW/m^2
conversation at 1 m		60 dB	1 μW/m^2
truck interior	noisy	70 dB	10 μW/m^2
city street		80 dB	100 μW/m^2
heavy truck	very noisy	90 dB	1 mW/m^2
shout at 1 m		100 dB	10 mW/m^2
jackhammer	intolerable	110 dB	100 mW/m^2
jet takeoff at 50 m		120 dB	1 W/m^2

Source: Adapted from Table 6.1 in *The Science of Sound*, 2nd Edition, by Thomas D. Rossing, Copyright 1990, by Addison-Wesley Publishing Company, Inc. Reprinted by permission.

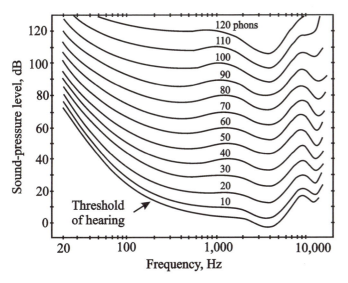

Figure 7.2. Loudness contours for people in the age range from 18 to 25. The bottom curve is the threshold of hearing. From Robinson and Dadson, *British Journal of Applied Physics*, volume 7, p. 166, 1956.

on each contour. The contours indicate combinations of pressure and frequency that are equally loud to the ear. The phon unit uses the sound pressure levels at 1 kHz as a reference. The top contour at 120 phons is the level where the sound begins to become painful. One interesting feature about the loudness contours is that they bunch together at low frequencies. This presents challenges in recording music, because we usually listen to music at a lower level than it is recorded at. In particular, you may need to boost the bass to compensate for this bunching.

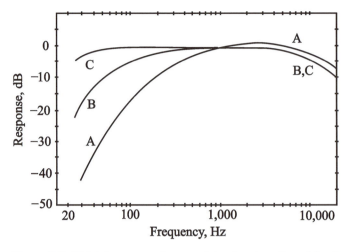

Figure 7.3. Weighting curves for sound level meters. Sound level meters usually allow a choice of A and C weighting. From Figure 1.11 in *Handbook of Recording Engineering,* by John Eargle, published by Van Nostrand Reinhold. Used by permission.

The loudness contours are also taken into account in designing offices and lecture halls. Greater sound pressure levels are allowed at lower frequencies than at high frequencies, because we are less sensitive to noise at low frequencies.

Sound-level meters take these loudness contours into account (Figure 7.3). The A curve is like the 40-phon loudness contour, but turned upside down. With A-weighting, a meter reading gives an indication of loudness. This is useful in evaluating background noise levels for offices and classrooms. Federal agencies also use the A-weighting to set workplace noise limits. For example, in the United States, OSHA (Occupational Safety and Health Act) regulations set the maximum continuous exposure level at 90 dB with A-weighting. In contrast, the C curve is quite flat. This is useful for comparing pressure levels at different frequencies in the lab.

7.3 Masking

How well a message is understood depends on the ratio of signal power to noise power. This power ratio is called the *signal-to-noise ratio,* and it is traditionally given in dB. Your transceiver is designed for receiving Morse Code at low power levels. Morse Code transmissions have a series of short and long pulses (*dits* and *dahs*). The receiver converts these pulses to 620-Hz audio tones. Your ear is quite good at rejecting noise and other interfering signals, as long as they fall outside of a 150-Hz band around 620 Hz. If the interfering signal is closer than this, your ears tend to perceive the two signals as a single tone. The bandwidth over which this fusing occurs is called the *critical* bandwidth. Noise within the critical bandwidth has the effect of hiding the signal. This is called *masking*. Masking occurs if the

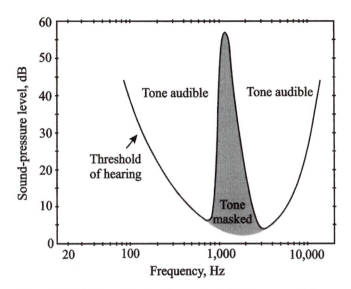

Figure 7.4. Masking of a signal tone by 1,200-Hz noise at $L_p =$ 60 dB. Adapted from Figure 3.6 in *Acoustic Noise Measurements* by Jens Broch, published by B & K Instrument Company.

noise power within the critical band is comparable to the signal power. The plot in Figure 7.4 shows the masking effect of 1,200-Hz noise at a pressure level of 60 dB. At frequencies that are far away from the noise, we can detect the signal tone right down to the normal threshold of hearing. As the frequency comes closer to 1,200 Hz, stronger and stronger tones are masked and cannot be heard. Right at 1,200 Hz, signal tones can be detected if they have sound pressure levels of 57 dB or larger. This is 3 dB below the noise. This means that Morse Code can be received when the signal power is comparable to the noise power within the critical bandwidth. By contrast, an AM or TV broadcaster aims at 50-dB signal-to-noise ratios over much larger bandwidths to provide a clear sound and picture. This means that the NorCal 40A can communicate at distances of thousands of miles with a transmitter power of 2 W, whereas a broadcaster might use 50 kW for coverage in a single city.

7.4 rms Voltages

In the next problem, you measure an AC voltage with a multimeter. The multimeter measures the *root-mean-square*, or rms, voltage rather than peak-to-peak voltage. The root-mean-square voltage V_{rms} is defined as the square root of the time-averaged value of $V^2(t)$. Mathematically, we write this as

$$V_{rms} = \sqrt{\overline{V^2(t)}}, \tag{7.23}$$

where the bar indicates the average value. For example, consider finding the rms voltage for the voltage given by $V = V_p \cos(\omega t)$, where V_p is the peak voltage. We can write V^2 in the form

$$V^2(t) = V_p^2 \cos^2(\omega t) = \frac{V_p^2}{2} + \frac{V_p^2}{2}\cos(2\omega t).$$ (7.24)

Because the average value of a cosine function is zero, we can write the average value of $V^2(t)$ as

$$\overline{V^2(t)} = \frac{V_p^2}{2}.$$ (7.25)

The rms voltage is given by

$$V_{rms} = \sqrt{\overline{V^2(t)}} = V_p/\sqrt{2}.$$ (7.26)

For a cosine voltage, the rms voltage is lower than the peak voltage by a factor of $\sqrt{2}$. This means that the rms voltage is lower than the peak-to-peak voltage by a factor of $2\sqrt{2}$. It is also convenient to use rms voltages for waveforms that do not have well-defined peak voltages, such as noise. Multimeters say "true rms" if they can also measure noise voltages. We can relate the rms voltage in a simple way to power. We write the average power P_a as

$$P_a = \overline{V^2(t)}/R = V_{rms}^2/R.$$ (7.27)

When we write power in terms of the rms voltage, we do not need to worry about remembering a factor of two. People often use rms voltages just because it makes power formulas simpler. This looks just like the formula for power for a DC voltage, and for this reason, the rms voltage is sometimes called the *effective* voltage.

FURTHER READING

Richard Feynman has an excellent discussion of the ideal gas law and the propagation of sound in the first volume of *The Feynman Lectures on Physics* (Addison-Wesley) in Chapters 39 and 47. The book, *Handbook of Recording Engineering*, by John Eargle, published by Van Nostrand Reinhold, is an excellent reference for recording music. A book with good coverage of the acoustics of musical instruments is *The Science of Sound* by Thomas Rossing, published by Addison-Wesley Longmans. A good book on the mathematical theory of sound is *Vibration and Sound*, by Philip Morse (McGraw-Hill).

PROBLEM 17 – TUNED SPEAKER

The NorCal 40A was originally designed for use with headphones, and for this reason the sound level is rather low for a loudspeaker. However, the transceiver is designed for output audio tones in the frequency range from 600 Hz to 650 Hz, and we can make a resonant speaker system that generates excellent sound levels in this frequency range. The response at other frequencies is reduced, and this helps suppress interference and noise. Figure 7.5 shows the construction.

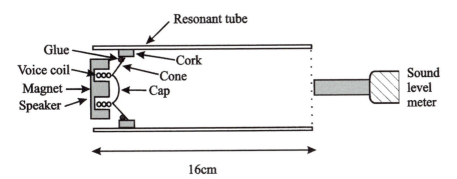

Figure 7.5. Tuned-speaker cross section. A speaker is mounted inside a 16-cm section of a cardboard mailing tube. The length is chosen to give a resonance in the 600-to-650 Hz range.

The speaker itself consists of a black paper cone that is attached to a coil of wire (the voice coil) that is mounted inside a ring magnet. When we put a current through the voice coil, the interaction between the coil and the magnetic field of the magnet produces a force on the paper cone. This gives the air next to the cone a velocity, which causes the sound. Our speaker is a small one, 2.25 inches in diameter, with a power rating of only a quarter watt.

To build the tuned speaker, slip a cork liner strip inside the tube about 1 cm from the end. Use a glue gun to attach the speaker to the cork liner. Start by plugging the glue gun into an outlet to heat up the glue. When the glue is hot, you can push it out of the tip with the trigger. Please be careful to get the glue at the joint between the speaker and the cork, and not on the tube. You should make a solid bead of glue completely around the edge of the speaker. Wait a few minutes for the glue to dry. The cork and speaker should still be able to slide inside the tube. Try to be neat with the glue.

Solder a pair of wires to the speaker, and add a stereo mini plug (Figure 7.6). You should do this carefully, because poor plug connections are a common cause of problems, and they are not easy to diagnose. One lead should be soldered to the outer ground strip, and the other should be soldered to one of the inner-conductor lugs. The inner-conductor contacts connect to leads that are shorted together on the circuit board, so that you do not have to connect to both. Take care that the pressure from the plastic hood does not cause the wires to short out. Also, do not crimp the plug ground strip until you have finished soldering the plug. Otherwise the insulation of the wires may melt through and

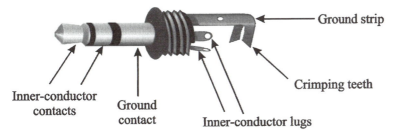

Figure 7.6. Stereo mini-plug connections.

short out the connection. Check your connections by measuring the resistance between the connector contacts with a multimeter.

The sound pressure level can be measured with a *sound level meter*. For the Tenma #72–860, the settings should be Lo, S, and C. The Lo (for low) position sets the measurement range from 35 dB to 100 dB. The S (for slow) setting gives a sound pressure level that is averaged over 1.5 seconds. By comparison, the F (fast) setting averages over 0.2 seconds. With the "slow" setting, the reading is more stable than with the "fast" setting, but the meter cannot respond to changes as quickly. The C weighting curve treats different frequencies the same, so that it is easy to make comparisons at different frequencies.

Mount the speaker and the meter on foam blocks so that the end of the meter with the microphone points into the tube, with the microphone centered in the mouth (Figure 7.5). The foam blocks should be oriented at angles so that sound reflections off the sides do not interfere with the measurement.

A.   Connect the function generator to the speaker. For these measurements set the function generator initially to a 600-Hz sine wave. The amplitude setting should be 25-mVrms. Do not use peak-to-peak voltages here, because we will be making a comparison with the AC voltage on a multimeter, which gives an rms voltage. The tube should have a resonance in the frequency range from 600 Hz to 650 Hz. The sound will peak at the resonant frequency. If you need to, you can lower the resonant frequency slightly by moving the speaker closer to the end, and you can raise it somewhat by pushing the speaker further into the tube. Find the resonant frequency and record L_p.

B.   Now measure f_l and f_u, where L_p drops by 3 dB from the maximum. Use these frequencies to calculate the Q.

C.   The speaker impedance is nominally 8 Ω, but it actually varies with frequency. Use an AC voltmeter reading at resonance to calculate the impedance. For this calculation, you may assume that the speaker is resistive.

D.   Now let us calculate the resonant frequency we would expect from a transmission-line equivalent circuit (Figure 7.7). In the circuit, pressure appears as voltage and velocity as current. The speaker controls the velocity of air, and so it is represented by a current source. The tube is a section of transmission line. The length is the distance from the cap in the center of the speaker to the mouth of the tube. At the open end of the tube, the sound pressure drops suddenly, because the air can spread out easily. For this reason, the open end is represented as a short circuit in the transmission-line model. There is a parallel resonance when the length

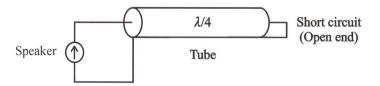

Figure 7.7. Transmission-line equivalent circuit for the tuned speaker.

is a quarter wavelength. Calculate this frequency, using 344 m/s for the speed of sound.

E. There is also a series resonance where the tube is a half wavelength long and the sound pressure is a minimum. Calculate the frequency where the tube length is a half wavelength long. Now measure the series resonant frequency, and note L_p. Measure the speaker impedance at this frequency.

F. There is another resonant frequency below 600 Hz that gives a pressure maximum. This is the resonant frequency of the speaker itself. Measure this frequency.

PROBLEM 18 – ACOUSTIC STANDING-WAVE RATIO

A. For this measurement, a section needs to be added to the tube to extend its length to more than a half wavelength. In addition, slip on a phenolic rod with ruler markings over the microphone of a sound level meter, so that sound-level measurements can be made down inside the tube. Insert the phenolic rod down the tube and measure the sound pressure level at intervals of 1 cm down the tube as far as you can go. The function generator should be set for a sine wave at 620 Hz, with the amplitude adjusted so that the maximum pressure level inside the tube is 114 dB, or 10 Pa. You should search carefully to make sure that you find the minimum pressure level and its position. Make a linear plot of the pressure in pascals versus the distance from the end in cm.

B. What is the ratio of the maximum pressure to the minimum pressure? This is the standing-wave ratio (SWR).

C. In the book, *Vibrations and Sound*, Philip Morse gives an approximate value for the SWR as

$$\text{SWR} = \frac{\lambda^2}{2\pi A} \tag{7.28}$$

where λ is the wavelength and A is the inside area of the tube. Calculate Morse's SWR value and compare it with your measurement.

D. The point of minimum pressure is actually less than a half wavelength from the mouth. It is as if the reflection actually occurs at a point beyond the end of the tube. The difference is written as δ (the Greek letter *delta*). Morse gives the formula for δ as

$$\delta = \frac{4d}{3\pi} \tag{7.29}$$

where d is the inner diameter of the tube. Calculate Morse's δ and compare with your measurement.

Transistor Switches

Now we begin the study of transistor circuits. Transistors have three terminals. Usually one of the terminals is the input, another is the output, and the third is a common connection that is shared between the input and the output. Transistor circuits can increase the power of a signal. For this they require an additional DC power source. Circuits that increase power are called *active* circuits. By comparison, a *passive* circuit has loss. The filters we covered in the earlier chapters are examples of passive circuits. We will study several different active circuits. An amplifier increases the power of a signal without changing the frequency. In an *oscillator*, an output sine wave is generated without any input signal. Transistors can also be used in passive circuits. In Problem 5, we saw that a transistor could act as a fast switch, with either a low resistance between the output terminals or a high resistance, depending on the input voltage. We will also use a transistor as a variable attenuator to control the signal level.

Manufacturers can combine many transistors on one chip of silicon. These circuits are called *integrated circuits*, or ICs. Many thousands of different integrated circuits are available. One common type of IC includes several amplifiers cascaded one after another, so that the output signal is much larger than the input. These circuits are called *op amps*, short for *operational amplifiers*. Our audio amplifier is a specialized kind of op amp. We also use integrated-circuit mixers to shift the frequencies in our transceiver. In addition, there is a *regulator* IC that provides a stable DC supply voltage that is close to 8 V over a wide range of currents.

A key component within transistors is a silicon diode (Figure 8.1). The two ends of a diode are traditionally labeled *p* and *n*. The *p* region is the anode, and the *n* region is the cathode. In the *n* region, current is carried primarily by electrons. Electrons have a negative charge (hence the label *n*). The *p* region is more difficult to describe. Physically, the current results from places in the silicon crystal where electrons are missing. These vacancies are called *holes*. Holes respond to an electric field just like electrons, except they act as if they have a positive charge rather than a negative charge. In addition, holes are usually slower. Whether a material is *p* type or *n* type depends on the kind of impurity introduced into the silicon during fabrication. Elements from the third column of the periodic table, such as boron or aluminum, make p-type material, whereas elements from column five, such as phosphorus or arsenic, produce material of n type. A diode is formed by a junction of p-type and n-type materials.

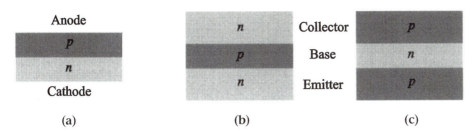

Figure 8.1. *p* and *n* regions in a pn diode (a). Layers for an npn transistor (b) and a pnp transistor (c).

There are two major families of transistors, *bipolar-junction* transistors, or BJTs, and *field-effect* transistors, or FETs. There are many different kinds of each transistor, and I defer to a book on solid-state devices for the details. We will think of a bipolar transistor as being controlled by an input current and a field-effect transistor as being controlled by an input voltage. Both kinds of transistors are found throughout electronics. We begin with BJTs.

8.1 Bipolar Transistors

A bipolar transistor is a three-layer sandwich and is called either *npn* (Figure 8.2a) or *pnp* (Figure 8.2b), depending on how the layers are arranged. The center layer is called the *base*, and the outer layers are the *collector* and *emitter*. In some respects, a transistor is a back-to-back connection of diodes. We can test a transistor like we test a diode by putting an ohmmeter across the terminals. We will find a diode between the base and the collector and between the base and the emitter. In practice, this is a good way to see if there is a problem with a transistor. If we connect the ohmmeter terminals to the collector and emitter, it will show an open circuit, because the diodes are connected back-to-back. This means that whichever way the voltage is applied, one of the diodes will be on, and one will be off, so that the current is blocked.

To an ohmmeter, there is no difference between a transistor and a pair of diodes connected back-to-back. But things become quite different if we have a current between the base and the emitter and a voltage applied to the collector. We

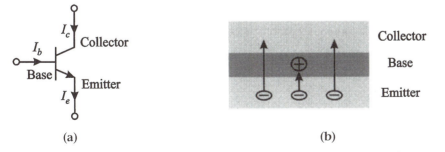

Figure 8.2. Schematic symbol and current directions (a) and electron flow (b) for an npn transistor.

start with npn transistors, which are more common than pnp transistors because they are faster, reflecting the fact that electrons generally move faster than holes. Figure 8.2a shows the schematic symbol for an npn transistor. The current flow in the base–emitter diode comes primarily from the flow of electrons from the emitter to the base (Figure 8.2b). Ordinarily, when current flows in a diode, an electron that crosses from the n region to the p region would meet with a hole and fill the vacancy. This is called *recombination*. However, in a transistor, the base region is made quite thin, a micron or less, so that the electron usually continues on to the collector, attracted by the voltage there. If the electron gets through the base without meeting up with a hole then it contributes to the collector current rather than to the base current.

The proportion of electrons from the emitter that make it to the collector is called the *collection efficiency*, and it is traditionally written as α. We write

$$I_c = \alpha I_e, \tag{8.1}$$

where I_c is the collector current and I_e is the emitter current. A typical value for α is 0.99, indicating that almost all of the electrons from the emitter end up in the collector rather than in the base. We can use Kirchhoff's current law to find the base current I_b by subtraction as

$$I_b = I_e - I_c = (1 - \alpha)I_e. \tag{8.2}$$

If α is close to one, then the base current is much smaller than the collector current. The ratio of collector current to base current is called the *current gain*, and we write it as β:

$$\beta = I_c/I_b. \tag{8.3}$$

We can divide Equation 8.1 by Equation 8.2 to get

$$\beta = \frac{\alpha}{1 - \alpha}. \tag{8.4}$$

If $\alpha = 0.99$, then β is about 100. In manufacturers' data sheets, you will usually see h_{FE} instead of β. The letter "h," short for *hybrid*, refers to a particular equivalent circuit model. So far, we have been thinking of the base current as a kind of leakage. However, we can turn this picture around and think of the base current as controlling the collector current. If we change the base current, the collector current follows. Because β is a big number, the collector current changes by a much larger amount. This is the basis for an amplifier with the base as the input and the collector as the output.

8.2 Transistor Models

We give an equivalent circuit for the transistor in Figure 8.3a. This circuit is not complete because only one diode is shown. The diamond denotes a current

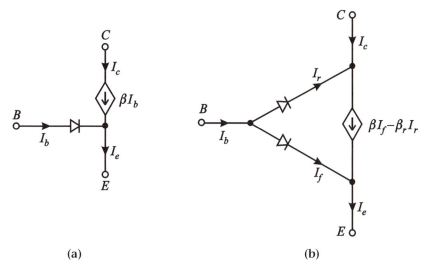

Figure 8.3. Transistor model with one diode (a), and the complete model with two diodes (b).

source βI_b. It is different from the current sources we have considered so far because it depends on another current in the circuit. It is called a *dependent* current source to contrast it with the previous *independent* sources. We will use diamonds for dependent sources and circles for independent sources.

We have considered the action of the base–emitter diode, but a similar picture holds for the base–collector diode. For this diode, we can define a reverse current gain β_r. Typically transistors are optimized for a large value of the normal current gain, and for this reason, β_r is much smaller, 10 or less. Figure 8.3b adds this effect to the circuit. Now I_b has two components, the forward base–emitter diode current I_f and the reverse base–collector diode I_r. The current generator has an additional component $-\beta_r I_r$.

We can use this model to understand how the collector current I_c behaves when we vary the collector–emitter voltage V_{ce} (Figure 8.4a). We will assume a constant

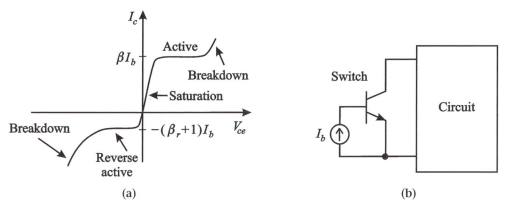

Figure 8.4. Collector current I_c for an npn transistor as a function of the collector–emitter voltage V_{ce}, with I_b constant (a). Transistor switch (b).

Table 8.1. Regions of Operation for an npn Transistor.

Region	V_{be}	V_{bc}	V_{ce}	I_c
Active	V_f	$<V_f$	$>V_s$	βI_b
Reverse active	$<V_f$	V_f	$<-V_s$	$-(\beta_r + 1)I_b$
On (saturated)	V_f	V_f	$>-V_s, <V_s$	$>-(\beta_r + 1)I_b, <\beta I_b$
Off	$<V_f$	$<V_f$	Any	0

Note: V_f is the forward voltage of the base–emitter and base–collector diodes. This is about 100 mV larger than in ordinary diodes. It is typically 0.7 to 0.8 V. V_s is the boundary between the saturation and active regions, typically 0.1 to 0.2 V.

base current I_b. When V_{ce} is positive, the base–emitter diode is conducting and the base–collector diode is off. The collector current is given by

$$I_c = \beta I_b. \tag{8.5}$$

We say the transistor is *active*. The active region is used for amplifiers. When V_{ce} is negative, the collector–emitter diode is on, and the base–emitter diode is off. This is the *reverse active region*. We can write the collector current as

$$I_c = -(\beta_r + 1)I_b. \tag{8.6}$$

The collector voltage that can be applied is limited by the reverse breakdown voltage of the diodes. In particular, one needs to be careful operating in the reverse active region, because breakdown usually occurs at only a few volts.

An interesting thing happens when V_{ce} is small, a few tenths of a volt or less. Current flows through both diodes at the same time. Small changes in the voltage shift the balance of current in the two diodes, and this causes large changes in the collector current. People call this the *saturation region* (Figure 8.4a), or they say the transistor is *on*. Saying that the collector current changes rapidly is another way of saying that the resistance is small. How small depends on the base current. The larger the base current, the steeper the slope in the saturation region, and the smaller the saturation resistance. Table 8.1 summarizes the regions of operation for an npn transistor.

8.3 Transistor Switches

We can use our model to understand how a transistor switch works. In Figure 8.4b, we connect a transistor to a circuit, and apply a control current I_b to the base. If the base current is zero, there is no collector current. We say the switch is *off*, and the transistor is effectively disconnected from the circuit. Now if we apply a base current, the resistance between the collector and emitter will be small, provided that the transistor stays in the saturation region. We say the switch is *on*. The transistor effectively shorts out the circuit.

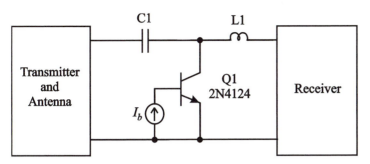

Figure 8.5. Operation of the Receiver Switch (Problem 19).

In the NorCal 40A, the Receiver Switch shorts out the RF Filter when we are transmitting (Figure 8.5). The transmitter and the antenna are connected to the receiver by a series resonant circuit. The transistor Q1 is the switch. In reception, the switch is off. Signals at the resonant frequency pass directly through from the antenna to the receiver. The switch is turned on when transmitting. This gives the switch a low resistance and blocks signals from the transmitter to protect the receiver.

To choose the base current to operate this switch, we need to know the relation between the saturation resistance and the base current. Figure 8.6 shows the saturation resistance for the transistor used in the Receiver Switch. Making this measurement is trickier than it might appear, because the collector current is

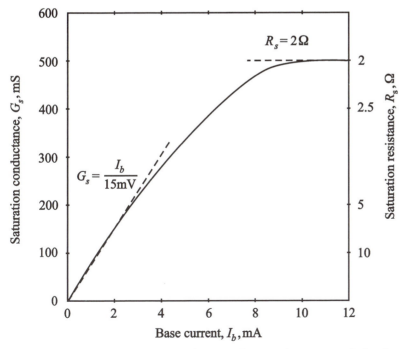

Figure 8.6. Measured saturation conductance G_s versus base current I_b for the 2N4124 transistor. The equivalent saturation resistance R_s is shown on the right.

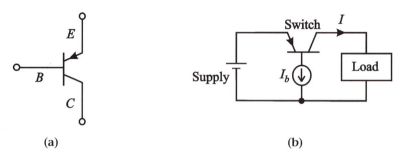

Figure 8.7. Circuit symbol for a pnp transistor (a), and a pnp switch for making a connection to a supply (b).

curved in the saturation region. The effective resistance depends on the value of the switch voltage and whether it is positive or negative. In the transmitter, the voltage is sinusoidal: It is positive half the time and negative the rest of the time. The plot shows the slope of the collector current versus collector voltage, evaluated at zero collector current. People call this a *small-signal conductance*, because it is appropriate for small voltages. There are two distinct regions in the plot. At small base currents, the saturation conductance G_s is proportional to I_b. We can write this relation as

$$G_s = I_b/15\,\text{mV}. \tag{8.7}$$

For large base currents, the conductance approaches a maximum value of 500 mS. Expressed as a resistance, we write

$$R_s = 2\,\Omega. \tag{8.8}$$

This is a *parasitic resistance* that is associated with the silicon itself. The Receiver Switch in Problem 19 is designed with a base current of about 5 mA, which gives a saturation resistance of about 3 Ω. Because this is much smaller than the reactances of C1 and L1, the transistor shorts out the signal.

An npn switch is convenient for providing a short to ground. A pnp transistor can connect a load to a voltage source. Figure 8.7a shows the circuit symbol for a pnp transistor. The arrow points *in*, in contrast to the npn symbol, where the arrow points out. For the pnp transistor, the models are identical except that the diodes and current generators change directions. It is a good idea to sketch these models to make sure that you understand them. Figure 8.7b shows a pnp switch for controlling the voltage applied to a load. A voltage source is connected to the emitter and a load is attached to the collector. The switch is controlled by base current. When the base current is zero, the transistor is off, and the load current and voltage are zero. We apply a base current to turn the transistor on, and to supply current to the load. In our transceiver, a pnp switch controls current to the transmitter circuits. This is the Transmitter Switch, and we build it in Problem 20.

Choosing the base current for this switch is simpler than for the previous one. The collector current does not change sign, and we can use the ordinary current gain β, provided we include a safety factor to make sure that the transistor saturates. We want the transistor to be saturated, so that the voltage drop is small and most of the supply voltage appears across the load. In the NorCal 40A, the Transmitter Switch load current is 7 mA. We need to look at the manufacturer's specification for the minimum current gain. The transistor in this switch is the 2N3906, and its data sheet in Appendix D specifies that the minimum value of h_{FE} (β) for a collector current of 10 mA is 100. Now we can calculate the required base current as

$$I_b = 2I/100 = 140\,\mu A, \tag{8.9}$$

where I is the load current and 2 is a safety factor. As a check, we can consult Figure 14 in Motorola's data sheet, which shows that the voltage drop between the collector and emitter should be about 200 mV. This would only knock the 8-V supply down to 7.8 V, which is acceptable. The figure also shows that this drop can be reduced by increasing the base current.

FURTHER READING

It is a good idea to study the chapters on transistor circuits in the *The Art of Electronics*, by Horowitz and Hill, published by Cambridge University Press. Horowitz and Hill have an excellent chapter on regulators, which we do not cover. *Device Electronics for Integrated Circuits*, by Muller and Kamins, published by Wiley, is a good book for information on transistors themselves. For an exhaustive survey at an advanced level on the behavior of different kinds of transistors, see *Physics of Semiconductor Devices*, by S. M. Sze, published by Wiley.

PROBLEM 19 – RECEIVER SWITCH

Transmitters produce much more power than receivers can handle. The NorCal 40A has an output power of about 2 W, which would destroy the Receive Mixer. The Receiver Switch (Figure 8.5) keeps the transmitter power out of the receiver. The receiver could be switched out by hand, but it is more reliable to have a transistor do this automatically. Figure 8.8 shows the detailed circuit. The transistor Q1 has its collector attached

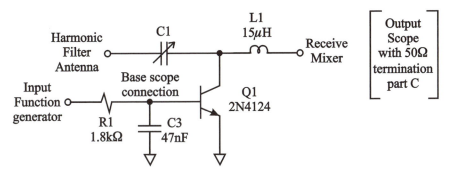

Figure 8.8. Details of the Receiver Switch and its connections. The triangles denote ground connections.

between the capacitor C1 and the inductor L1, and the emitter goes to ground. The base is connected to an RC delay circuit (R1 and C3). When transmitting, 8 V is applied to the delay circuit. This connection is called 8 V TX in the transceiver schematic. (TX is an abbreviation from telegraphy for "transmit.") The voltage produces a current in the base and turns the transistor on. This shorts out the filter, blocking the transmitter signal. In receiving, the input goes to zero volts. This stops base current and turns the transistor off, effectively removing it from the circuit. The filter can now operate normally.

The Receiver Switch prevents the receiver from being destroyed by the transmitter, but even more blocking is needed. The transmitter would still produce loud, annoying tones. In early radios, operators slipped off their headphones when they wanted to transmit. Modern transceivers have attenuators to reduce the signal before it gets to the audio amplifier. In the NorCal 40A, this job is handled by the AGC circuit.

Solder R1, C3, and Q1 into the circuit. For R1, leave enough room to connect the scope and the function generator. For the transistor, use the white outline to orient the package.

Set the function generator to give a 1-kHz square wave with an open-circuit high voltage of 8 V and low voltage of 0 V. This means that a 50-Ω function generator should be set to 4 Vpp with a DC offset of 2 V. These settings work because the open-circuit voltage is twice the indicated voltage. It is a good idea to check the voltage on an oscilloscope.

Connect the function generator to the input end of R1 and the scope to the other end (Figure 8.8). The base voltage should alternate between zero and the forward voltage of the base–emitter diode, with rising and falling transitions in between.

A. Consider the rising part of the base voltage waveform. Measure the initial slope. You will find it convenient to set the scope trigger for a positive slope, so that you can zoom in on the rising part of the waveform. Calculate approximately what the slope should be.

B. Now consider the falling part. It is best to set the slope trigger for a negative slope. Measure the time t_2 that it takes for the voltage to drop by a factor of two. At first the base–emitter diode will be on, and this causes the voltage to drop much faster than it does later. For this reason, you should make the measurement over a part of the curve where the voltage is below 0.6 V. Calculate what t_2 should be.

In the following sections, we need to measure small signals, and you will find it convenient to turn on the scope's low-pass filter, if one is available. This reduces high-frequency noise and makes the traces sharper. Attach the function generator to the Antenna jack J1 and the scope at the output as shown in Figure 8.8. Use a 50-Ω termination on the scope. The function generator should be set for a 1-Vpp, 7-MHz sine wave.

C. Now we measure the attenuation of the switch. First consider the signal with the Receiver Switch off. Adjust C1 for maximum scope voltage and record the voltage.

D. Now measure the voltage with the Receiver Switch on. You can turn on the switch by connecting a 12-V power supply to the input at R1. You should increase the function-generator amplitude setting to 10 Vpp, to make it easier to see the signal.

You will need to account for this voltage setting in the calculations. Measure the output voltage, and calculate the on–off rejection ratio R in dB from the expression

$$R = 20 \log(V_{off}/V_{on}).\qquad(8.10)$$

E. Find an approximate formula for the attenuation in terms of the saturation resistance R_s. The easiest way to do this is to think of the circuit as a pair of cascaded voltage dividers.

F. Now calculate the attenuation that we would expect. To start, find the base current from the voltage drop across R1. Then find R_s from Figure 8.6, and apply your attenuation formula.

G. Simulate the attenuator with *Puff* from 0 MHz to 14 MHz.

H. The designer chose the 2N4124 for this switch because of its low off capacitance. This capacitance causes loss even when the transistor is off. Use your *Puff* simulation to find the loss. For this calculation use $C_{obo} = 3.5$ pF (Figure 1 in the data sheet). You will need to adjust C1 slightly for best transmission, just as in your measurements. For comparison, repeat the calculation for the 2N2222A transistor that you used in Problem 5, using $C_{cb} = 10$ pF (Figure 9 in the data sheet).

PROBLEM 20 – TRANSMITTER SWITCH

In the NorCal 40A, the Transmitter Switch uses a pnp transistor to provide the 8 V TX line that drives the Receiver Switch (Figure 8.9a). The Transmitter Switch also supplies the Transmit Mixer, the Buffer Amplifier, and the RIT circuit that shifts the frequency of the VFO during transmission. We will discuss these circuits later. The switch is controlled by a line to the Key jack J3. In operation, the jack is connected to a telegraph key. When the key is down, the line shorts to ground, and the radio transmits. When the key is up, the line opens, and the radio receives. We will use a relay in place of the key.

First consider what happens when the key is down. The capacitor C57 discharges through the diode D11, dropping the capacitor voltage down to the forward voltage

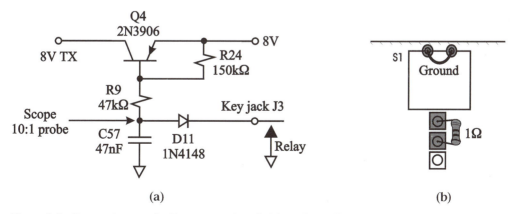

Figure 8.9. Connections to the Transmitter Switch (a), and installing a 1-Ω resistor in the S1 holes (b).

of the diode. A current flows in R9, and this pulls current through the base to turn on the transistor Q4, providing 8 V to the Receiver Switch. When the key is up, the base current stops flowing in the diode and begins to charge C57. As the capacitor charges, the base current drops. For a while nothing happens to the collector voltage, because the base current is still sufficient to turn the transistor on. However, eventually the base current drops enough to make the transistor *active*, where the collector current is given by $I_c = \beta I_b$. The collector voltage begins to fall now. Finally the capacitor voltage gets so high that the base–emitter diode turns off, and the collector current stops. The resistor R24 keeps the capacitor charging even after the base–emitter diode is off.

Install the parts in Figure 8.9: Q4, R24, R9, D11, C57, and J3. Leave a couple of millimeters of lead length on the diode so that you can attach a probe. Also, install the regulator circuit consisting of U5 (78L08), C42 (10 μF), and C43 (47 nF). Refer to the endpaper to see how these components fit into the circuit. The regulator provides a nearly ideal 8-V source for currents up to 100 mA. The capacitors prevent the regulators from oscillating and help filter out signals from other parts of the circuit and noise from the power-supply connection to J2. The large capacitor provides protection at low frequencies, and the small capacitor at high frequencies. We will not study regulators but detailed information about this regulator is given in Appendix D. You do need to be careful with the electrolytic capacitor C42. This type of capacitor can provide very large capacitances from 1 μF to 100 mF. However, they have limitations. The capacitance values are not precise. Typically the tolerance is only ±20%. The standard values are usually limited to multiples of 10, 22, 33, 47, and 68. They are also *polarized*, which means that they cannot take large negative voltages. You must get the wires in the right holes. You can recognize the positive lead because it is longer than the negative lead. The positive hole is marked on the circuit board with a plus sign. In addition, the minus lead is marked on the can with a minus sign. If you install the capacitor incorrectly, there will be a large current, and the capacitor will heat up. Often you can smell that something is wrong. Eventually there will be a small explosion, or a big explosion if it is a large capacitor, and material will pop out of the capacitor can.

In addition, we need to install the Supply jack J2 and the Schottky diode D7 (1N5817). Refer to the endpaper schematic for this. The diode prevents damage if the power supply is hooked up backwards. Install a 1-Ω resistor in the S1 holes as shown in Figure 8.9b. This is a good place to attach probes to measure the supply voltage and current. The third hole farthest from the edge of the board is not connected to the circuit; it should be left open. You should add a loop of wire between the two holes in the S1 outline on the edge of the board to provide a ground connection.

Attach a relay to the Key jack J3. A relay is a switch that is operated by an electromagnet. See Appendix A for more information. We use the Magnecraft W171DIP-7, which requires 5 V to close the relay. The switching time is 200 μs. It has a coil resistance of 500 Ω and includes a snubber diode.

Connect the function generator to the relay, and use a 20-Hz square wave with an amplitude of 5 Vpp. This gives an open-circuit positive voltage of 5 V, which is the proper voltage for the relay. This speed of 20 pulses per second is about as fast as the best operators can receive Morse code. A more typical speed would be two to three times slower. Slower speeds are not convenient to observe on an oscilloscope without storage because the trace fades. A 10:1 scope probe should be attached at the anode of D11 to monitor the voltage of the capacitor C57. We do need to use the 10:1 probe, for otherwise the scope resistance drains too much charge from the capacitor.

A. Sketch the voltage on C57, indicating the key-down time when the relay is closed and the key-up time when the relay is open. Measure the time that it takes the capacitor to charge halfway from its minimum voltage to its maximum voltage. Now calculate approximately what this time should be.

B. Calculate approximately the collector current I_c when Q4 is on. Assume that the base–emitter voltage of Q1 is 700 mV. You can neglect the saturation voltage of Q4. Calculate the base current I_b that is required to produce this collector current, assuming the manufacturer's minimum β value of 100.

C. For comparison, calculate I_b at key down, assuming a 700-mV drop in the base–emitter diode of Q4 and a 600-mV drop in D11.

D. Now sketch the collector voltage for Q4, showing where the transistor is saturated. What is the delay in going active? This delay is useful because it allows the Power Amplifier to shut down gradually over a time period of 1 to 2 ms. If a transmitter turns off too quickly, it causes an annoying clicking sound on nearby frequencies that interferes with other operators.

E. The time at which the transistor goes active is interesting because we can use it to infer β. Measure the voltage across R9 at this time, and use it to calculate I_b. Compare this with the collector current I_c that you calculated previously to find β.

Transistor Amplifiers

When a transistor is active, the current gain β is large, in the range of 100 or more. This means that we can use the transistor as an amplifier to increase the power of a signal. The amplifier may be considered the single most important device in communications electronics, and it is key to both receivers and transmitters. Developing amplifiers has been a central focus of electrical engineering from the days of the first vacuum tubes, and it is just as important today. There are many issues to consider in designing an amplifier. In transmitters, we are very interested in efficiency. High efficiency makes it easier to dissipate the heat and allows long battery life in portable transmitters. In receivers, it is important to add as little noise as possible to the signal. In this chapter, we study *linear* amplifiers, where the amplitude of the output tracks the amplitude of the input. In the next chapter, we consider saturating amplifiers, where only the frequency of the output follows the input.

9.1 Common-Emitter Amplifier

The basic transistor amplifier is shown in Figure 9.1a. It uses an npn transistor with a load resistor R at the collector. The supply voltage is written as V_{cc}. It is traditional to double the subscript of a supply voltage to distinguish it from an AC voltage. This circuit is called a *common-emitter* amplifier. You do have to be on your guard with amplifier names. In this example, the emitter is grounded, and so the name *common emitter* is straightforward enough. The idea is that the input is made up of the base and emitter, and the output is made up of the collector and emitter, so that the emitter is *common* to both the input and the output. However, there will usually be other components connected to the emitter, so that it may not be easy to make a distinction. Later we will study a common-collector amplifier, the emitter follower.

At the top of Figure 9.1b is the base-voltage waveform. The positive voltage is limited to the forward voltage V_f of the base–emitter diode. Base current will only flow when the source voltage V_o is sufficiently positive to turn on the base–emitter diode. When V_o goes negative, the base voltage also goes negative, and the base current will cease. The collector-voltage waveform is shown beneath the base voltage. When current flows in the base, the transistor is active, and there will be a large collector current. This collector current causes a voltage across the

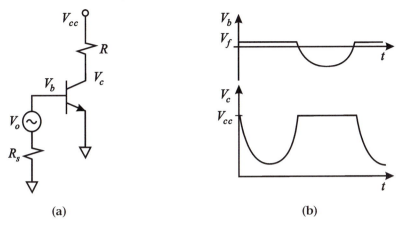

Figure 9.1. (a) Common-emitter amplifier. The triangles indicate ground connections. (b) Base voltage V_b (top) and collector voltage V_c (bottom).

load resistor, and the collector voltage drops. If the source voltage is large enough, we will turn the transistor completely on, and drive the collector voltage to the saturation level, as in the Receiver Switch in Problem 19.

This amplifier has a drawback. The transistor is off half the time, and the output is only half a cosine. To get the entire cosine, we must offset the base voltage to keep the transistor from shutting off. We should note, however, that this half-cosine circuit is not as dumb as it looks, and we will consider it again when we discuss Class-B amplifiers in the next chapter. For now, we add a DC voltage V_{bb} to the base circuit (Figure 9.2a). V_{bb} is called a *bias*, and it is usually provided by a resistive voltage divider between the supply and ground. In the lab, function generators often allow us to add this offset voltage. The idea of the bias voltage is to keep the base conducting continuously. Thus the collector voltage will be a full cosine wave (Figure 9.2b). The largest collector voltage we can get is a swing from V_{cc} when the collector current I_c is zero to near ground when the transistor is on the edge of saturation. We can get this large swing by adjusting the bias for a collector voltage near $V_{cc}/2$.

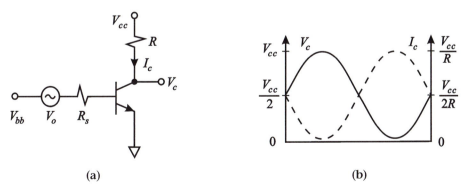

Figure 9.2. Common-emitter amplifier with a bias voltage V_{bb} (a), and collector waveforms for a maximum peak-to-peak voltage (b).

We distinguish between different classes of amplifiers by the bias. If we bias so that the base–emitter diode is always on, it is a *Class-A* amplifier. In the NorCal 40A, the Driver Amplifier is Class A. In the next chapter, we define other letter classes of amplifiers where the transistor is off during part of the cycle. The advantage of Class A is that the output is a good replica of the input with little distortion. The disadvantage of Class A is poor efficiency. The other classes distort the waveform but are much more efficient than Class A.

9.2   Maximum Efficiency of Class-A Amplifiers

We can calculate the maximum efficiency for a Class-A amplifier, assuming that the output voltage varies from zero to V_{cc}. The collector current will vary from zero to V_{cc}/R. We define the efficiency η (the Greek letter *eta*) as

$$\eta = P/P_o, \tag{9.1}$$

where P is the AC load power and P_o is the DC supply power. We can write P_o as

$$P_o = V_{cc}I_o, \tag{9.2}$$

where I_o is the average collector current, given by

$$I_o = \frac{V_{cc}}{2R}. \tag{9.3}$$

This gives us the supply power

$$P_o = \frac{V_{cc}^2}{2R}. \tag{9.4}$$

Now we can write the AC load power P as

$$P = \frac{V_{pp}I_{pp}}{8} = \frac{V_{cc}^2}{8R}. \tag{9.5}$$

We can see that $P = P_o/4$, so that the efficiency η is 25%. This is the maximum efficiency for a Class-A amplifier with a resistive load.

It is interesting to track the power flow. In addition to the AC load power P, there is a DC load power P_{rdc}. We can write this as

$$P_{rdc} = \frac{V_{cc}^2}{4R}. \tag{9.6}$$

This means that half the power from the supply is lost as DC power in the resistor. There is also DC power in the transistor. Because the average voltage and current across the transistor are the same as for the resistor, the DC power is also the same. We write this as P_{tdc}, given by

$$P_{tdc} = \frac{V_{cc}^2}{4R}. \tag{9.7}$$

However, this is not the end of the story. The AC peak-to-peak voltage and current for the transistor are also the same as for the resistor, except Figure 9.2b shows

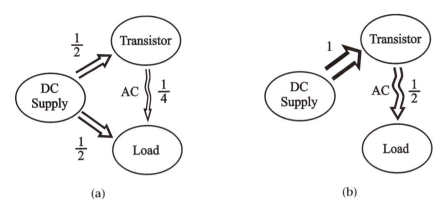

Figure 9.3. Power flow in a Class-A amplifier with maximum efficiency for a resistive load (a) and for a transformer-coupled load (b).

that they are 180° out of phase. Thus we can write the transistor AC power P_{tac} as

$$P_{tac} = -\frac{V_{cc}^2}{8R},$$ (9.8)

where there is a minus sign because the voltage and current are out of phase. The interesting part about this power is that it is negative, indicating that the transistor produces AC power rather than absorbing it like a resistor. Half the DC power delivered to the transistor is converted to AC power, which is consumed in the load. Figure 9.3a summarizes this power flow. The DC power from the supply splits between resistor and the transistor. The transistor power further splits between the power that is absorbed and the AC power that is delivered to the load.

There are two major disadvantages of this Class-A circuit. First, half the power is lost as DC power in the resistor. Second, many loads cannot be connected directly to the supply. If the load is the base of another npn transistor, for example, we will want a voltage that is near ground. We can use a transformer to solve these problems (Figure 9.4a). Because DC power does not couple through a transformer, this eliminates the DC power in the load. In addition, the DC resistance between the supply and the transistor is zero, and so the average collector voltage is V_{cc}

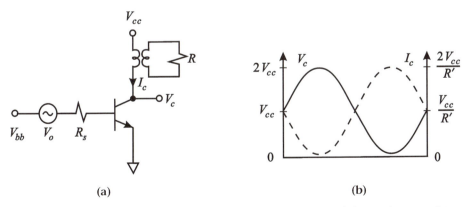

Figure 9.4. Class-A amplifier with a transformer-coupled load (a), and the maximum peak-to-peak voltage and current waveforms (b).

instead of half V_{cc} (Figure 9.4b). This means that the peak-to-peak voltages and currents can be twice as large as before. The Driver Amplifier in Problem 21 uses a transformer. We can write the supply power as

$$P_o = V_{cc}I_o = V_{cc}^2/R', \tag{9.9}$$

where R' is the effective load resistance, given by

$$R' = n^2 R, \tag{9.10}$$

where n is the turns ratio. The average output power P can be written as

$$P = \frac{V_{pp}I_{pp}}{8} = \frac{V_{cc}^2}{2R'}. \tag{9.11}$$

We can see that $P = P_o/2$, and thus the maximum efficiency is 50%. The maximum efficiency is twice as high as for a resistive load because the transformer prevents DC power from being lost in the resistor. Figure 9.3b shows the power flow. The efficiency in practical Class-A amplifiers is typically between 30 and 40%. One important point is that a Class-A amplifier dissipates even more power when there is no output. This is a major disadvantage, and it is in sharp contrast to the other classes of amplifiers, where power dissipation is small when there is no output.

The transformer turns ratio controls the peak-to-peak current. Manufacturers specify a collector current limit in their data sheets. We would aim for a maximum current I_m under this limit. From Figure 9.4b, we can write

$$I_m = \frac{2V_{cc}}{R'} = \frac{2V_{cc}}{n^2R}. \tag{9.12}$$

This lets us choose n. To get the full current swing, the bias current should be set to half this maximum.

9.3 Amplifier Gain

In addition to efficiency, we characterize an amplifier by its gain, which is the ratio of output power to input power. Figure 9.5 shows a circuit with an amplifier,

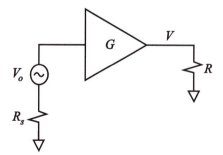

Figure 9.5. Amplifier with a source and a load.

a source, and a load. We write the gain G in dB as

$$G = 10\log(P/P_+)\,\mathrm{dB}. \tag{9.13}$$

In the lab, we will calculate the output power P from a measurement of the peak-to-peak voltage across the load. We write

$$P = \frac{V_{pp}^2}{8R}. \tag{9.14}$$

The available power P_+ is the maximum power that would be delivered by the source to a matched load. For a matched load, the voltage is half the open-circuit voltage of the source. We write the matched-load voltage as V_+, given by

$$V_+ = V_0/2. \tag{9.15}$$

We then write P_+ as

$$P_+ = \frac{V_{+pp}^2}{8R_s}. \tag{9.16}$$

In the lab, V_{+pp} is the amplitude setting on a properly calibrated function generator. These formulas show that the gain depends on both voltages and resistances.

9.4 IV Curves

One thing that you may notice when you build the Driver Amplifier in Problem 21 is that the output looks somewhat distorted on the oscilloscope (Figure 9.6a). The collector voltage V_c is wider on the bottom and narrower on the top. We can understand what is happening if we check the base voltage V_b. It is distorted in the same way as the collector voltage, except that the widening is on the top and the narrowing is on the bottom. If we take into account the fact that the

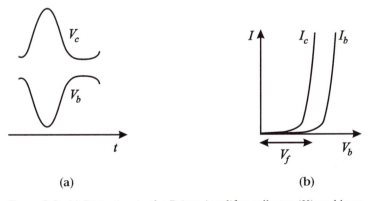

Figure 9.6. (a) Distortion in the Driver Amplifier collector (V_c) and base (V_b) voltage waveforms, as seen on an oscilloscope. The scales are adjusted to compensate for gain. (b) IV curves for the base–emitter diode when the transistor is active.

amplifier inverts the output, it means that at least to the eye, the circuit is acting as a good voltage amplifier, in that the output voltage is a faithful replica of the input voltage. It is interesting to check this on an oscilloscope by inverting one channel and adjusting the voltage scale to compensate for the gain of the amplifier. If you do this, you will see that the input and output waveforms match quite well.

We can understand the distortion in the base voltage curve if we consider IV curves for the base–emitter diode (Figure 9.6b). There are two currents to consider, the base current I_b and the collector current I_c. They have a similar shape, except that the base current is lower by a factor of β. The effect is to shift the base-current curve to the right of the collector-current curve. We leave the details to a book on solid-state devices, but we can say that when the base voltage is near V_f, the currents are given to a close approximation by

$$I_b = I_{bs} \exp(V_b/V_t), \tag{9.17}$$

$$I_c = I_{cs} \exp(V_b/V_t), \tag{9.18}$$

where I_b is the base current, I_c is the collector current, and V_b is the voltage between the base and emitter. A similar expression can be used for the current and voltage in a pn diode. V_t is called the *thermal voltage*, and it is given by

$$V_t = kT/q, \tag{9.19}$$

where k is Boltzmann's constant, 1.38×10^{-23} J/K, T is the absolute temperature in kelvins, and q is the electronic charge, 1.60×10^{-19} C. At a typical room temperature of 295 K, or 22°C, we have

$$V_t = 25\,\text{mV}. \tag{9.20}$$

This formula gives a hint that the underlying physics of current flow in diodes and transistors is thermal excitation. The currents I_{bs} and I_{cs} are proportionality constants. They are called *saturation currents*, and they are related by the current gain β via

$$I_{cs} = \beta I_{bs}. \tag{9.21}$$

Both I_{cs} and I_{bs} increase strongly with temperature, and β varies with both temperature and current level. Hence the saturation currents are not really constants.

Another thing that you should notice is that in Equations 9.17 and 9.18, I_b and I_c are always positive, even when the applied voltage is negative. This must be incorrect. Otherwise, we would have a perpetual power source, because the power for negative voltages would be negative. Often you will see a term I_{bs} and I_{cs} subtracted from these equations to make the currents go to zero when the voltage is zero, and to make the current negative when the voltage is negative. The actual currents at low and negative voltages are more complicated. In particular, the effective value of V_t changes. For this reason I prefer to give a simpler formula that holds where we apply it, at voltages near V_f.

The equations tell us that diode and transistor currents increase by a factor of e for every 25-mV increase in voltage. This is a steep increase. For a factor of ten, we need only change the voltage by

$$\Delta V = V_t \ln(10) \approx 60 \text{ mV}. \tag{9.22}$$

This means that in a device with a β of 100, the I_b curve is shifted to the right of the I_c curve by 120 mV. This is the reason that the forward voltages for the base–emitter diodes tend to be between 700 mV and 800 mV, rather than the 600 mV to 700 mV for ordinary diodes.

9.5 Base Resistance

The slope of the base current gives us the conductance of the base–emitter diode. This is a small-signal conductance, and we write it as g_b. Because Equation 9.17 is exponential, this has a simple form:

$$g_b = \frac{dI_b}{dV_b} = \frac{I_b}{V_t}. \tag{9.23}$$

You should go through the details to verify this formula. It is more common to use the base resistance r_b, which is the inverse of g_b. At room temperature, we write

$$r_b = \frac{25 \text{ mV}}{I_b}. \tag{9.24}$$

This expression can also be used for a diode. The formula is similar to the one we found for the resistance of a switch in saturation, except that the voltage constant is different (Equation 8.7). As an example, a typical base bias current for the Driver Amplifier in Problem 21 might be 250 μA. This gives us $r_b = 100 \ \Omega$. We can use this base resistance to draw a small-signal AC model of an active transistor (Figure 9.7a) and to write the base voltage as

$$v = i_b r_b. \tag{9.25}$$

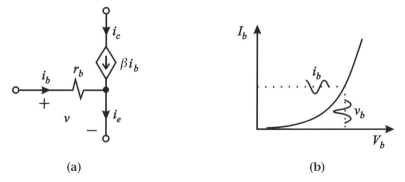

(a)   **(b)**

Figure 9.7. Small-signal model of an active transistor, including the base resistance r_b (a). Understanding the distortion in the base voltage waveform (b).

We will use lower-case letters to distinguish this signal voltage and current from the DC bias voltages and currents.

The slope of the collector-current curve is written in a similar way as

$$g_m = \frac{dI_c}{dV_b} = \frac{I_c}{V_t},$$

(9.26)

where g_m is called the *transconductance* to indicate that the current and voltage are for different terminals. We will not use the transconductance in our calculations for bipolar transistors, but we will need it for field-effect transistors.

The IV curves explain the distortion in the base waveforms (Figure 9.6a). Consider applying a sine-wave current i_b in addition to the bias current (Figure 9.7b). The resulting voltage v_b has bigger negative peaks than positive peaks because the graph is curved. This accounts for the distortion we see. We can improve the distortion by lowering the source impedance. The source impedance for the Driver Amplifier in Problem 21 is rather high, around 600 Ω. In the complete transceiver, the Buffer Amplifier provides a much lower source impedance for the Driver Amplifier, and this reduces the distortion.

There is another thing that you might think about from the current plots. A small change in voltage causes a large change in current, and this can make it difficult to set the base current with a voltage source. Since β is usually not well known, we will have even more uncertainty in the collector current. Moreover, the forward voltage V_f shifts downward by 2 mV for every degree that the temperature rises. We will not dwell on this, but it serves as a warning that biasing can be a tricky business. We can make biasing much easier if we add an emitter resistor.

9.6 Emitter Degeneration

In Figure 9.8a, we add an emitter resistor R_e. This is called *emitter degeneration*. This resistance makes it easier to set the bias and control the gain. We consider the bias first. Figure 9.8b shows a simplified model for the amplifier input. The base–emitter diode is represented by a voltage source. This is because a diode has a low resistance when it is on and a voltage that normally varies over only a small range. Since this is just what a voltage source does, it is convenient to use it in the equivalent circuit. In addition, we have shown the emitter resistor R_e, together with the collector current I_c. The base current is much smaller than the collector current, so that $I_c \approx I_e$. If we assume that we have a particular collector current goal in mind for the bias, we can write the required base bias voltage V_{bb} as

$$V_{bb} \approx V_f + I_c R_e.$$

(9.27)

Notice that this voltage does not depend on β as long as it is large. This makes the voltage much easier to set.

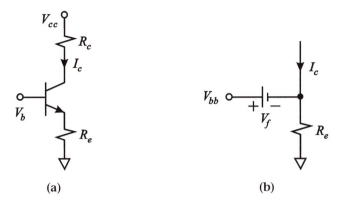

Figure 9.8. Adding an emitter resistor to the common-emitter amplifier (a), and equivalent circuit for calculating the bias voltage (b).

The emitter resistor also allows us to set a voltage gain that is determined by resistances that are under much better control than β. To see how this works, let us define the voltage gain G_v as the ratio

$$G_v = v/v_i. \tag{9.28}$$

In the formula, v is the output AC voltage and v_i is the input AC voltage. Working in terms of AC voltages and currents simplifies things, because the bias voltages have no effect on AC signals. We can replace the bias voltages by shorts in the equivalent circuit (Figure 9.9a). We assume that β is large so that $i_b \ll i_e$ and $i_e \approx i_c$, and we write the input voltage v_i as

$$v_i = i_b r_b + i_e R_e \approx i_c R_e. \tag{9.29}$$

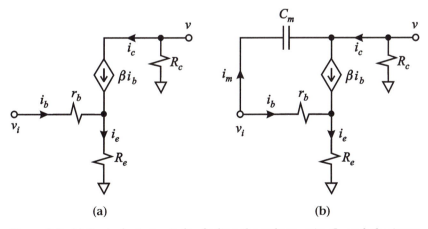

Figure 9.9. (a) Equivalent circuit for finding the voltage gain G_v and the input impedance Z_i of a common-emitter amplifier with an emitter resistor. (b) Adding the Miller capacitance.

The output voltage is a little trickier, because the current flows up through R_c. This gives the output voltage a minus sign. We can write the output voltage v as

$$v = -i_c R_c. \tag{9.30}$$

Notice that since v has the opposite sign from v_i, the amplifier inverts. We can divide these formulas to get the gain as

$$G_v = -R_c/R_e. \tag{9.31}$$

Now we have a gain that does not depend on the value of β, but rather on resistors that have precise values. Note that the gain can be less than 1 if $R_e > R_c$.

We can also use Figure 9.9a to find the AC input impedance. This time, we cannot neglect the base current, because it is the input. We write the input impedance as

$$Z_i = v_i/i_b. \tag{9.32}$$

From Equation 9.29, we have

$$v_i \approx i_c R_e = \beta i_b R_e. \tag{9.33}$$

This gives us our input impedance:

$$Z_i = \beta R_e. \tag{9.34}$$

Here the factor of β enters directly to increase the input impedance. This helps if we want to see the full open-circuit voltage of the source. In Problem 22, however, we will find that we may not actually achieve this high impedance, because the input is shunted by capacitance from the base to the collector. This capacitance is called the *Miller capacitance*. It arises because the layers in a transistor lie on top of one other, so that a voltage in one layer induces charges in another layer. The capacitance is small, usually only a few picofarads. However, the effect is magnified by the gain of the amplifier. To see this, consider the circuit in Figure 9.9b, where we have added a capacitor at the input. The voltage on one side of the capacitor is v_i, and the voltage on the other side is the collector output voltage $v = G_v v_i$. We can write the capacitor current i_m in terms of the total voltage across the capacitor:

$$i_m = j\omega C_m(v_i - v) = j\omega C_m(v_i + |G_v|v_i) = j\omega C_m(|G_v| + 1)v_i. \tag{9.35}$$

The effect of the capacitance is multiplied by a factor of $|G_v| + 1$. We can write the input impedance in the form

$$Z_i = \beta R_e \parallel (|G_v| + 1)C_m. \tag{9.36}$$

Now we calculate the output impedance that we would use in writing a Thevenin or Norton equivalent circuit for our amplifier. So far we have assumed that the collector voltage does not affect the current of an active transistor. However, there is a tilt to the IV curves (Figure 9.10). The reason the lines slope is that the effective thickness of the base becomes smaller at high voltages, and this increases β. An

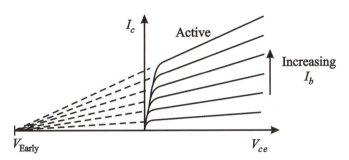

Figure 9.10. Collector current and voltage, showing the slope in the active region. If we extend these lines backwards, they intersect at the Early voltage.

interesting thing about the lines is that they intersect at one point on the voltage axis. This point is called the *Early voltage*. The Early voltage is usually not quoted in manufacturer's data sheets, and it varies greatly among transistors, because it depends strongly on the thickness of the base. For the 2N2222A transistor that we use in the Driver Amplifier in Problems 21 and 22, the Early voltage is 145 V. Because this is much larger than any voltage we apply to the collector, we can write the collector resistance r_c approximately as

$$r_c \approx V_{\text{Early}}/I_c. \tag{9.37}$$

Like the emitter resistance r_b, the collector resistance varies inversely with the current. For a collector bias current of 50 mA, we get a collector resistance r_c of 3 kΩ. We can draw a small-signal model for the transistor that includes a shunt collector resistance r_c (Figure 9.11a).

In Figure 9.11b, we show a small-signal model for the common-emitter amplifier. In the model, we use a combined source resistance R'_s, given by

$$R'_s = R_s + r_b. \tag{9.38}$$

The collector also has capacitance associated with it, and this should be included, because its reactance is often comparable to the collector resistance. We define a

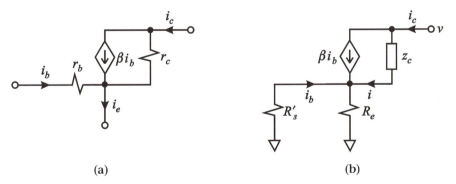

(a) **(b)**

Figure 9.11. Small-signal transistor model that includes the collector resistance r_c (a). Small-signal model for the common-emitter amplifier for calculating the output impedance Z_o (b).

collector impedance z_c given by

$$z_c = r_c \parallel C_c, \tag{9.39}$$

where C_c is the output capacitance that is specified in manufacturer's data sheets. For the 2N2222A used in the Driver Amplifier, C_c is 8 pF, which gives a reactance of 2.8 kΩ at 7 MHz.

We write the output impedance of the amplifier as

$$Z_o = v/i_c. \tag{9.40}$$

The key to the calculation is finding the current i through the collector impedance z_c. We use Kirchhoff's current law to write it as

$$i = i_c - \beta i_b. \tag{9.41}$$

We can use the current-divider formula to write i_b as

$$i_b = -\frac{i_c R_e}{R'_s + R_e}. \tag{9.42}$$

When we substitute into the previous formula, we get

$$i = i_c \left(1 + \frac{\beta R_e}{R'_s + R_e}\right). \tag{9.43}$$

Now we write the voltage v as

$$v = i z_c + i_c R'_s \parallel R_e \tag{9.44}$$

and the output impedance as

$$Z_o = \frac{v}{i_c} = z_c \left(1 + \frac{\beta R_e}{R'_s + R_e}\right) + R'_s \parallel R_e. \tag{9.45}$$

Usually $|z_c| \gg R_e$, and we can write

$$Z_o \approx z_c \left(1 + \frac{\beta R_e}{R'_s + R_e}\right). \tag{9.46}$$

Ordinarily z_c is multiplied by a large factor, giving a large output impedance. Hence a common-emitter amplifier is an excellent current source.

9.7 Emitter Follower

We have found that the emitter resistor couples the effect of the collector current back to the input to affect the bias, gain, and input impedance. This is called *feedback*. In the emitter follower, we take this to an extreme, eliminating the collector resistor altogether (Figure 9.12a) and using the emitter resistor as the load.

When the base–emitter diode is conducting, the output voltage is the same as the input voltage, except for the base–emitter diode drop V_f. This means that for

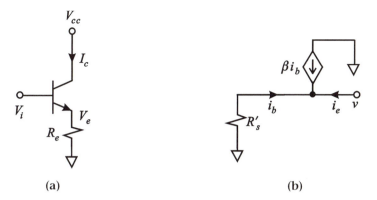

Figure 9.12. Emitter follower (a), and the small-signal circuit model for finding the output impedance Z_o (b).

AC voltages, the voltage gain is 1. Because the output voltage has the same sign as the input voltage, the emitter follower does not invert. A follower does not have any voltage gain, but it does have full current gain, and for this reason it can increase the power. It is useful as a buffer amplifier because the input impedance is very high. We can insert an emitter follower between a source and load when we have a source such as an oscillator or filter that might be affected by load changes. The formulas that we developed for biasing (Equation 9.27) and input impedance (Equation 9.34) for the common-emitter amplifier still hold, because we are using the base for the input as before. The Miller capacitance is not a problem because the collector voltage is constant. However, the output impedance is quite different, because the output is taken from the emitter, which is a low-impedance point. The small-signal equivalent circuit is given in Figure 9.12b. We write the output impedance as

$$Z_o = v/i_e. \tag{9.47}$$

We can write the output voltage as

$$v = -i_b R_s', \tag{9.48}$$

where R_s' is given by

$$R_s' = R_s + r_b. \tag{9.49}$$

We write the current i_e as

$$i_e \approx -\beta i_b. \tag{9.50}$$

If we divide v by i_e, we get the output impedance

$$Z_o \approx R_s'/\beta. \tag{9.51}$$

In practical amplifiers, this is quite small, as low as an ohm, making the emitter follower a nearly ideal voltage source. Also notice that in the small-signal model, the collector is an AC ground. The emitter follower is classified as a common-collector amplifier for this reason, even though the collector is actually connected to the supply, not to ground. In the NorCal 40A, there are emitter followers in the Audio Amplifier and in the oscillator circuits in the Product Detector and the Transmit Mixer.

9.8 Differential Amplifier

A differential amplifier amplifies the difference between two signals. Differential amplifiers appear in many places in electrical engineering, because they allow interference to be canceled. In addition, thermal drifts can be reduced. The mixers and the Audio Amplifier in the NorCal 40A are based on differential amplifiers. A representative differential amplifier is shown in Figure 9.13. The actual mixer and Audio Amplifier circuits are considerably more complicated than this, and we will wait until the later chapters for a discussion. The amplifier is made up of a pair of identical common-emitter amplifiers with the emitter resistors tied together. The resistor R_t connects the emitter resistors to ground. This design is called a *long-tailed pair*. The "tail" is the common resistor R_t, and "long" means that $R_t \gg R_e$. We will see why this is useful shortly.

Consider what happens if we apply equal and opposite input voltages,

$$v_{i1} = -v_{i2}. \tag{9.52}$$

This sign change is also reflected in the emitter currents. We write

$$i_{e1} = -i_{e2}. \tag{9.53}$$

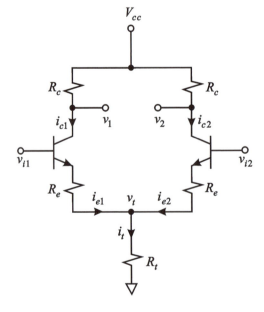

Figure 9.13. The long-tailed pair differential amplifier. The voltages and currents are AC signals. Bias voltages and circuits are not shown.

The tail current i_t is given by

$$i_t = i_{e1} + i_{e2} = 0. \tag{9.54}$$

Thus the tail voltage v_t is zero, and so each amplifier is effectively grounded. We write the collector voltages as

$$v_1 = -(R_c/R_e)v_{i1}, \tag{9.55}$$
$$v_2 = -(R_c/R_e)v_{i2}. \tag{9.56}$$

The output voltage v_d is the difference between the collector voltages. We write

$$v_d = v_1 - v_2 = -(R_c/R_e)(v_{i1} - v_{i2}) = -(R_c/R_e)v_{id}, \tag{9.57}$$

where v_{id} is the differential input voltage given by

$$v_{id} = v_{i1} - v_{i2}. \tag{9.58}$$

The gain for a differential input voltage is the same as the gain of one of the individual amplifiers in the pair. We write the differential gain G_d as

$$G_d = -R_c/R_e. \tag{9.59}$$

In calculating the output resistance, we assume that the internal transistor collector resistance r_c is large and can be neglected. This means that the only path between the two output terminals is the series connection of the two collector resistors. This gives us the differential output impedance

$$Z_d = 2R_c. \tag{9.60}$$

Now consider what happens if we apply the same voltage to each input, so that

$$v_{i1} = v_{i2}. \tag{9.61}$$

We call this a *common-mode* voltage to distinguish it from the differential voltage. By symmetry, the emitter currents are the same, and we write

$$i_{e1} = i_{e2}. \tag{9.62}$$

The tail current i_t is given by

$$i_t = i_{e1} + i_{e2}. \tag{9.63}$$

We can write the input voltages as

$$v_{i1} = R_e i_{e1} + R_t i_t = (R_e + 2R_t)i_{e1}, \tag{9.64}$$
$$v_{i2} = R_e i_{e2} + R_t i_t = (R_e + 2R_t)i_{e2} \tag{9.65}$$

and the output voltages as

$$v_1 = -R_c i_{c1} \approx -R_c i_{e1} = -\frac{R_c v_{i1}}{R_e + 2R_t}, \tag{9.66}$$
$$v_2 = -R_c i_{c2} \approx -R_c i_{e2} = -\frac{R_c v_{i2}}{R_e + 2R_t}. \tag{9.67}$$

The outputs are equal. We call the common-mode input voltage v_{ic} and the common-mode output voltage v_c, and we write

$$v_c = -\frac{R_c v_{ic}}{R_e + 2R_t}. \tag{9.68}$$

The common-mode gain G_c is given by

$$G_c = \frac{v_c}{v_{ic}} = -\frac{R_c}{R_e + 2R_t}. \tag{9.69}$$

In practice, R_t is usually much larger than R_e, which may be small, or even omitted. This means that $G_c \ll G_d$. This is useful because interference often appears on both inputs at the same time, so that it is a common-mode signal.

In the NorCal 40A mixers, we only apply the signal v_i to one input, and the other input is bypassed to ground. In this case, we consider that we have both a differential input voltage v_{id}, given by

$$v_{id} = v_i, \tag{9.70}$$

which is the difference between the two inputs, and a common-mode voltage v_{ic}, which is the average of the two inputs:

$$v_{ic} = \frac{v_i + 0}{2} = \frac{v_i}{2}. \tag{9.71}$$

We can then find the differential output voltage as

$$v_d = G_d v_{id} = G_d v_i \tag{9.72}$$

and the common-mode output voltage as

$$v_c = G_c v_{ic} = G_c v_i/2. \tag{9.73}$$

This is just the average of the two output voltages and would be much smaller than the differential output.

9.9   Field-Effect Transistors

In addition to bipolar transistors, there is another major device family called field-effect transistors. We found that bipolar transistors are controlled by an input current. In contrast, field-effect transistors are controlled by an input voltage. There are many types of field-effect transistors. The major use is in computer circuits, but they are important in other areas as well. The field-effect transistor in our transceiver is called a junction field-effect transistor, or JFET for short. JFETs come in two varieties: n-channel, where the charge carriers are electrons, and p-channel, where the carriers are holes. Generally speaking, electrons move faster than holes, and for this reason, n-channel JFETs are more common than p-channel JFETs. In

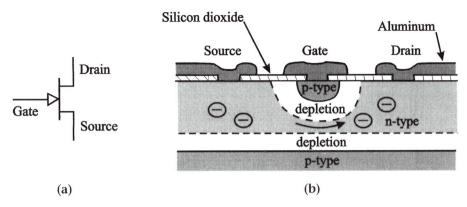

Figure 9.14. Schematic symbol for an n-channel JFET (a), and a cross section (b).

the NorCal 40A, there are three JFET circuits: the Buffer Amplifier in Problem 23, the VFO in Problem 26, and the Automatic Gain Control in Problem 32.

The circuit symbol and construction of an n-channel JFET is shown in Figure 9.14. The silicon wafer itself is p-type. The top part of the wafer is converted to n-type by putting it in a furnace with n-type impurities. This forms a diode that is called the body diode. Next a pn diode is made in the n-layer. The connection to this pn diode is called the *gate*. The gate is the input terminal for the device. In addition, two metal contacts are added on each side of the gate. These are called the *source* and *drain*. In a circuit, the drain is usually given a positive bias and electrons flow from the source to the drain. The gate, source, and drain correspond to the base, emitter, and collector in a bipolar transistor, and most people probably wish that the same names had been kept for both.

The voltage on the gate controls the current. A JFET is operated with the gate in reverse bias, so that the gate diode does not conduct. This gives JFET amplifiers a very high input impedance. Associated with the gate diode and body diode are regions where there are no charge carriers and the silicon effectively becomes an insulator. These insulating regions are called *depletion* regions. A depletion region arises because carriers diffuse across the junction between the p and n regions and recombine, leaving charged impurity atoms that expel charge carriers. The electrons flow between the depletion regions of the gate diode and the body diode. This region is called the *channel*. The thickness of the depletion region depends on the diode voltage. The p-type substrate is typically connected to the source, so that the thickness of its depletion region is fixed. As the gate voltage is made more negative, its depletion width increases. This reduces the channel height and decreases the current. This characteristic is shown in Figure 9.15 for the J309 JFET that is used in the NorCal 40A. If we make the gate voltage sufficiently negative, the channel closes off entirely, and no current flows. This voltage is called the *cut-off voltage* V_c.

A field-effect transistor has two different regions of operation, depending on whether the drain voltage is small or large. These correspond to the saturation region and the active region in a bipolar transistor. The low-voltage region in a

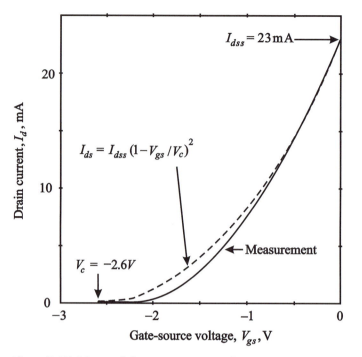

Figure 9.15. Measured drain current I_d as a function of gate-source voltage V_{gs} for the J309 JFET in the NorCal 40A. The drain-source voltage is fixed at 5 V. The dashed line is a plot of Equation 9.74 with $I_{dss} = 23$ mA and $V_c = -2.6$ V.

field-effect transistor is called the linear region, and we will study it in Chapter 13 when we consider the Automatic Gain Control. In this chapter, we consider only the active region where the drain voltage is large. The JFET is relatively difficult to analyze precisely, but we can write the drain current I_d with tolerable accuracy in the active region as

$$I_d = I_{dss}(1 - V_{gs}/V_c)^2, \tag{9.74}$$

where I_{dss} is the current at zero gate-source voltage. The abbreviation *dss* denotes *drain-source* current with the gate *shorted* to the source. This formula is plotted in Figure 9.15 for comparison.

The slope of the drain-current curve dI_d/dV_{gs} is the transconductance g_m. It corresponds to the current gain β in a bipolar transistor, because it tells us how much change in output current we get for a change in the input voltage. We write

$$g_m = \frac{dI_d}{dV_{gs}}. \tag{9.75}$$

Figure 9.16 shows a plot of the transconductance. It is large near zero volts and decreases to zero at the cut-off voltage V_c. This characteristic is used in the VFO start up in Problem 26. The oscillation begins with a zero-voltage bias, where the transconductance is large. This assures that the oscillation starts properly. As the

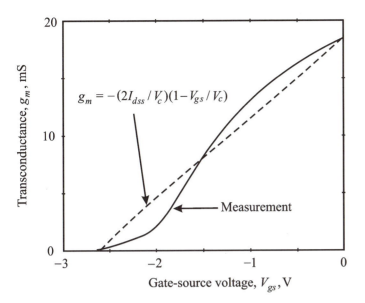

Figure 9.16. Measured transconductance g_m as a function of gate-source voltage V_{gs} for the J309 JFET. The drain-source voltage is fixed at 5 V. The dashed line is a plot of Equation 9.76 with $I_{dss} = 23$ mA and $V_c = -2.6$ V. The data sheets in Appendix D give further information.

oscillation reaches a steady state, the bias shifts to near the cut-off voltage, where g_m is much smaller. This gives a stable operating condition. We can differentiate Equation 9.74 to find an approximate formula for g_m:

$$g_m = -(2I_{dss}/V_c)(1 - V_{gs}/V_c). \tag{9.76}$$

This relation is a straight line, and it is also shown on Figure 9.16.

We can draw a simple small-signal equivalent circuit for the JFET (Figure 9.17a). The gate input draws no current; thus we leave it open-circuited. We represent the

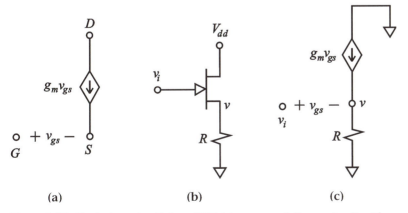

Figure 9.17. Equivalent circuit for a JFET (a), a source-follower circuit with a load R (b), and the equivalent AC circuit for a source follower (c).

control of the drain current by a dependent current source $g_m v_{gs}$. This model is appropriate for the source follower in Problem 23 and the VFO start up in Problem 26. At higher frequencies, capacitances would need to be added to the model. These are given in the data sheets in Appendix D.

9.10 Source Follower

The source-follower circuit (Figure 9.17b) is similar to the bipolar emitter follower. In our transceiver, a source follower is used to isolate the Transmit Mixer from the Driver Amplifier, so that changes in the Driver Amplifier impedance do not affect the Transmit Mixer. One attractive feature of the source-follower circuit is that the load resistor R determines the bias voltage and current – no additional bias components are needed. Let us see how this works. We assume that the gate has a DC connection to ground, so that its DC voltage is zero. For example, in the Buffer Amplifier there is an inductor (L6) that makes this connection. We can write the source voltage as $I_b R$, where I_b is the bias current. This gives us a relation between gate-source bias voltage V_b and the current. We can write

$$V_b = -I_b R. \tag{9.77}$$

This is the equation of a straight line through the origin. It is called a *load line*. In addition, the bias voltage and current must satisfy the relationship given by Figure 9.15. We can find the solution graphically by drawing the load line on this plot and finding the intersection. This is shown in Figure 9.18.

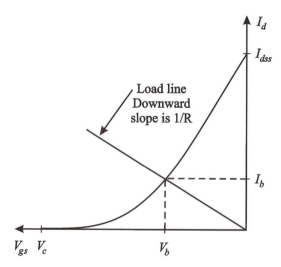

Figure 9.18. Finding the bias voltage V_b and current I_b for the JFET source follower. The bias point is at the intersection of the load line and the drain-current plot.

In the equivalent AC circuit, we replace the drain supply with a short circuit to ground, because the supply does not affect an AC current (Figure 9.17c). From this circuit, we can write an expression for the output voltage v in terms of the gate-source voltage v_{gs}:

$$v = R\, g_m v_{gs}. \tag{9.78}$$

The gate-source voltage is written as the difference between the input voltage v_i and the output voltage v:

$$v_{gs} = v_i - v. \tag{9.79}$$

Between these two equations, we can eliminate v_{gs}, and write

$$v = \frac{R v_i}{1/g_m + R}. \tag{9.80}$$

This is the same formula we get for a potential-divider circuit with source resistance $1/g_m$ and load resistance R. This means that the output impedance of the follower is

$$Z_o = 1/g_m. \tag{9.81}$$

We can write the voltage gain as

$$G_v = \frac{v}{v_i} = \frac{1}{1 + \frac{1}{g_m R}}. \tag{9.82}$$

The gain is near 1 if $R \gg 1/g_m$.

FURTHER READING

A good introductory textbook for amplifiers is *Microelectronic Circuits and Devices*, by Mark Horenstein, published by Prentice-Hall. The classic advanced textbook is *Analysis and Design of Analog Integrated Circuits*, by Paul Gray and Robert Meyer, published by Wiley. This book emphasizes integrated-circuit amplifiers. *Device Electronics for Integrated Circuits*, by Muller and Kamins, published by Wiley, gives a derivation of the current characteristics for a JFET. *Radio Frequency Design*, by Wes Hayward, published by the American Radio Relay League, has a detailed discussion of JFET amplifiers.

PROBLEM 21 – DRIVER AMPLIFIER

Figure 9.19 shows the Driver Amplifier and the measurement connections. The transformer T1 and the resistor R14 were installed in Problem 15. You should solder in the transistor Q6 and the emitter resistor R12, leaving room on each to attach probes. We use R12 to measure the bias current. Measure the resistance of R12, and make a note of it for later.

Solder in R13. R13 is a variable resistor, or potentiometer ("pot" for short). It is wired so that the resistance can be adjusted from near zero to 500 Ω. R13 controls the gain of

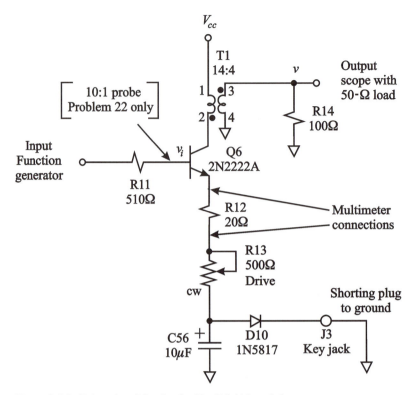

Figure 9.19. Driver Amplifier for the NorCal 40A and the measurement connections.

the amplifier. You should adjust the pot so that it is fully clockwise, leaving it at its lowest resistance. Install C56 and D10, and insert a shorting plug into the Key jack J3. Install the end of R11 that is connected to the base of Q6, and leave the other end free for attaching the function generator. Plug in the power supply and hook up the function generator at the free end of R11 as shown in Figure 9.19. Attach the oscilloscope to R14 with a 50-Ω termination.

The next job is to adjust the amplitude and offset of the function generator. You should work for a large sine-wave output at 7 MHz with high efficiency. A good starting point is an amplitude of 2 V and an offset of 0.5 V. If the offset is too small, the output will clip, because the transistor will turn off during part of the cycle (Figure 9.20a). However, be careful when you increase the offset, because the current grows quickly, reducing the efficiency. It is a good idea not to let the emitter current exceed 50 mA to avoid burning out the transistor. You may feel it getting hot. If the amplitude is too high, the waveform will bottom out when the transistor saturates (Figure 9.20b).

A. Once the amplitude and offset are adjusted, measure the output voltage and calculate the output power P. The load resistance is the parallel combination of R14 and the scope termination.

B. Now we calculate the power delivered by the power supply. Record the DC voltage across R12, and calculate the emitter current. Now use the multimeter to measure

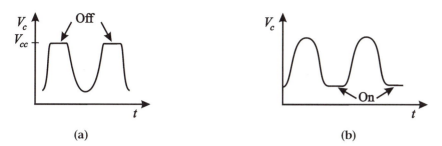

Figure 9.20. Output waveforms for setting the offset and amplitude of the function generator. In (a), the offset is too small, and in (b), the amplitude is too large.

V_{cc}. It is best to measure this voltage at the end of the 1-Ω resistor across S1 that connects to T1. Calculate the supply power P_o and the efficiency η.

C. Next we find the gain. Calculate the available power P_+ from the function generator, using the amplitude setting on the function generator for V_+. Take the source resistance to be the sum of R11 and the function-generator resistance. Calculate the gain G in dB.

PROBLEM 22 – EMITTER DEGENERATION

This is a continuation of the previous problem. This time, we will concentrate on the voltage gain and the input impedance. In the Driver Amplifier, there are two emitter resistors in series, R12 and R13. R12 is a fixed 20-Ω resistor, and R13 is a variable resistor that can be adjusted anywhere from a very low resistance to a resistance of about 500 Ω. This allows the output power to be controlled. The fixed resistance R12 sets the maximum voltage gain.

Make the connections for the previous problem, and add a 10:1 scope probe at the base end of R11. This allows us to measure the AC input voltage v_i. You should use a 7-MHz sine wave with the same offset as before, but use a smaller amplitude of 1 Vpp to reduce the distortion in the voltage waveforms.

A. Measure the voltage gain $G_v = v/v_i$ with R13 set fully clockwise (maximum gain) and fully counterclockwise (minimum gain).

B. Calculate what you expect the voltage gain to be for each setting. You need to take the turns ratio of the transformer into account. Use the multimeter to measure the total emitter resistance for R12 and R13 for each setting. Turn off the function generator and disconnect the power supply when you make resistance measurements. Otherwise there will be errors in the readings.

C. Measure the magnitude of v_i at the maximum gain setting. We can use this to deduce the Miller capacitance, with the equivalent circuit in Figure 9.21 as a guide. In the circuit, we have a 560-Ω source resistance and an open-circuit voltage $v_o = 2$ V, twice the amplitude setting. We will assume that the input impedance of the amplifier is dominated by the Miller capacitance and ignore the rest. Use the equivalent circuit to find an expression for the magnitude of v_i in terms of the Miller capacitance C_m.

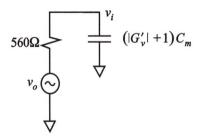

Figure 9.21. Equivalent circuit for extracting the Miller capacitance.

You will need to consider the effect of the transformer carefully. The voltage gain G'_v that is needed is the ratio of the effective collector resistance to the emitter resistance, not the gain you measured at the load. Solve for C_m. When you are finished, solder in the other end of R11, and remove the shorting plug from J3.

PROBLEM 23 - BUFFER AMPLIFIER

The Driver Amplifier has an input impedance that is dominated by Miller capacitance. The effective input capacitance varies over a large range, from 10 pF at low gain to 105 pF at high gain. In the transceiver, the input for the Driver Amplifier comes from the Transmit Mixer, after passing through the Transmit Filter. A large change in load capacitance would upset the filter resonance if we put the filter output into the Driver Amplifier directly. In order to isolate the filter from changes in the driver amplifier, the transceiver has a JFET Buffer Amplifier (Figure 9.22). The Buffer Amplifier is a source-follower circuit, which has a high input impedance but no voltage gain.

To start, solder in the JFET Q5, C36, and R10 for the amplifier. Leave R10 a few millimeters off the board, so that you can get a scope probe on it. C36 is a *bypass* capacitor. It ensures that the AC impedance that the drain sees is low, less than an ohm, to prevent oscillations. The resistor R10 has the same purpose. It is dangerous to put a resonant circuit at the input of a transistor unless there is a resistance to damp out oscillations.

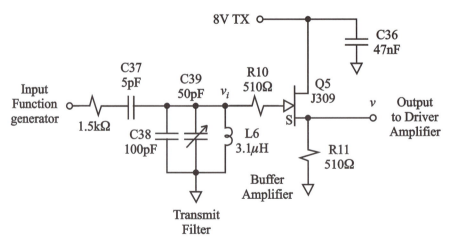

Figure 9.22. Buffer Amplifier for isolating the Driver Amplifier from the Transmit Filter.

In addition, you should solder one end of a 1.5-kΩ resistor into the hole #4 of U4. This provides a connection to C37 for the function generator. We use the 1.5-kΩ resistor to play the role of the Transmit Mixer. The Buffer Amplifier requires 8V TX to operate, and we must put a shorting plug into the Key jack J3 for this.

A. The source resistor R11 determines the bias of both the JFET and the Driver Amplifier. Initially set the gain of the Driver Amplifier to its minimum value, with the R13 pot set fully counterclockwise. This gives the Driver Amplifier a high impedance. Connect the power supply and turn it on. Measure the DC voltage of the source of the JFET with the multimeter. It is convenient to do this measurement by hooking the multimeter lead onto R11. Calculate the drain bias current.

B. For comparison, use Figure 9.15 to calculate the source voltage and drain current that you should expect for a source resistor of 510 Ω.

C. Next we measure the voltage gain of the Buffer Amplifier. This is tricky, because the scope probe affects the resonance of the filter, and the filter must be tuned each time the probe moves. The gain of the Driver Amplifier should be left at its minimum value. Connect the function generator to the circuit through the 1.5-kΩ resistor. Initially the 10:1 scope probe should be attached to the filter end of R10 (Figure 9.22) to measure the input voltage v_i. Set the frequency to 7 MHz and the amplitude to 1 Vpp. The probe capacitance affects the filter resonance, and you should adjust C39 to give a maximum probe voltage. The adjustment is coarse, and the capacitance and pressure of the screwdriver affect the capacitance. You can find the maximum voltage by varying the frequency a few kilohertz. Record the maximum input voltage. Now move the probe to R11 to measure the output voltage v. You will need to adjust C39 again. Find the voltage gain G_v, given by

$$G_v = v/v_i. \tag{9.83}$$

Use this value of G_v to deduce the transconductance g_m.

D. For comparison, use Figure 9.16 to find the transconductance you would expect for the bias voltage you measured.

E. In measuring the input voltage, you attached the 10:1 probe at the filter end of R10, rather than at the gate end. To see why, measure the ratio of these two voltages. Make sure that you retune C39 for these measurements. Now calculate the ratio that you expect, assuming that the impedance of the Buffer Amplifier is very high.

F. Calculate the available power P_+ from the function generator through the 1.5-kΩ resistor and the power P delivered to the 510-Ω load. Make sure that C39 is tuned. Calculate the power gain G in dB.

10

Power Amplifiers

Class-A amplifiers produce outputs with little distortion because the transistors are biased and driven so that they are always active. However, when a transistor is active, the voltage and current are large at the same time, so that the dissipated power is substantial and the efficiency is poor, in the range of 35%. In addition, the amplifier dissipates power even when there is no output. These are severe limitations for even modest output power levels; consequently, few power amplifiers run Class A. To eliminate the power drain when there is no signal, we can leave the transistor unbiased, so that it does not dissipate power when it is off. In addition, if we drive the transistor clear to saturation, using the transistor as a switch, the dissipated power can be greatly reduced because the saturation voltage is low. This is Class-C amplification, which achieves excellent efficiencies, in the range of 75%. We will also see variations of Class C, the Class D, E, and F amplifiers, that achieve even higher efficiencies. The disadvantage of operating Class C is that the output amplitude no longer follows the input level. There is significant distortion at both low and high levels. We say the amplifier is *nonlinear*, and this presents challenges in amplifying signals that vary in frequency and amplitude at the same time, such as music in stereo amplifiers. However, Class C is quite suitable for signals that simply turn on and off, such as Morse Code in the NorCal 40A, or signals that only vary in frequency, such as FM transmissions. Class C also works for signals that vary in amplitude alone, such as AM broadcasts, because the amplitude can be controlled by the supply voltage.

Class-B amplifiers are active for half of a cycle. They usually have a small bias to reduce low-level distortion, and they operate at output levels below saturation to stay linear. They represent a compromise between the poor efficiency of Class A and the distortion of Class C. The bias is usually small enough that the power drain is not a problem. They achieve efficiencies that are substantially better than Class A, but not as good as Class C, in the range of 60%. The output stage of the Audio Amplifier in the NorCal 40A is a Class-B amplifier. Figure 10.1 shows the relationships among the different amplifier classes.

Even efficient transistor amplifiers are often limited in their output power by heat. Transistors are small, and this makes it difficult to get the heat out. In addition, transistors are limited to modest operating temperatures, and they are made of semiconductors that are only fair conductors of heat. This is in contrast

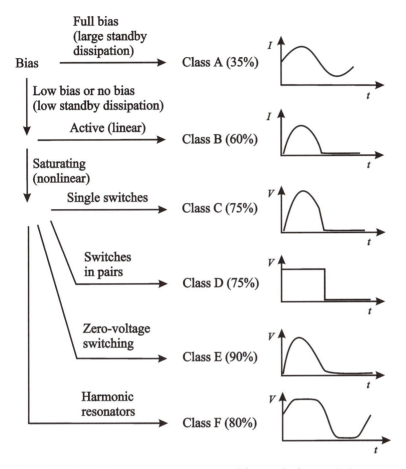

Figure 10.1. Classes of transistor power amplifiers with characteristic waveforms. Representative efficiencies at radio frequencies are given in parentheses. The classification can be quite confusing, partly because transistor and vacuum-tube amplifiers operate differently, and partly because usage has changed over time. Transistor amplifiers achieve high efficiency by saturating, whereas vacuum-tube amplifiers achieve high efficiency by being active only over a small part of the cycle. Traditionally Class C refers to amplifiers that conduct over less than half of the cycle. This is appropriate for tube amplifiers, but difficult to apply to the transistor Power Amplifier in the NorCal 40A, where the transistor conducts for a full half cycle. We will call saturating amplifiers Class C even if they conduct over more than half a cycle. We will also call amplifiers Class C if they conduct for less than half a cycle, even if they never saturate. We will see that the oscillators we study in the next chapter operate in this mode. Class D uses transistors as switches in pairs to increase both efficiency and power. Class E uses a network that allows switching when the voltage is low, achieving spectacular efficiencies, as high as 90%. Class-F amplifiers add harmonic resonators to shape the voltage waveforms to reduce the peak voltage.

to vacuum tubes, which are large structures of metal, glass, and ceramic that can operate at high temperatures and dissipate large amounts of heat. Transistor limitations make it important to understand their thermal behavior. We can make a thermal model for a power amplifier that is the analog of an RC circuit.

10.1 Class-C Amplifiers

Figure 10.2 shows the Class-C Power Amplifier in the NorCal 40A. It is a common-emitter amplifier, with no emitter resistor at all. The collector voltage is supplied through a large inductor called an *RF choke*. The choke gives the supply a large RF impedance, so that it effectively acts as a DC current source. The load current is taken through a large capacitor called a *DC block*. The blocking capacitor has a small RF impedance that does not affect AC currents, but it prevents DC current from the supply from getting to the antenna. Many antennas have transformers that make them DC short circuits. The Harmonic Filter removes the harmonic components (Problem 13).

We represent the transistor by a switch that opens and closes at the operating frequency (Figure 10.3a). When the switch is open, the transistor is off. When it is closed, the transistor is on. A voltage source V_{on} is included to take the on-voltage into account. This is an approximation, because the on-voltage varies with the current. When the switch turns off, there is a large ringing voltage caused by the

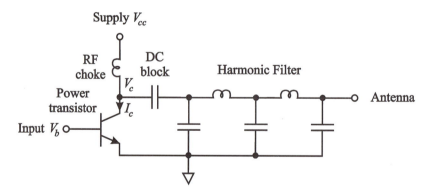

Figure 10.2. Class-C Power Amplifier in the NorCal 40A.

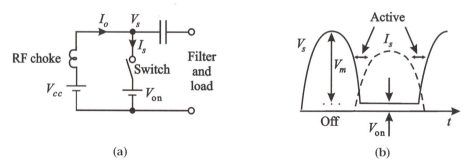

Figure 10.3. Switch model for the Class-C amplifier (a), and switch voltage V_s and current I_s waveforms (b).

Harmonic Filter. We saw a voltage like this in the transistor switch in Problem 5. As an approximation, the switch voltage V_s is a rectified cosine wave superimposed on V_{on} (Figure 10.3b). The switch current I_s is zero when the switch is off. When the switch turns on, current flows. There is some overlap with the cosine voltage, because the transistor is active during the transition from off to on. There is another active period when the switch turns off.

First we can relate the supply voltage V_{cc} to the switch voltage V_s. We can write V_s in the form

$$V_s(t) = \begin{cases} V_{\text{on}} + V_m \cos(\omega t), & \text{switch off,} \\ V_{\text{on}}, & \text{switch on,} \end{cases} \qquad (10.1)$$

where V_m is the peak value of the rectified cosine wave. Because the choke has no DC resistance, the DC or average value of V_s must be the same as the supply voltage. The average value of the rectified cosine over a full cycle is V_m/π. Therefore we can write

$$V_{cc} = V_{\text{on}} + V_m/\pi. \qquad (10.2)$$

We can write V_m in terms of the supply voltage as

$$V_m = \pi(V_{cc} - V_{\text{on}}). \qquad (10.3)$$

This formula indicates that we can interpret $V_{cc} - V_{\text{on}}$ as the effective supply voltage.

We can write the power from the supply P_o as

$$P_o = V_{cc} I_o, \qquad (10.4)$$

where I_o is the DC supply current. We can also write an expression for the switch loss. Because the blocking capacitor passes no DC current, the DC switch current must also be I_o. For now, we assume that the time that the transistor is active is small, and we neglect it. Therefore we can write the power dissipated in the switch P_d as

$$P_d = V_{\text{on}} I_o. \qquad (10.5)$$

Now that we have accounted for the power lost in the transistor, the remaining power must be the output P. We can write

$$P = P_o - P_d = (V_{cc} - V_{\text{on}})I_o. \qquad (10.6)$$

The efficiency is given by

$$\eta = P/P_o = (V_{cc} - V_{\text{on}})/V_{cc}. \qquad (10.7)$$

The efficiency is the ratio of the effective supply voltage to the actual supply voltage. This formula tells us that to make efficient amplifiers, we need to make the on-voltage small and the supply voltage large. We can reduce the on-voltage by reducing the DC current, but this also reduces the output power. Moreover, we have to be careful not to exceed the manufacturer's limit for peak voltage.

To go further with the analysis, we need to interpret the voltage waveform as a sum of harmonic frequency components. This is called a Fourier series. In

Appendix B, Section 3, we derive a formula for the Fourier coefficients. They are given by

$$V_s(t) = V_{cc} + \frac{V_m}{2}\cos(\omega t) + \frac{2V_m}{\pi}\left(\frac{\cos(2\omega t)}{3} - \frac{\cos(4\omega t)}{15} + \frac{\cos(6\omega t)}{35} - \cdots\right).$$

(10.8)

The first term in the sum is just the DC voltage. The next term is the component at the transmitter frequency. This is called the *fundamental*. The peak value of the fundamental component is $V_m/2$. This makes sense, because we have only half a cosine. The other components are even harmonics, the second, the fourth, and so on. Because the filter greatly reduces the harmonics at the load, we will not consider the power in them. We will assume that the input impedance of the filter at the fundamental is real and given by R. Since the peak voltage of the fundamental component is $V_m/2$, we can write the output power as

$$P = \frac{V_m^2}{8R} = \frac{\pi^2(V_{cc} - V_{on})^2}{8R}.$$

(10.9)

We can increase the power by increasing the supply voltage, as long as we do not exceed the maximum voltages, currents, and temperatures specified for the transistor. Since the output power varies inversely with the input impedance of the filter, if we want to increase the power, we should reduce the impedance. We considered this in Problem 13. There are limits to how much the power can be increased, however, because lower impedances raise the current, and this increases V_{on}. This reduces both the effective supply voltage and the efficiency. In addition, at high current levels in bipolar transistors, I_c is proportional to $\sqrt{I_b}$ rather than I_b. This is shown in Figure 10.4a. This limits the current that can be driven through the transistor.

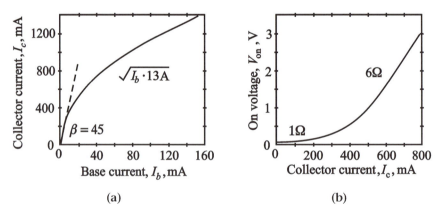

(a)   (b)

Figure 10.4. Effects of large currents in the 2N3553 bipolar transistor used in the Nor-Cal 40A Power Amplifier. (a) Collector current I_c as a function of base current I_b. The collector-emitter voltage is 2.5 V. The relationship at low currents is linear, with a β of 45, but for collector currents above 200 mA, the measurements follow the curve $I_c = \sqrt{I_b \cdot 13\,A}$ quite closely. (b) On-voltage V_{on} as a function of the collector current I_c. The base current is 20 mA. There is a kink in the curve, where the slope changes from $1\,\Omega$ to $6\,\Omega$.

For Class-A amplifiers, we derived an amplifier output impedance, which is useful for writing a Thevenin equivalent circuit and for matching for maximum power transfer. For transistor Class-C amplifiers, the output impedance is usually small, and it is harder to pin down. In a Class-A amplifier, the output impedance tells us how much the output voltage will drop as current is drawn. In a Class-C amplifier, V_{on} plays this role, because it increases with current, and this reduces the effective supply voltage. At the large currents in a power amplifier, the collector voltage makes a gradual transition from saturation to active, so that there is not a simple relation between V_{on} and I_c (Figure 10.4b).

10.2 NorCal 40A Power Amplifier

We can use the formulas we developed for Class-C amplifiers to make a quick but optimistic prediction of the performance of the NorCal 40A Power Amplifier and to compare these predictions with measured values. Figure 10.5 shows the measured collector and base voltage waveforms. The supply voltage is $V_{cc} = 12.8$ V, and $V_{on} \approx 2$ V. The on-voltage of 2 V is much larger than we have seen in the low-level switches in Problems 19 and 20. This large voltage is caused by the large current that flows in the power transistor when it is on. The input impedance of the filter R is close to 50 Ω. From Equations 10.9 and 10.7, we predict a power of 2.9 W and an efficiency of 84%, which are somewhat higher than the 2.5 W and the 78% that was measured.

In both Class-C and Class-D amplifiers, there is additional loss when the transistor is active. If you look carefully at Figure 10.5a, you can see a slope change where the active regions begin. The active transition from off to on is the easier one to see, and it is 17 ns long. The transition from on to off is not as distinct and lasts only 8 ns. For comparison, the on time is 50 ns, and the off time is 68 ns. The off time is actually somewhat less than half of the cycle, which is 143 ns long. The off time corresponds closely to the time that the base voltage in Figure 10.5b is below 0.6 V. It seems surprising that the base–emitter diode conducts through more than half the cycle, because the base is driven with a sine wave. The explanation is that a saturated transistor requires additional time to turn off, because there are a large number of electrons in the base that must recombine or leave first. This is called a *charge-storage* delay. It also occurs in pn diodes, but not in Schottky diodes, which have a metal anode instead of a silicon one.

It is interesting to account for the transistor loss P_d. We can calculate it as

$$P_d = P_o - P = 3.2 \text{W} - 2.5 \text{W} = 700 \text{mW}. \tag{10.10}$$

We start with the active period loss. The active transition from off to on has a larger loss than the one from on to off, because it is longer and the voltages are higher. Physically we account for the loss as a capacitive discharge through the transistor (Figure 10.6a). The input capacitor in the Harmonic Filter C45 (330 pF) is connected through a DC block to the collector of the power transistor Q7. The transistor itself, the zener diode D12, and the RF Filter also have capacitance, but

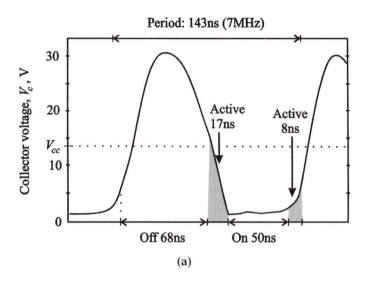

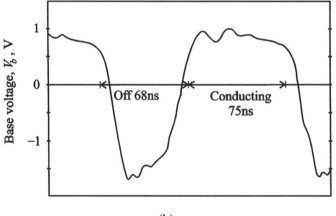

Figure 10.5. Voltages for the NorCal 40A Power Amplifier at 7 MHz. The output power is 2.5 W, with a supply voltage of 12.8 V and a supply current of 250 mA. The efficiency is 78%. Collector voltage (a), and base voltage (b).

all are small, and we will neglect them. The capacitor is charged to 15 V at the beginning of the active period. We can write the capacitive energy E as

$$E = CV^2/2 = 37 \, \text{nJ}, \tag{10.11}$$

where C is the capacitance of C45, 330 pF, and V is 15 V. During the active period, this energy dissipates in the transistor. At the end of the active period, the voltage has dropped to 2 V, and the capacitive energy is less than a nanojoule. We can

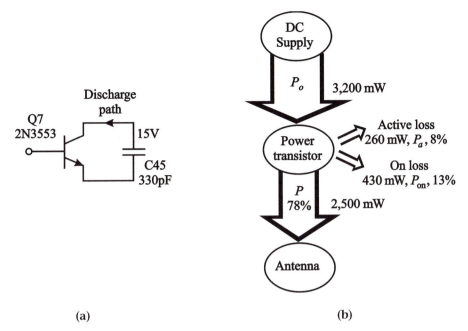

Figure 10.6. (a) Capacitive discharge through the collector of the power transistor. The DC block (C44) is not shown. C45 is charged to 15 V when the active transition begins. (b) The power flow in the NorCal 40A Power Amplifier when the output is 2.5 W.

convert this dissipation to a power P_a by multiplying by the frequency to get

$$P_a = Ef = 260 \, \text{mW}, \tag{10.12}$$

where f is the frequency, 7 MHz.

We can find the current I_c associated with the capacitive discharge by multiplying the capacitor charge Q by f. We write

$$Q = CV = 5.0 \, \text{nC} \tag{10.13}$$

so that

$$I_c = Qf = 35 \, \text{mA}. \tag{10.14}$$

We can calculate the on-period current I_{on} by subtracting I_c from the DC current $I_o = 250$ mA. This gives us

$$I_{on} = I_o - I_c = 215 \, \text{mA}. \tag{10.15}$$

We can find the loss during the on period P_{on} by multiplying I_{on} by $V_{on} = 2$ V:

$$P_{on} = V_{on}I_{on} = 430 \, \text{mW}. \tag{10.16}$$

We get the total dissipated power P_d by adding P_a and P_{on}. This gives us

$$P_d = P_a + P_{on} = 690 \, \text{mW}, \tag{10.17}$$

which is close to the measured value. Figure 10.6b summarizes the power flow in the amplifier.

10.3 Class D

We can extend the idea of a switching amplifier by employing a pair of transistor switches that alternately connect a load to a voltage source and to ground. This is a Class-D amplifier, and a simplified circuit is shown in Figure 10.7a. The switch pair is called a *push–pull* circuit, in contrast to the Class-C amplifier, which is said to be *single-ended*. The supply voltage V_{cc} is connected directly to a switch without a choke. There is a band-pass filter to prevent DC and harmonic currents from reaching the load. We can simplify the circuit by representing the pair of switches by a single double-throw switch at the operating frequency (Figure 10.7b). This gives us a square-wave voltage superimposed on the on-voltage of the transistor (Figure 10.7c). We write the switch voltage V_s as

$$V_s(t) = \begin{cases} V_{cc} - V_{on}, & S_1 \text{ on, } S_2 \text{ off,} \\ V_{on}, & S_2 \text{ on, } S_1 \text{ off.} \end{cases} \tag{10.18}$$

The difference between the maximum and minimum voltages is given by

$$V_m = V_{cc} - 2V_{on}. \tag{10.19}$$

For a Class-C amplifier, we interpreted the rectified cosine in terms of its frequency components. We can also write a square wave this way. The coefficients are derived in Appendix B, in Section 2. The voltage $V_s(t)$ is given by

$$V_s(t) = \frac{V_{cc}}{2} + \frac{2V_m}{\pi}\left(\cos(\omega t) - \frac{\cos(3\omega t)}{3} + \frac{\cos(5\omega t)}{5} - \cdots\right). \tag{10.20}$$

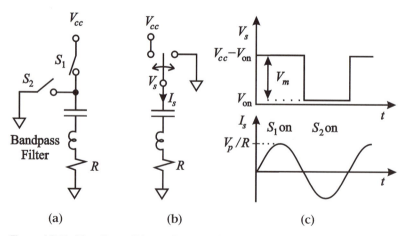

(a)   (b)   (c)

Figure 10.7. Class-D amplifier with a pair of transistor switches (a), and a simplified circuit with a single double-throw switch (b). The switch voltage and current waveforms (c).

There are only odd harmonics. The peak voltage of the fundamental component V_p is given by

$$V_p = 2V_m/\pi. \tag{10.21}$$

This means that we can interpret $V_m = V_{cc} - 2V_{on}$ as the effective supply voltage. Since the filter stops all of the current components except the fundamental, the output power P is given by

$$P = \frac{V_p^2}{2R} = \frac{2V_m^2}{\pi^2 R}. \tag{10.22}$$

The output power is inversely proportional to the load resistance. The input power P_o is given by

$$P_o = V_{cc}I_o, \tag{10.23}$$

where I_o is the average supply current. The supply current is a rectified cosine with a peak value I_p of

$$I_p = V_p/R, \tag{10.24}$$

so that we can write

$$I_o = \frac{I_p}{\pi}. \tag{10.25}$$

We can rewrite this in terms of the effective voltage V_m by substituting from Equation 10.21. This gives us

$$I_o = \frac{2V_m}{\pi^2 R}, \tag{10.26}$$

and the output power P is

$$P = V_m I_o. \tag{10.27}$$

We get the efficiency by dividing Equation 10.27 by Equation 10.23 to get

$$\eta = P/P_o = V_m/V_{cc}. \tag{10.28}$$

The efficiency is the ratio of the effective supply voltage to the actual supply voltage. In practice, the efficiencies of Class-D amplifiers are comparable to Class-C amplifiers. The square-wave voltage of a Class-D amplifier is a major advantage in many situations because the maximum voltage is low, the same as the supply voltage. By comparison, the Class-C amplifier peak voltage is π times larger than the DC voltage. However, the switches must be carefully synchronized, and this limits Class D to lower frequencies. They are common in AM transmitters and power supplies.

10.4 Class E

In the NorCal 40A Power Amplifier, 40% of the loss comes from capacitive discharge when the transistor turns on. One thing that you might notice in Figure 10.5a is that the voltage is already coming down when the transistor goes active. If we substitute a series resonant circuit for a series inductor in the harmonic filter (Figure 10.8a), the voltage can be made to come all the way down to zero before the transistor turns on (Figure 10.8b). This is called zero-voltage switching, and it eliminates capacitive discharge loss. Gerald Ewing demonstrated this idea in 1964, and it is the basis for the Class-E amplifier defined by Nathan and Alan Sokal in 1975. In the Class-E amplifier, both the voltage and its slope are zero when the transistor goes active. This means that even if the switching point is mistimed, or the switching is slow, the loss is still low. Class-E amplifiers are the most efficient amplifiers known, with efficiencies in the 90% range. The disadvantage is that the peak voltages for Class-E amplifiers are even higher than those for Class-C amplifiers.

You might wonder why it is important to work to raise the efficiency from 75% to 90%. There would be a modest improvement in the electric bill for a transmitter or battery life for a portable transceiver. The real answer, however, is heat – the loss is two and a half times lower for 90% efficiency than for 75% efficiency. If an amplifier is limited by the heat it can dissipate, we can calculate the maximum power output as a function of efficiency. We can write the output power P in terms of the supply power P_o as

$$P = \eta P_o \tag{10.29}$$

and the dissipated power P_d as

$$P_d = (1 - \eta)P_o. \tag{10.30}$$

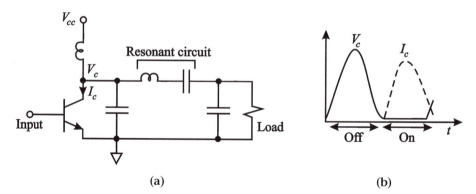

(a)   (b)

Figure 10.8. Class-E amplifier (a), and collector voltage and current waveforms (b). The components in the resonant circuit are adjusted so that the collector voltage comes all the way down to zero before the transistor turns on.

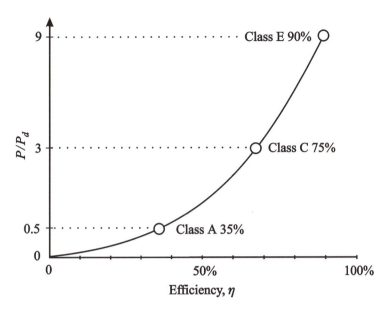

Figure 10.9. Output power versus efficiency for amplifiers that are limited by heat.

We can eliminate the input power between these two formulas and get

$$\frac{P}{P_d} = \frac{\eta}{1 - \eta}. \tag{10.31}$$

The maximum output power increases greatly at high efficiencies (Figure 10.9). As an example, a large power transistor might be able to dissipate 50 W safely. The maximum power that we would be likely to get out of a Class-A amplifier with this transistor would be 27 W. For a Class-C amplifier, we would expect 150 W, and for a Class E, we could get 450 W.

An interesting Class-E amplifier was developed by Caltech undergraduate students that uses the NorCal 40A as a driver. This amplifier allows the output power to be increased from the few watts available from the NorCal 40A to 500 W. The circuit is shown in Figure 10.10a. It uses a field-effect transistor called a MOSFET. MOS is short for *metal-oxide-semiconductor*. In the circuit, C_d sets the current. For higher current, we would increase C_d. L_s and C_s form the resonant circuit. The resonant frequency for L_s and C_s is set somewhat below the operating frequency. L_2 and C_2 transform the antenna impedance of 50 Ω to about 10 Ω, which is appropriate for this amplifier. In addition, L_2 and C_2 have a resonance at the second harmonic. This circuit is called a *trap*, and it acts as a band-stop filter to prevent the second harmonic from getting to the output.

The drain voltage is shown in Figure 10.10b. The peak voltage is 400 V, more than ten times larger than the peak voltage in the NorCal 40A. The efficiency is extremely high, about 90%, and this makes it possible to dissipate the power in an aluminum heat sink without a fan.

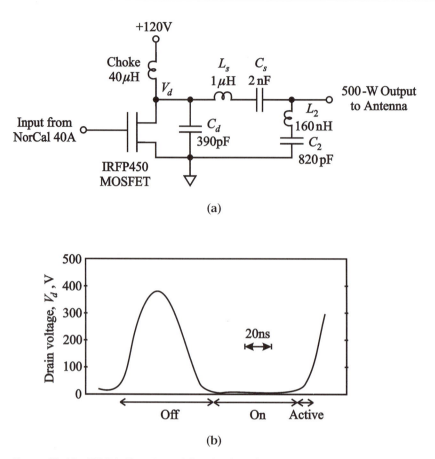

Figure 10.10. 500-W Class-E amplifier developed at Caltech. Circuit diagram (a), and drain voltage (b). For more information, see "High-Efficiency Class-E Power Amplifiers," by Eileen Lau, Kai-Wai Chiu, Jeff Qin, John Davis, Kent Potter, and David Rutledge, in *QST* magazine, Part 1, May 1997, pp. 39–42, and Part 2, June 1997, pp. 39–42.

10.5   Class F

Class-D amplifiers give a square-wave voltage that is attractive because it keeps the peak voltage low. However, Class-C amplifiers are simpler, because they need only one transistor switch. The Class-F amplifier uses a single switch like the Class C, but it adds a third-harmonic trap to produce a flattened voltage as in Class D. We can think of it as a single-ended Class D. The circuit is shown in Figure 10.11a. L_3 and C_3 have a resonance at the third harmonic that causes a third-harmonic component to be added to the collector voltage to flatten it (Figure 10.11b).

The efficiency of Class-F amplifiers is typically higher than that of Class C but lower than that of Class E. Which design is appropriate depends on whether the transistor output is limited by collector voltage or heat. It is not easy to generalize about this, because the answer depends on the frequency and power levels. Also, the amplifier may only operate a fraction of the time. The fraction of the time

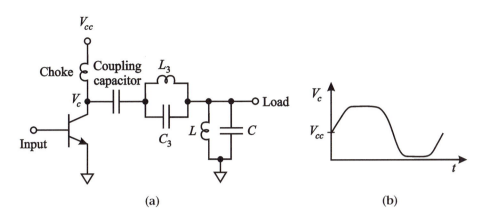

Figure 10.11. Class-F amplifier (a), and collector voltage waveform (b).

that an amplifier is on is called the *duty cycle*. A small duty cycle will reduce the heat load, but the peak voltages are not affected. At higher frequencies in the GHz range, the maximum voltages for transistors may only be a few volts, and a Class-F design may be better. However, for higher-power transmitters at lower frequencies, Class E may be appropriate.

10.6 Class B

The high-efficiency classes C, D, E, and F all are nonlinear, and this means that they are not suitable for amplifying signals that have both frequency and amplitude changes. In mathematical terms, an amplifier is *linear* if we can relate the output voltage V and the input voltage V_i by a scalar multiplier:

$$V = \alpha V_i. \tag{10.32}$$

Notice that this is different from the definition we use in mathematics, where

$$y = ax + b \tag{10.33}$$

would be considered linear. In Figure 10.12, I plot output voltage against input voltage for the NorCal 40A Power Amplifier in Class C. The relationship is not at all linear, and you can see distortion at both low and high levels.

The curve shows a *threshold*, with very little output when the input is less than 1.2 Vpp. Lower voltages are not large enough to turn on the base–emitter diode. We can eliminate threshold distortion by biasing for Class A. However, Class A is inefficient and there is large DC power consumption even when there is no input signal. However, there is a way to reduce the threshold distortion, and to keep high efficiency, if we bias enough to shift the threshold near zero (Figure 10.12). This is Class B, and it combines efficiency with low threshold distortion. You may see it called Class AB to indicate that there is a bias. There is a tradeoff between the amount of bias power and the reduction in distortion.

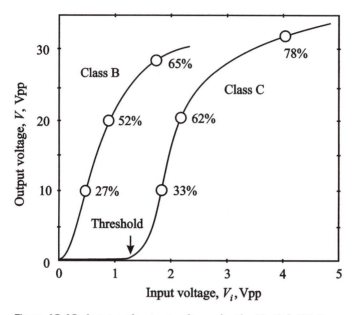

Figure 10.12. Input and output voltages for the NorCal 40A Power Amplifier, with the efficiencies noted on the curves. The curve marked Class C is normal operation, while the curve marked Class B is for a base bias of 500 mV (calibrated for 50-Ω source and load). The base bias greatly reduces the threshold so that the amplifier can be used for linear operation. At a given output, the efficiency for Class-B operation is lower than that for Class C.

Class-B amplifiers are more complicated that either Class-A or Class-C amplifiers, but for many applications they represent the best of both worlds. There is high-level distortion in both Class B and Class C where the transistors saturate. The efficiency rises as we push further into saturation, and this tells us that we must trade off distortion for power and efficiency.

Figure 10.13 shows the collector-voltage waveform during Class-B operation. The transistor is active for only half the cycle, but the voltage is approximately sinusoidal because of the ringing of the harmonic filter. Residual threshold distortion can be seen, as well as the onset of saturation.

We can derive a formula for the maximum possible efficiency for a Class-B amplifier by analyzing idealized waveforms (Figure 10.14a). The voltage is a cosine with a maximum of twice the supply voltage and a minimum of zero. The current is a rectified cosine. The DC current I_o can be written as

$$I_o = I_m/\pi, \tag{10.34}$$

where I_m is the maximum value of the rectified cosine. The supply power is

$$P_o = V_{cc}I_o = V_{cc}I_m/\pi. \tag{10.35}$$

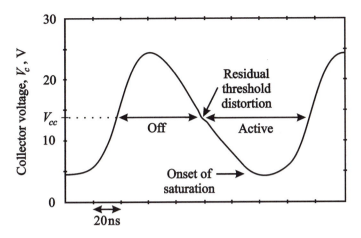

Figure 10.13. Collector voltage for the NorCal 40A Power Amplifier in Class-B operation at 7 MHz. The output is 1 W into a 50-Ω load. The base RF voltage is set for 880 mVpp, with a DC offset of 500 mV, calibrated for a 50-Ω source and load.

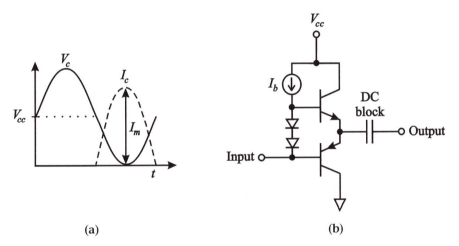

(a)   **(b)**

Figure 10.14. Idealized collector voltage and current waveforms for a Class-B amplifier (a). A push–pull Class-B audio amplifier (b).

Section B.2 shows that for a rectified cosine, the peak value I_p of the fundamental component of the current is given by

$$I_p = I_m/2 \tag{10.36}$$

so that the output power is

$$P = V_{cc}I_p/2 = V_{cc}I_m/4 \tag{10.37}$$

and the maximum efficiency is given by

$$P/P_o = \pi/4 = 79\%. \tag{10.38}$$

This is the ideal efficiency, and 60% would be typical of practical Class-B amplifiers. This is much better than for Class A, but falls short of Class-C efficiencies.

One problem with a single-ended Class-B amplifier is that the harmonic filter limits the bandwidth. We cannot amplify a signal at a frequency and its harmonic at the same time, because the harmonic will be taken out by the filter. This poses a problem in an audio amplifier, where we would like to amplify over a very wide frequency range. To solve this problem, the Audio Amplifier in the NorCal 40A uses a push–pull Class-B amplifier. In a push–pull amplifier, there are two transistors, and each transistor provides current for half the cycle. This amplifier comprises the final stage of the LM386N-1 integrated circuit. There are two stages of Class-A amplification, followed by a final Class-B power amplifier. A simplified version of the power amplifier is shown in Figure 10.14b. It is an emitter follower. When the input voltage is high, the top npn transistor is active, and the bottom pnp transistor is not. When the input voltage is low, the situation is reversed, and the bottom pnp transistor is active. The two diodes set a voltage difference between the two bases that minimizes threshold distortion.

10.7   Thermal Modeling

Controlling the temperature is critical for high-power amplifiers. Manufacturers specify a maximum operating temperature for their devices. This is typically between 150° and 200°C. Because the failure rate for transistors increases dramatically as the temperature increases, it is a good idea to be well below this limit. We will use an electrical analogy to make a mathematical model that relates the dissipated power to the temperature rise. In the analogy, the temperature T corresponds to voltage, and the dissipated power P_d corresponds to current:

$$T \Longleftrightarrow V, \tag{10.39}$$

$$P_d \Longleftrightarrow I. \tag{10.40}$$

We call this a *dual* relationship. In our power amplifier, we clip a piece of metal called a *heat sink* onto the transistor that allows the heat to escape into the air. The power that is lost to the air is proportional to $T - T_0$, where T is the temperature of the heat sink and T_0 is the ambient temperature. This means that we can characterize the heat sink by the ratio of $T - T_0$ to the dissipated power P_d. This ratio is analogous to resistance, and it is called the *thermal resistance* R_t. We write it as

$$R_t = (T - T_0)/P_d. \tag{10.41}$$

The units of thermal resistance are °C/W. In addition to the heat lost to the air, thermal energy is stored in materials as they heat up. The temperature rise is proportional to the stored energy, which is the integral of the power. The integral of power is dual to charge, and thus the ratio of heat energy to temperature rise is like capacitance. It is called the thermal capacitance C_t, and we write

$$C_t T' = P_d, \tag{10.42}$$

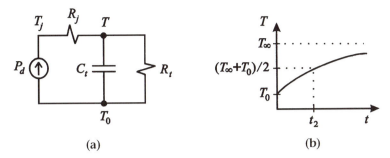

Figure 10.15. Thermal model for the transistor and heat sink (a), and plot of heat-sink temperature T versus time (b).

where the prime denotes a time derivative. The units of thermal capacitance are J/°C.

Figure 10.15a shows a thermal circuit model for the transistor and heat sink. We draw a current source for the power, and include the thermal resistance to air and the thermal capacitance. We need to consider an additional effect. There is a large thermal resistance between the transistor and its package, because the transistor is small, and this makes it difficult to get the heat out. It is traditional to write this resistance as R_j, where j stands for junction. We can write

$$R_j = (T_j - T)/P_d, \tag{10.43}$$

where T_j is the transistor temperature. R_j is in series with the other components. We can understand this if we consider that the thermal power from the transistor must pass to the package and heat sink before we can consider the effect of thermal capacitance and thermal resistance to air. This means that the transistor will have a greater temperature than the sink.

The equations for this circuit are like the ones for RC circuits we studied in Chapter 2. We start with the first-order differential equation

$$f(t) + \tau f'(t) = 0, \tag{10.44}$$

where τ is a time constant. This equation has a decaying solution

$$f(t) = f_0 \exp(-t/\tau), \tag{10.45}$$

where f_0 is the initial value of f. If the right side in Equation 10.44 is zero, the equation is *homogeneous*. We want to solve the *inhomogeneous* equation, where the right side is equal to some value x:

$$f(t) + \tau f'(t) = x. \tag{10.46}$$

In the long term, the derivative will approach zero, and this means that x is just the final value for $f(t)$. We relabel x as f_∞ to show this, obtaining

$$f(t) + \tau f'(t) = f_\infty. \tag{10.47}$$

To solve this equation, define a new variable g, which is the difference between $f(t)$ and the final value f_∞:

$$g(t) = f(t) - f_\infty. \tag{10.48}$$

The derivative of g is the same as the derivative of f:

$$g'(t) = f'(t). \tag{10.49}$$

The equation for g is homogeneous, and it looks the same as Equation 10.44, with gs replacing fs:

$$g(t) + \tau g'(t) = 0. \tag{10.50}$$

This gives us the solution

$$g(t) = g_0 \exp(-t/\tau), \tag{10.51}$$

or in terms of f,

$$f(t) = f_\infty - (f_\infty - f_0) \exp(-t/\tau). \tag{10.52}$$

Now let us apply this approach to find the transistor temperature from the thermal circuit. We start by finding the heat-sink temperature T. We can write the dissipated power P_d as a sum of two parts: a resistive component $(T - T_0)/R_t$ and a capacitive component $C_t T'$. Notice that R_j does not affect the heat-sink temperature. We have

$$P_d = \frac{T(t) - T_0}{R_t} + C_t T'(t). \tag{10.53}$$

We can multiply by R_t and rearrange to find

$$T(t) + R_t C_t T'(t) = P_d R_t + T_0. \tag{10.54}$$

This is in the same form as our inhomogeneous equation (Equation 10.46), and we can rewrite it as

$$T(t) + \tau T'(t) = T_\infty, \tag{10.55}$$

where

$$\tau = R_t C_t \tag{10.56}$$

and

$$T_\infty = P_d R_t + T_0. \tag{10.57}$$

The heat-sink temperature is given by

$$T(t) = T_\infty - P_d R_t \exp(-t/\tau). \tag{10.58}$$

The transistor temperature T_j is given by

$$T_j = T(t) + R_j P_d. \tag{10.59}$$

FURTHER READING

Power amplifiers raise entirely different issues from small-signal amplifiers, and a good place to start is *Radio Frequency Transistors, Principles and Practical Applications*, by Norm Dye and Helge Granberg, published by Butterworth and Heinemann. This book is full of the details of construction and measurements. *Solid State Radio Engineering*, by Herbert Krauss, Charles Bostian, and Frederick Raab, published by Wiley, is an excellent reference. The chapters on power amplifiers were written by Frederick Raab, a pioneer in Class-F power-amplifier development. Raab also gives a list of all the letter classes of amplifiers as they have been defined and redefined in the literature.

Puff is a linear circuit simulator, and therefore it cannot handle nonlinear circuits such as power amplifiers and oscillators. The standard program for simulating nonlinear circuits is called SPICE. This program was developed at the University of California at Berkeley and there are many commercial versions. A popular one is PSPICE, and it is available free from the OrCad Company in a demonstration version that is fine for many problems. Consult the company web site at http://www.orcad.com for more information.

PROBLEM 24 – POWER AMPLIFIER

The Power Amplifier is shown in Figure 10.16. The transistor is the 2N3553, and the manufacturer's data sheet in Appendix D specifies the maximum collector voltage as 40 V.

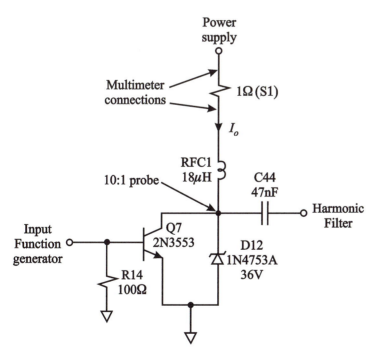

Figure 10.16. Power Amplifier.

D12 is a zener diode across the output that conducts at 36 V to prevent excessive collector voltages. R14 is a 100-Ω resistor across the input. This limits the reverse voltage on the base–emitter diode, which should not exceed 4 V.

To start, solder in the transistor Q7. You should slip a plastic spacer over the leads to keep the can from touching the board. The can is connected to the collector, and it will short out the circuit if it touches the base or emitter solder pads. Then install C44, D12, and RFC1 ("RFC" stands for radio-frequency choke). Leave the cathode of the diode high enough above the board that you can get a probe under it. Slip a heat sink onto the transistor for cooling.

We will use the 1-Ω resistor across S1 for monitoring the current. We will need to measure its resistance accurately. There is a problem in measuring small resistance, however, because the test leads add appreciable resistance. Clip the test leads from the multimeter together to measure the lead resistance. Now measure the resistance of the 1-Ω resistor, and subtract the lead resistance. Record this corrected resistance for future reference.

Connect the scope to the Antenna jack J1 and use a 50-Ω termination. The function generator should be connected across R14. It should be set for a 1-Vpp, 7-MHz sine wave, with no offset. In addition, a 10:1 probe should be connected to the cathode of D12 to monitor the collector voltage.

A. Calculate the peak-to-peak voltage across the 50-Ω load that is required for an output power of 2 W. Gradually increase the function-generator voltage until the output power is 2 W. Sketch the collector voltage. What is the available power from the function generator? Calculate the gain G in dB. Use the multimeter to measure V_{cc} and record this value. A good place to make a connection for measuring V_{cc} is the choke side of your 1-Ω current-sensing resistor.

B. Next make a series of measurements for peak-to-peak output voltages of 5, 10, 15, 20, 25, and 30 V. For each output voltage, use the multimeter to measure the DC voltage across the 1-Ω current-sensing resistor. Calculate the DC supply current I_o, subtracting 2 mA for regulator current. Now using your previous measurement of V_{cc}, calculate the supply power P_o.

C. Calculate the output power for each output voltage. Plot the efficiency η given by

$$\eta = P/P_o, \tag{10.60}$$

with output power P on the x axis. Now plot the power that is dissipated in the circuit P_d, given by

$$P_d = P_o - P, \tag{10.61}$$

with output power P on the x axis.

D. At sufficiently high output voltages the efficiency begins to drop. Extend your plots until you can see this drop off.

PROBLEM 25 – THERMAL MODELING

We will use the thermal model in Figure 10.15a to characterize our amplifier. R_j can be difficult to measure, because the transistor is inside the can. However, since R_j depends only on the characteristics of the transistor and the package, and not on the heat sink, a manufacturer can do this measurement for us. Motorola specifies that for the 2N3553 transistor, $R_j = 25°C/W$.

Thermal time constants are much longer than electrical time constants, and this means that you can take the data by hand as the temperature rises. However, this requires planning, because you cannot hurry things. If something goes wrong during the measurements, you need to wait for things to cool down before you start over. You will be making measurements over a twenty-minute period.

Coat the end of a thermometer with heat-sink compound. This helps to keep the thermal resistance between the thermometer and the heat sink low. Try not to get the compound on anything except the thermometer and the heat sink, because it is difficult to get off.

A. Place the thermometer bulb on the heat sink. Some heat sinks have clips with enough tension to hold the thermometer upright. Make sure that the thermometer is oriented so that you can read the temperature easily. Also make sure that you take each reading with your head in the same place to avoid parallax error. Take an initial temperature reading with the power off. This is the ambient temperature T_0.

B. The connections are the same as in the previous problem, except that you do not need the 10:1 probe. Turn up the function-generator voltage until the output across the 50-Ω scope termination is 30 Vpp. This is an output power of 2.25 W. Take a temperature reading each minute for the first ten minutes. You may need to adjust the function generator from time to time if the output voltage changes.

C. At the end of ten minutes, use the multimeter to measure the voltage across the 1-Ω sense resistor and to measure V_{cc}. From the multimeter measurements, deduce the power being dissipated by the transistor. You should allow for a 2-mA regulator current.

D. Take a final temperature reading at twenty minutes to get T_∞. Use this measurement to calculate R_t and the final value of T_j.

E. Make a plot of heat-sink temperature versus time, and from the plot calculate the time t_2 that it takes for the temperature to come a factor of two closer to the final temperature. A good starting point for this calculation is the temperature after one minute. Often the first minute of a thermal measurement is complicated by the time that it takes to adjust the equipment. Use the measurement of t_2 to calculate C_t. You should wipe the thermal compound off the thermometer after you are done with it.

Now we consider the entire transmitter amplifier chain from the Buffer Amplifier through the Driver Amplifier and the Power Amplifier. Install C48 (10 nF). This should complete the circuits shown on Figure 10.17. The capacitor C48 is next to the Key jack J3. It has a low impedance at the 7-MHz operating frequency, only about 2 Ω, and its purpose is to keep RF voltages on the Keyline low, so that they do not cause mischief.

We can also understand the effect of the Schottky diode D10. This diode only lets current leave the Driver Amplifier circuit; it does not allow it to flow back in. This prevents other circuits from charging up C56.

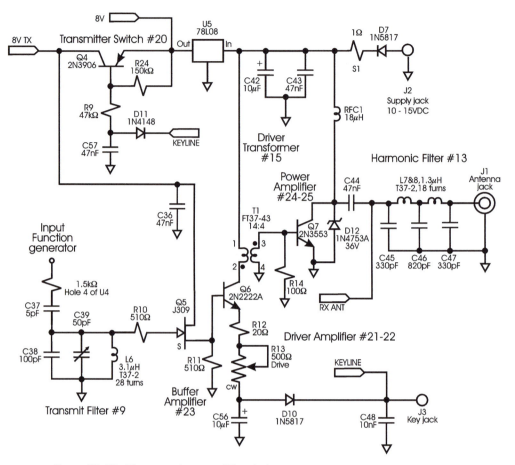

Figure 10.17. The transmitter amplifier chain.

So far, when we have tested the Power Amplifier, we have adjusted the function-generator voltage to set the output power. In the actual transmitter, however, the input signal that comes from the Transmit Mixer is at a fixed level, and the output is set with the Drive pot R13, which controls the gain of the Driver Amplifier.

To start, plug in the power supply, attach leads for the function generator, and connect the Antenna output to the scope with a 50-Ω termination. Insert a shorting plug into the Key jack J3. Set the function generator to a frequency of 7 MHz and an amplitude of 600 mVpp. Check that C39 is tuned for maximum output, and set the Drive pot R13 to give an output of 2 W. You should leave the pot in this position for the rest of the lab.

Now we investigate the bias for the Drive Amplifier. You will need to turn off the signal from the function generator. The bias circuit is simple and elegant. It comes directly from the JFET Buffer Amplifier. The resistor R11 (510 Ω) controls the bias for both the JFET Buffer and the BJT Driver. The value of R11 is chosen by balancing off the bias currents of the Buffer and the Driver. Making the resistance larger reduces the JFET bias current but increases the BJT current.

F.   Find the emitter resistance (R12 and R13 together). In making an ohmmeter reading, you will need to pull out the shorting plug. Make sure you put the shorting plug back after you take the reading, or else the amplifiers will have no bias. Use a multimeter to find the following: the base–emitter diode drop for Q6, the drop across R12 and R13 together, and the diode drop for the Schottky diode D10. What is the emitter bias current I_e?

G. You should notice that the Schottky diode has a lower forward voltage than a pn diode. A Schottky diode is made as a contact between metal and silicon. Usually the silicon will be n-type, and many metals can be used. Gold and platinum are common. What would I_e become if the designer had used a pn diode for D10 instead of a Schottky? You may assume that a pn diode has a forward voltage of 0.6 V. What problem would this new emitter current cause?

The Driver capacitor C56 is important in turning the transmitter signal off. Transmitters should not be turned off suddenly, because the sudden change creates spurious frequency components that are heard by operators on other frequencies as annoying clicks. Typically transmitters are designed to turn off gradually over a period of several milliseconds. The same consideration applies in turning a transmitter on. This timing is controlled by the Transmit Mixer, and we will look at it later.

When the Key jack J3 is short-circuited (key down), the emitter bias current for the Driver flows through the emitter resistors, and out through the Schottky diode D10. The Driver capacitor C56 is charged to the forward voltage of the diode. When the Key jack is open-circuited (key up), the current through D10 stops. This leaves C56 charging through the emitter resistance. The emitter bias current decays exponentially. As the emitter current drops, this reduces the peak-to-peak current that the amplifier can produce, giving the gradual reduction in output power that we need.

H. Measure the time t_2 that it takes the emitter current to drop by a factor of 2, and compare it with theory. To do this measurement, you will need to cycle repeatedly through key up and key down. Use a keying relay cable for this. You should set the frequency to 20 Hz.

I. Finally, we will look at the gain of the entire amplifier and filter chain from the Band-pass Filter through the Buffer, Driver, Power Amplifier, and Harmonic Filter. To start, the function generator should be set to an amplitude of 250 mV and a frequency of 7.03 MHz. This is a little higher than the frequency in previous measurements, and it is the actual frequency that we will use for the transceiver tests. The gain of the Driver Amplifier should be set to its maximum value, with the R13 pot set fully clockwise. Adjust the filter capacitor C39 for maximum output voltage. Vary the frequency a few kilohertz on each side to confirm that the peak is within 10 kHz of 7,030 kHz. Now adjust the amplitude setting on the function generator until the output power P is 2 W. Calculate the power available P_+ from the function generator with the 1.5-kΩ resistor. Calculate the gain in dB of the entire chain. Find the 3-dB bandwidth. You can remove the 1.5-kΩ resistor from hole 4 of U4 when the measurements are completed.

11

Oscillators

It is common in public-address systems to hear a loud tone when someone moves the microphone too close to a speaker. People call this *feedback,* and it is caused by sound from the speaker getting back into the microphone. When this happens, the sound is amplified again and again until the amplifier overloads. It is perhaps surprising that we hear a single tone rather than a broad range of frequencies, and it suggests that we can use feedback to make a sine-wave oscillator. We can distinguish between two kinds of feedback. The public-address oscillations are an example of *positive feedback,* where the output adds to the signal. In *negative feedback,* the output cancels part of the input, reducing the gain. We have had two examples of negative feedback so far, in the emitter resistor of the Driver Amplifier and in the source resistor of the Buffer Amplifier. In both cases, the output generates a voltage across a resistor that cancels part of the input. We will see two more examples of negative feedback when we consider the Automatic Gain Control and the Audio Amplifier. Positive feedback increases the gain of an amplifier. This was used in early radios in a circuit called a *regenerative* receiver, where positive feedback brought the receiver to the brink of oscillation. This gave a large gain and allowed a receiver to operate with only a single stage of amplification. However, regenerative receivers were difficult to adjust, and they tended to break into oscillation. Eventually they were superseded by the more stable superheterodyne receiver. Today we use positive feedback for oscillators.

11.1 Criteria for Oscillation

We start by establishing general criteria for oscillation. In Figure 11.1, we show an amplifier with gain and a linear feedback network. We will call the input to the amplifier x and the output y. We characterize the amplifier by a gain G that is the ratio of the output to the input. In the NorCal 40A oscillators, the amplifier is a transistor with an output current and input voltage, and so the gain is a transconductance. For the feedback network, the roles of x and y reverse, so that y is the input and x is the output. We characterize the feedback network by a loss L, which is the ratio of input to output. We can write the following equations:

$$y = Gx, \tag{11.1}$$
$$y = Lx, \tag{11.2}$$

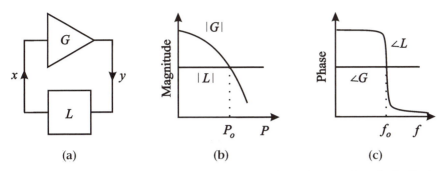

Figure 11.1. Oscillator network consisting of an amplifier with a gain G, and a feedback network with a loss L (a), satisfying the magnitude (b) and phase (c) criteria.

where all the quantities are complex numbers. If $G \neq L$, x and y must be zero, and there is no oscillation. However, if $G = L$, then we can have an output. Since G and L are complex numbers, this is actually two conditions, one for magnitudes and one for phases. We can write the oscillation criteria as

$$|G| = |L|, \tag{11.3}$$

$$\angle G = \angle L. \tag{11.4}$$

Physically, the gain of the amplifier must compensate for the loss of the feedback network, and the phase shift of the feedback network must offset the phase shift of the amplifier.

The key to satisfying the magnitude criterion is to recognize that the gain of practical amplifiers falls at high power levels. This can be from overload, or from a gain-limiting circuit. This is shown in Figure 11.1b, where the gain and loss are plotted as a function of the output power P. The loss $|L|$ is shown as a flat line, because the feedback network is linear. We meet the magnitude criterion at the intersection of these two curves:

$$|G(P_o)| = |L|, \tag{11.5}$$

where P_o is the oscillator output power. For the phase criterion, we consider the frequency variation in the feedback network. In public-address system feedback, this is provided by the acoustic delay between the speaker and the microphone. In our oscillators, it is provided by a resonant circuit that gives a rapidly varying phase near the resonant frequency. This is shown in Figure 11.1c, which gives the phase as a function of the frequency f. The amplifier phase $\angle G$ usually varies much more slowly than the phase of the resonator. We meet the phase criterion at the intersection:

$$\angle L(f_o) = \angle G. \tag{11.6}$$

This determines the oscillation frequency f_o. In our oscillator models, there is no phase shift in the amplifier; therefore, the phase criterion is satisfied at the resonant frequency of the feedback network, where there is no phase shift. In summary, the power characteristics of the amplifier determine the output power, and the frequency characteristics of the feedback network determine the frequency. In general, high-Q networks have a fast phase variation that gives precise control of the oscillation frequency. This means that crystal oscillators are more stable than LC oscillators.

These formulas determine the power and frequency of an oscillation, but we also have to think about how the oscillation starts. This is actually a more serious mathematical problem than we can explore here, and we will give criteria that are sufficient for starting but do not cover all cases. Physically, oscillations either build up from noise or start from an external signal. If

$$|G| > |L| \tag{11.7}$$

at low powers, noise at a frequency that meets the phase criterion is repeatedly amplified until it reaches the output level that satisfies the magnitude criterion. This gives us *starting* criteria:

$$|g| > |L|, \tag{11.8}$$

$$\angle L(f_o) = \angle g, \tag{11.9}$$

where g is the small-signal gain. Some systems that oscillate well at high power levels will not start by themselves. For example, consider using a Class-C amplifier in an oscillator. Class-C amplifiers have a threshold that gives them low gain at low power levels. This is shown in Figure 11.2. I have drawn a line to indicate the response of the feedback network (L). For power levels below the starting power P_s, signals are not amplified, and there will be no oscillation. Ordinarily this is a good thing, and it means that Class-C amplifiers will not oscillate on their own. To make one oscillate, we need an external signal that is larger than P_s for a kick start. After the oscillation is started, we can remove the external signal. This is the basis of operation of the *super-regenerative receiver*, which is used in garage-door openers and inexpensive walkie-talkies. In these circuits, the received signal triggers an oscillation. The oscillators in the NorCal 40A begin in Class A; thus they are self starting. However, as the oscillation builds up, operation shifts to Class C.

The figure raises another question. There are actually two powers that meet the magnitude criterion – P_s and P_o. At which level will the oscillator settle? It

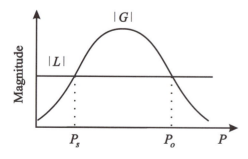

Figure 11.2. Starting problem for an oscillator with a Class-C amplifier.

turns out that P_s is not a stable operating point. If we drop slightly below P_s, then $|G| < |L|$, and the signal will die out. However, above P_s, $|G| > |L|$, and the signal increases until the output power reaches the stable operating point P_o.

11.2 Clapp Oscillator

Many, many different oscillators have been invented, and by tradition they carry the inventor's name. You may run across Hartley, Colpitts, Clapp, Pierce, Gouriet, Hansen, Harris, Butler, Lampkin, Seiler, Miller, Meissner, Vackar, Gunn, and Wien. This can be quite confusing, because some of the designs are improvements on a previous circuit. One can, however, think in terms of families of oscillators and variations within a family. One oscillator may be called by more than one name, depending on whether one is referring to the family or to a variation within a family. Each type usually has both field-effect and bipolar versions. It is often difficult to say that one design is better than another, because much depends on how well the components for the oscillator are chosen and how well the oscillator is isolated from the other circuits in the receiver. In general, each oscillator has a resonator to determine the frequency and a divider network to feed back part of the output to the input. We can classify oscillators in two major families depending on whether the inductor or the capacitor in the resonant circuit is used for the divider. The Colpitts oscillator uses a capacitive divider (Figure 11.3a), and the Hartley oscillator uses the inductor as the divider (Figure 11.3b).

The oscillators in the NorCal 40A are Clapp oscillators. The Clapp oscillator is in the Colpitts family, with a capacitor divider for feedback. Consequently, you may also hear them referred to correctly as Colpitts oscillators. In the Clapp oscillator, the inductor is replaced by a resonant circuit. This allows us to have large voltages in the resonator elements, which helps in keeping the frequency stable, without overloading the transistor. We use this circuit with an LC resonator in the variable-frequency oscillator, or VFO (Figure 11.4a). The VFO uses a JFET source follower

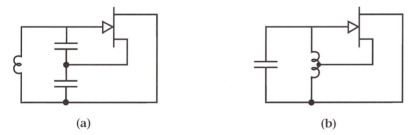

(a)   (b)

Figure 11.3. Colpitts oscillator, with a capacitive-divider feedback network (a), and Hartley oscillator, with inductor feedback (b). The inductor network is usually made by a connection part way down the coil called a *tap*. The amount of feedback is controlled by the position of the tap. In both figures, the bias and load networks are omitted for simplicity.

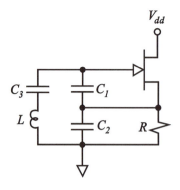

Figure 11.4. Clapp oscillator circuit that is used for the variable-frequency oscillator (VFO) in the NorCal 40A. C_1 and C_2 form the divider network, and R is the load. The gate and source bias networks and the tuning are complicated, and we omit them for now.

for the amplifier. The BFO and Transmit Oscillators use a crystal resonator with a BJT emitter follower.

We start with a small-signal model for the JFET VFO. This will give us the phase condition for starting and for oscillation and the magnitude condition for starting. The final oscillations are large voltages, and we will need a large-signal analysis for the oscillation condition for magnitudes. We replace the JFET in Figure 11.4 with the model in Figure 9.17a, and we substitute a ground connection for the supply voltage V_{dd}. We can write the small-signal equivalent circuit shown in Figure 11.5.

In this analysis, the gate-source voltage v_{gs} corresponds to the input x in Equation 11.1, and the drain current i_d is the output y. For the JFET, we write

$$i_d = g_m v_{gs}. \tag{11.10}$$

We can therefore write the small-signal gain g as

$$g = g_m. \tag{11.11}$$

There is no phase shift, and so we meet this phase condition at the resonant frequency ω_0 for the network. To see this, consider that in resonance, the reactances of the two arms of the feedback network must cancel, and

$$-\frac{1}{j\omega_0 C_2} = j\omega_0 L + \frac{1}{j\omega_0 C_3} + \frac{1}{j\omega_0 C_1}. \tag{11.12}$$

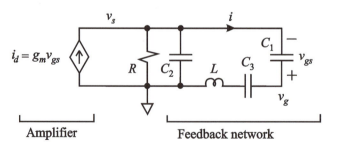

Figure 11.5. Small-signal equivalent circuit for the Clapp VFO. It takes quite a bit of rearrangement to turn Figure 11.4 into this circuit; so you should go through it carefully.

We can rewrite this as

$$\omega_0 = \frac{1}{2\pi\sqrt{LC}}, \tag{11.13}$$

where C is the series combination of the three capacitors, given by

$$C = \frac{1}{1/C_1 + 1/C_2 + 1/C_3}. \tag{11.14}$$

At resonance, the currents in the arms of the feedback network cancel, and we can write

$$i = -j\omega_0 C_2 v_s \tag{11.15}$$

and thus

$$v_{gs} = -\frac{i}{j\omega_0 C_1} = \frac{v_s C_2}{C_1}. \tag{11.16}$$

At resonance, we can write the source voltage v_s as

$$v_s = R i_d, \tag{11.17}$$

so that we can rewrite v_{gs} as

$$v_{gs} = i_d R C_2 / C_1. \tag{11.18}$$

This means that the loss L is given by

$$L = \frac{i_d}{v_{gs}} = \frac{C_1}{R C_2}. \tag{11.19}$$

Because both g and L have zero phase shift, we meet the phase criterion (Equation 11.9). The starting condition becomes

$$g_m > \frac{C_1}{R C_2}. \tag{11.20}$$

In measurements, it is convenient to measure the gate voltage v_g and the source voltage v_s. We can write

$$v_g = v_{gs} + v_s = v_s(1 + C_2/C_1). \tag{11.21}$$

This formula only depends on the feedback network, which is linear, and so it is valid for AC quantities large and small. In the NorCal 40A VFO, $C_1 = C_2$, and the circuit voltages are related by

$$v_{gs} = v_s \tag{11.22}$$

and

$$v_g = 2v_s. \tag{11.23}$$

Equation 11.19 becomes

$$L = 1/R \tag{11.24}$$

so that the starting condition (Equation 11.8) becomes

$$g_m > 1/R. \tag{11.25}$$

The oscillation starts if the transconductance is larger than the inverse of the load resistance. In practice, the required transconductance will be higher than this because of the output resistance of the JFET and losses in the resonator. In a circuit, g_m is determined by the bias voltage. In the VFO, we begin with a bias voltage near zero, so that g_m is large and the oscillation starts easily.

11.3 Variable-Frequency Oscillator

The VFO is a key component in a transceiver, because it sets the operating frequency. The VFO is shared between the transmitter and receiver, so that they track in frequency. The challenge in a VFO is that we would like to be able to adjust the frequency easily, but once we set the frequency, we would like it to stay put. If the frequency drifts, it forces the other operator to retune, and there is the danger of interfering with others on nearby frequencies. The frequency shift that can be tolerated is different for each radio service, but for the NorCal 40A, a reasonable limit is 100 Hz.

There are two basic approaches to making variable oscillators. One can start with a crystal oscillator, and use dividing and multiplying circuits to create sine waves at other frequencies. This is called a *synthesized* oscillator, and it gives good stability over a wide range of frequencies. However, the circuits are complex. The other approach is to build an LC oscillator and to minimize the frequency drift. Our VFO is an LC oscillator. There are several things that can be done to stabilize the VFO. We work at as low a frequency as possible, because frequency drifts are usually proportional to frequency. In our transceiver, the VFO is at 2.1 MHz, well below the operating frequency, 7 MHz. To vary the frequency, we use a varactor, which is a reverse-biased diode that acts as a variable capacitor. A major factor that causes drift is temperature change from heat produced by circuit components, particularly the Power Amplifier, and by changes in the outside temperature. The VFO is isolated to some extent from the Power Amplifier because it is at the other end of the board, but we also need to select the capacitors and inductors to minimize the frequency shift.

We can predict the temperature stability of the oscillator from the temperature coefficients of the components. Mathematically we write a temperature coefficient α of a quantity x in the form

$$\alpha = \frac{1}{x} \cdot \frac{dx}{dT}, \tag{11.26}$$

where T is the temperature. Usually we multiply α by a million to state it as parts per million. We can rewrite this formula in terms of logarithms as

$$\alpha = \frac{d \ln(x)}{dT}. \tag{11.27}$$

Now consider a resonant frequency given by

$$f = \frac{1}{2\pi\sqrt{LC}}.$$ (11.28)

We can write the logarithm of the frequency as

$$\ln(f) = -\ln(2\pi) - \frac{\ln(L)}{2} - \frac{\ln(C)}{2}.$$ (11.29)

We can write the frequency temperature coefficient α_f as

$$\alpha_f = \frac{d\ln(f)}{dT} = -\frac{\alpha_l + \alpha_c}{2},$$ (11.30)

where α_l and α_c are the inductor and capacitor temperature coefficients. This formula shows that for good temperature stability, we must either choose stable components or find components whose temperature coefficients cancel. Appendix D gives the temperature coefficients for the resistors, capacitors, and inductors in the transceiver. The capacitors in the VFO are made of polystyrene, which has a small negative temperature coefficient. The inductor is made of a carbonyl iron powder with a small positive temperature coefficient to cancel the capacitive shift. In addition, instead of a ceramic trimmer capacitor, we set the VFO with an air-variable capacitor, which is quite stable.

11.4 Gain Limiting

The JFET in the VFO does not actually overload. Instead, the gain is limited by a detector diode, which keeps the output well below the supply voltage, giving a clean sine wave. Figure 11.6 shows the circuit. The starting resistor ensures that initially the gate voltage begins at ground, giving a large initial g_m so that the oscillation starts easily. The detector diode conducts on positive peaks of the gate

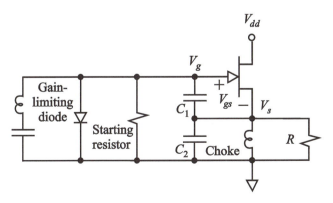

Figure 11.6. The VFO circuit in the NorCal 40A. The choke sets the DC source voltage to zero. The diode limits g_m to give a clean sine wave.

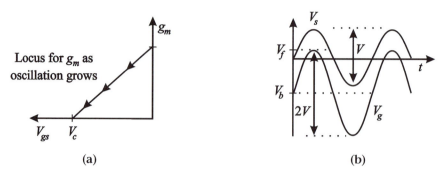

Figure 11.7. Effect of the gain-limiting diode on g_m as the oscillation builds up (a). V_g and V_s waveforms (b).

voltage. When the diode conducts, it pulls charge through the capacitors, and this gives the gate a negative bias, reducing g_m. As long as g_m is greater than $1/R$, the value required by the oscillation condition, the oscillation grows. This causes more current through the diode, and the bias voltage drops further, taking g_m with it (Figure 11.7a). Eventually an equilibrium is reached where the oscillation condition is satisfied, and the oscillation does not grow any more.

Figure 11.7b shows the final gate and source voltage waveforms V_g and V_s. They are both sinusoidal. We let V be the peak-to-peak value of the output voltage V_s. For our VFO, $C_1 = C_2$, and Equation 11.23 tells us that the peak-to-peak value of V_g is twice that of V_s. In addition, V_g and V_s have different DC values. V_s has no DC offset because of the choke. However, the peak voltage of V_g is limited to V_f, the forward voltage of the gain-limiting diode. This gives V_g a large negative DC offset V_b. From Figure 11.7b we can write

$$V_b = V_f - V. \tag{11.31}$$

To find the gate-source voltage V_{gs}, we subtract V_s from V_g. This is shown in Figure 11.8a. The peak-to-peak value of V_{gs} is V. V_{gs} is always negative, and the DC bias is V_b. The maximum value of the gate-source voltage V_m is given by

$$V_m = V_b + V/2. \tag{11.32}$$

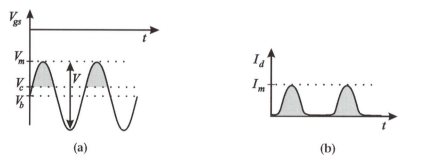

Figure 11.8. VFO gate-source voltage V_{gs} (a), and drain current I_d (b).

We can substitute for V_b from Equation 11.31 to get

$$V_m = V_f - V/2. \tag{11.33}$$

In practice, the bias voltage V_b is below the cut-off voltage V_c. This means that the JFET is active for less than half a cycle and that it is operating in Class C. Power is added to the oscillator in current pulses, rather than continuously (Figure 11.8b). One might liken this to a person pushing a child in a swing. We calculated the start oscillation condition in terms of g_m, which only applies when the JFET is active and the signals are small. Now we have large signals and a JFET that is off more than half the cycle. We define a *large-signal transconductance* G_m given by

$$G_m = I/V, \tag{11.34}$$

where I and V are the peak-to-peak values of the fundamental components of the drain current and gate-source voltage. G_m is also called a *describing function*. In the VFO, it is the amplifier gain; thus we can combine Equations 11.5 and 11.24 to get the oscillation condition:

$$G_m = 1/R. \tag{11.35}$$

G_m can be calculated approximately from JFET characteristics. We start with Equation 9.74 for the drain current:

$$I_d = I_{dss}(1 - V_{gs}/V_c)^2. \tag{11.36}$$

As an approximation, we assume that the current is on for nearly half a cycle, and this gives the current a cosine-squared shape. Therefore the average value of the current over the half cycle will be half its peak value, and the average current over the entire cycle will be a quarter of its peak value. We can write the DC drain current I_o approximately as

$$I_o = I_m/4, \tag{11.37}$$

where I_m is the maximum drain current. The fact that the current is proportional to the cosine squared gives the current a narrow pulse shape. In Section B.4, the Fourier series for a pulse train is derived, and it is shown that the peak-to-peak value of the fundamental component is four times the DC component. This means we can write

$$I \approx I_m. \tag{11.38}$$

Now we can divide by V to get

$$G_m = I_m/V. \tag{11.39}$$

Figure 11.9 is a plot of G_m versus V, calculated using this formula and Equation 11.33 and Figure 9.15. The oscillation condition is $G_m = 1/R$, and I have added

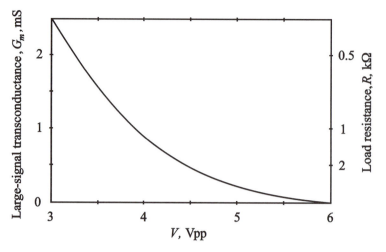

Figure 11.9. Large-signal transconductance G_m of the J309 JFET in a Clapp oscillator with $C_1 = C_2$. V is the peak-to-peak value of both the gate-source voltage and the output voltage. The right axis gives the load resistance R on an inverted scale to allow prediction of the output voltage.

an inverted scale for R to show this relation. The plot also allows us to predict the output voltage because V is equal to the output voltage.

11.5   Crystal Oscillators

The crystal oscillators in the NorCal 40A are also Clapp oscillators (Figure 11.10). These circuits are included in the SA602AN mixer ICs. The amplifier is an emitter

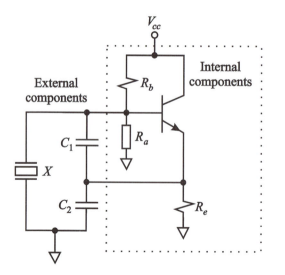

Figure 11.10. Clapp crystal oscillator in the SA602AN integrated circuit that is used for the Transmit Oscillator and the Beat Frequency Oscillator.

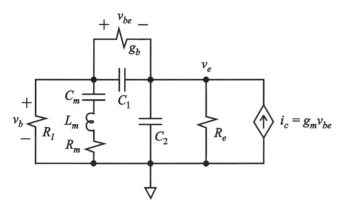

Figure 11.11. Equivalent circuit for the Clapp crystal oscillators.

follower, and R_b (18 kΩ) and R_e (25 kΩ) are internal bias resistors. R_a represents the internal load, which is a buffer amplifier that drives the mixer. We neglect the reactance of the internal load. C_1 and C_2 are external capacitors that form the divider network. X is an external crystal resonator.

The analysis can be simplified if we use a small-signal BJT model with a transconductance that is like that for the JFET. The circuit shown in Figure 11.11. The collector current i_c is given by

$$i_c = g_m v_{be}. \tag{11.40}$$

In the oscillator analysis, we consider v_{be} as the amplifier input and i_c as the output, so that we can write the small-signal gain g as

$$g = g_m, \tag{11.41}$$

The crystal is represented by its motional inductance L_m and capacitance C_m. Provided the Q is large, we meet the phase condition again at the resonant frequency ω_0 given by

$$-\frac{1}{j\omega_0 C_1} - \frac{1}{j\omega_0 C_2} = j\omega_0 L_m + \frac{1}{j\omega_0 C_m}. \tag{11.42}$$

We can rewrite this in terms of the resonant frequency as

$$\omega_0 = \frac{1}{\sqrt{L_m C}}, \tag{11.43}$$

where C is the series combination of the three capacitances, given by

$$C = \frac{1}{1/C_1 + 1/C_2 + 1/C_m}. \tag{11.44}$$

Finding the starting condition is complicated by the fact that there are many sources of loss in the oscillator that may be important. In Figure 11.11, we include the effect of the load as R_l, the crystal resistance as R_m, the base–emitter diode as a conductance g_b, and the emitter bias resistor as R_e. R_l itself represents the parallel combination of the base bias resistor R_b and the buffer amplifier impedance R_a. The crystal resistance R_m varies considerably from crystal to crystal.

For the base–emitter conductance we use Equation 9.23,

$$g_b = I_b / V_t, \tag{11.45}$$

where I_b is the DC base-bias current and V_t is the thermal voltage, 25 mV at room temperature. We have to be careful in using this formula, because it is a small-signal formula, and the base current varies greatly during a cycle. We can think of it as the average of the conductance over a cycle.

We transform all the resistances to an equivalent resistance R parallel to the current source, assuming the Q is high in each case. There are successive transformations from parallel to series and then series to parallel. We skip the details and write

$$\frac{1}{R} = \frac{(C_1 + C_2)^2}{R_l C_1^2} + R_m (\omega_0 C_2)^2 + g_b \left(\frac{C_2}{C_1} \right)^2 + \frac{1}{R_e}. \tag{11.46}$$

Now we can use Equation 11.20 that we developed for the JFET VFO, provided we identify "base" with "gate" and "source" with "emitter." The starting condition becomes

$$g_m > \frac{C_1}{RC_2} = \frac{(C_1 + C_2)^2}{C_1 C_2 R_l} + (\omega_0 C_1)(\omega_0 C_2) R_m + \frac{C_2}{C_1} \frac{I_b}{V_t} + \frac{C_1}{C_2 R_e}. \tag{11.47}$$

Previously we found that g_m is proportional to the collector bias current I_c (Equation 9.26). If the oscillation does not start, it may be possible to start it by adding an external resistor in parallel to R_e to increase the bias. Also, it should be noticed that although each resistance pushes up the required value of g_m, making it harder to start, the effect of the divider capacitors depends on where the loss in the circuit comes from.

For a large-signal analysis, we need to consider that the emitter current comes in narrow pulses. This is because the relationship between the base voltage and collector current is exponential, and small changes in voltage cause large spikes in current. This is shown for a Clapp crystal oscillator in Figure 11.12. The fact that the emitter current comes in narrow pulses allows us an easy way to predict the output voltage, because we know that the peak-to-peak value of the fundamental component is four times the DC current (Appendix B, Section 4). We write

$$I = 4I_o, \tag{11.48}$$

where I is the peak-to-peak value of the emitter current and I_o is its DC value. Equation 11.21 lets us write the peak-to-peak value of the output voltage V across the load R_l as

$$V = IR \frac{C_1 + C_2}{C_1}. \tag{11.49}$$

It is hard to say much about this formula because R is itself a function of the various

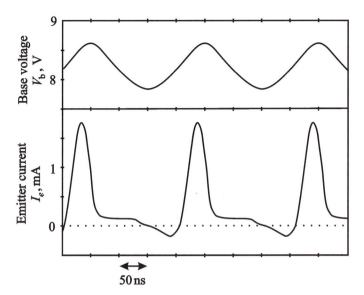

Figure 11.12. Measured emitter current I_e and base voltage V_b for a breadboard Clapp crystal oscillator. The component values are the same as those used in SA602AN and in the NorCal 40A. A 2N4124 transistor like the one in the Receiver Switch is used in place of the internal transistor in the SA602AN to make it easy to measure the emitter current through a 100-Ω series resistor.

resistances and capacitances. However, if the load resistance R_l is dominant, we can use Equation 11.46 to write

$$V = \frac{4I_o R_l C_1}{C_1 + C_2}.$$
(11.50)

This formula predicts that the output voltage increases as C_1 increases. In the NorCal 40A, C_1 is a trimmer capacitor, and this effect is quite noticeable during tuning.

11.6 Phase Noise

In addition to long-term frequency drifts due to temperature changes, oscillators have random phase shifts due to fluctuations in the oscillator current (Figure 11.13a). This is called *phase noise*. Phase noise causes power to spread away from the carrier frequency to nearby frequencies. Phase noise is a problem because receivers treat it like an ordinary signal, and we hear increased noise at the receiver output. Phase noise is also a problem in circuits such as multimeters that convert analog signals to digital ones, because it causes *jitter* in the timing circuits.

We can get a feeling for the effect of current fluctuations by considering an LC resonator with a charge q injected (Figure 11.13b). The injection phase is

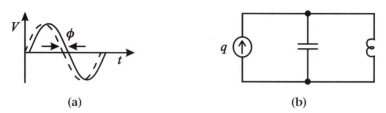

Figure 11.13. (a) Phase noise in an oscillator. The dashed line indicates a sine wave at the carrier frequency, while the solid line shows the output of an oscillator with a phase shift ϕ. (b) Model for phase noise with a charge q injected into an LC resonator.

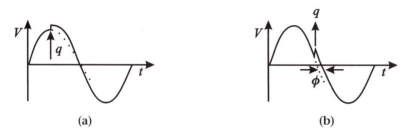

Figure 11.14. (a) Charge injection at the voltage peak (a) and at the zero crossing (b). Charge injection at the peak causes no phase shift, whereas charge injection at the zero crossing causes a large phase shift.

crucial. The charge goes directly into the capacitor, because the current in an inductor cannot change suddenly. This means that the voltage on the capacitor jumps, while the inductor current remains the same. If the charge comes at the peak of a cycle, where the capacitor voltage is maximum and the inductor current is zero, there is no change in the oscillator phase, only a change in amplitude (Figure 11.14a). The amplitude change will be smoothed out by the gain-limiting action of the oscillator, and it is not of much consequence. However, if the charge is injected at the zero crossing where the capacitor voltage is zero and the inductor current is large, the charge causes a large change in phase with only a small change in amplitude (Figure 11.14b).

The fluctuations in transistor currents are much larger when the transistors are active than when they are off. For this reason, it is important that the transistor be active near the peak voltage, rather than near the zero crossing, where the fluctuations would cause phase noise. In the Clapp oscillator, the transistor is active at the peak and off during the zero crossing, and this gives it low phase noise.

FURTHER READING

This chapter owes much to *The Design of CMOS Radio-Frequency Integrated Circuits,* by Thomas Lee, published by Cambridge University Press. The research by Hajimiri and Lee on phase noise has rendered previous writing on oscillators obsolete. Many practical designs and construction details for VFOs are given in *The ARRL Handbook*, published by the American Radio Relay League. Synthesized oscillators are thoroughly covered in *Communications Receivers, Principles and Design,* by Ulrich Rhode and T.T.N. Bucher, and published by McGraw-Hill.

PROBLEM 26 – VFO

The VFO shown in Figure 11.15 is more complicated than the previous circuits we have built. It is a good idea to check all the connections carefully. Oscillator circuits can be frustrating, because one bad connection can prevent the oscillation from starting, and it is difficult to find the problem because there is no signal to trace.

We will consider each component in turn. C52 and C53 form the capacitor divider network. L9 and C51 form the series resonator. D8 is a *varactor* diode. It acts as a variable capacitor, controlled by the reverse bias voltage. The voltage determines the capacitance by controlling the thickness of the depletion layer. Our diode is the MVAM108. These diodes are intended for tuning AM radios, and that is where the "AM" in the type number comes from. "M" is for Motorola, "V" is for varactor, and "8" is for 8 V, the top voltage at which this diode should be used. The MVAM108 gives a wide range of capacitance from 600 pF at low voltages to 30 pF at 8 V. In the circuit, we control the bias on the varactor diode with a potential divider network formed by R17 and R20. R17 is a large potentiometer that sets the bias on the varactor. This is our tuning knob for the radio. R19 is a large series resistor that keeps the divider network from loading the resonator, and C49 prevents L9 from shorting out the divider.

C50 is an air-variable capacitor in parallel with L9. The capacitance is a maximum, about 25 pF, when the leaves are fully meshed, and a minimum, about 2 pF, when the leaves are unmeshed. We use C50 to set the range of frequencies that can be covered by the VFO. RFC2 is a choke that gives a DC ground connection to the source. C54 is an RF bypass capacitor that keeps the RF impedance between the drain and ground low. D9 is the detector diode that controls the gain of the JFET. R21 sets the initial gate bias

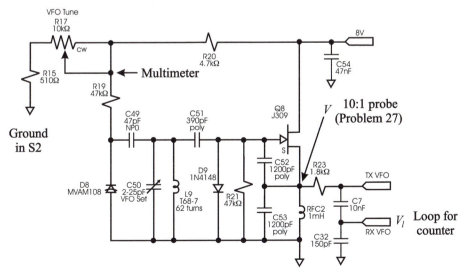

Figure 11.15. The VFO in the NorCal 40A.

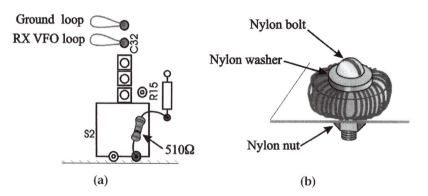

Figure 11.16. Special solder connections for the VFO (a), and mounting L9 (b).

at zero volts to ensure that the oscillation starts. R23 is the load. R23 and C32 act as a low-pass filter that sets the input voltage levels for the Transmit Mixer and the RF Mixer. C7 provides a DC block between the mixer inputs.

To start, install all of the components in Figure 11.15. Be careful not to overheat the polystyrene capacitors C51, C52, and C53 when you are soldering them in. If the plastic melts, they short out. After you have installed the R17 pot, take a pair of pliers and twist off the locking tab to the left of the shaft. This tab is meant to fit in a slot in the front panel, and we will not need it. If you do not take it off, it will interfere with the front panel. Also make sure that the nut and washer stay on the pot. You will need them later. There are some special connections that are shown in Figure 11.16a. The 510-Ω resistor should be soldered between the R15 hole and one of the ground holes in S2 at the edge of the board. There are two additional wire loops to solder in to make it easy to hook on probes to check the output of the VFO. Stick both ends of a loop in the same hole and solder them.

The L9 inductor uses a 68-7 core. The number "68" indicates the size of the core (0.68″ diameter), and "7" is the number of the iron-powder mix. #7 cores are painted white. In Problem 9 we used the red 37-2 cores for the Transmit Filter. #7 cores are advertised to be more stable than #2 cores, but in our measurements they have similar temperature coefficients. See Appendix D for more information. For the coil, you should start with 1.5 m of #28 wire. Wind 62 turns tightly and neatly around the core without overlapping the wires. The turns should be spread as evenly as possible. Check your coil carefully. If the turns bunch or the count is wrong, you may have to add or subtract a turn later. Solder the coil to the board. L9 is also secured to the board by a nylon bolt, washer, and nut to keep the core from moving (Figure 11.16b). It also helps pin down the turns of the winding. The stepped side of the washer should face the core to help it fit in neatly.

Now we are ready to test the oscillator. With a screwdriver, set the air-variable capacitor (C50) to half-mesh. Plug in the power supply, and turn on the power. Attach a 10:1 probe to the probe loops (RX VFO and the ground loop in Figure 11.16a). The oscilloscope should be set for an internal trigger. Verify that you see a sine wave.

A. Attach a multimeter probe to measure the DC voltage of the wiper of the R17 pot. This is the multimeter connection shown in Figure 11.15. Record the wiper voltage when R17 is fully counterclockwise and when it is fully clockwise.

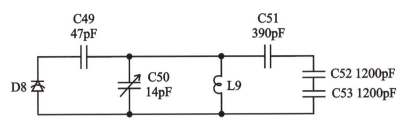

Figure 11.17. Simplified circuit diagram for calculating the resonant frequency, showing L9 and the important capacitances.

B. Now we will try to calculate these numbers. From Figure 11.15, what would you expect the multimeter voltage to be for the counterclockwise setting? For the clockwise setting?

C. Study the data sheets in Appendix D to find the capacitance of the MVAM108 for these voltages. You may need to extrapolate a bit.

D. Replace the scope probe with counter leads. What is the frequency when R17 is counterclockwise? How much does the frequency increase when we rotate R17 fully clockwise?

E. Now we are ready to calculate the oscillation frequency. Our experience has been that the inductance constant for the 68-7 core varies considerably. A reasonable value to use is $A_l = 5.0$ nH/turn^2. Calculate the inductance for L9 with 62 turns. I have drawn a simplified circuit diagram for the important reactive elements in Figure 11.17. We have neglected the transistor Q8, the diode D9, the choke RFC2, and the load circuits. You can take C50 to be the average of its fully meshed and unmeshed values, 14 pF. Calculate the resonant frequency for the counterclockwise setting. Now calculate how much the frequency should increase for the clockwise setting.

PROBLEM 27 – GAIN LIMITING

A. Measure the peak-to-peak source voltage V. You should connect the 10:1 probe at the source end of R23 for this (Figure 11.15).

B. Now use Figure 11.9 to predict the output voltage V that satisfies the large-signal oscillation condition. Your prediction may be high because we have not considered circuit losses.

C. In deriving the oscillation condition for our VFO, we neglected the inductor resistance and the drain-source resistance of the JFET r_d. How does the oscillation condition change if we add these effects? If L9 has a Q of 250, and the JFET output resistance is $r_d = 5$ kΩ, what is the new prediction for V?

D. The voltage is smaller at the RX VFO loop. Measure the loop voltage V_l with the 10:1 probe. You should remove the counter probe for this measurement so that it does not load the circuit. Find the loss ratio $|V/V_l|$. For comparison, calculate the ratio that you would expect.

E. Now measure the temperature dependence of the oscillator. Reattach the counter probe to the RX VFO loop so that you can measure the frequency. Rest the bulb of a thermometer on the screw hole in the circuit board near R6. Heat the board with a hair drier until the temperature reaches 50°C. This is best done with the board in a plastic box with holes that direct the flow of air. See Appendix A for more information. Switch off the hair drier. Record the temperature and the frequency as the temperature drops. From your measurements, find the frequency sensitivity in Hz/°C. Find the temperature coefficient in ppm/°C. How large a temperature change would be necessary to cause the frequency to shift 100 Hz?

F. There is an advantage in having a high VFO frequency, because this makes it easy for filters to reject the image. However, there is a big disadvantage: less stable oscillators. Assume we reduce the capacitance and inductance by a factor of six to increase the VFO frequency by the same factor. This would leave us with the same IF frequency as before. The radio would work with no modification, and the image of the VFO would be rejected more easily. To see the difficulty with this approach, calculate the temperature change that would be needed to cause the frequency to drift 100 Hz, assuming the same temperature coefficient as before.

G. Now calculate the temperature coefficient that you expect, assuming that the polystyrene capacitors have a temperature coefficient of −150 ppm/°C and the 68-7 core has a temperature coefficient of +50 ppm/°C. You may neglect the effect of the other components.

H. Calculate the change in the oscillation frequency that you would expect from taking a turn off the inductor. Now set R17 fully counterclockwise, and adjust the air-variable capacitor (C50) until the oscillation frequency is 2,085 kHz. This sets the lower end of the tuning range for the radio. If you cannot reach this frequency, you will need to change the number of turns.

The next step is to install the Receiver-Incremental-Tuning (RIT) circuit. This is our first chance to use one of the eight-pin integrated-circuit (IC) packages. You will also hear these called DIP (for "dual-in-line package," because the pins come in two lines). In digital circuits, it is common to mount the ICs in sockets, so that they can be easily removed if they burn out. In radio circuits, however, people usually solder in the ICs. This reduces problems of noise pickup and helps prevent unwanted oscillations. The problem with soldering an IC is that it is difficult to remove, because you have to get all the solder off each lead before it will budge. Check that you have the IC oriented correctly before you solder! Manufacturers make a notch or a dimple at one end, and there is a matching notch in the board outline. It is important to be able to identify each pin by number. The pins are numbered in a counterclockwise direction from the notch (Figure 11.18a). You should realize that the numbering is only counterclockwise if you look from the top of the package. From the bottom side, it is clockwise. This can be confusing, particularly when the IC is mounted on the board.

The RIT allows us to offset the receive frequency from the transmit frequency. Some operators may have transmitters and receivers that are not aligned, or their transmitters

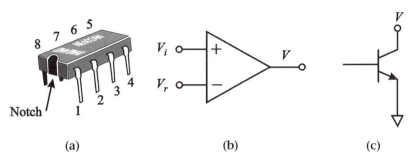

Figure 11.18. Identifying the pins on an eight-pin DIP package (a). Schematic symbol for a comparator (b). Open-collector circuit for one of the LM393N comparators (c).

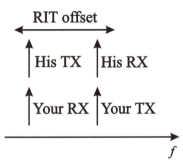

Figure 11.19. Using the RIT to adjust your receive frequency (RX) when the other operator's receive and transmit (TX) frequencies are not aligned.

may drift from heating, while their receiver does not. The RIT shifts the receive frequency without changing the transmit frequency. For example, assume that the other operator's transmit frequency is lower than his receive frequency. This means that you need an RIT offset to align your receiver with his transmitter (Figure 11.19). In our measurements, the RIT is useful for fine adjustments of the receiver frequency.

The RIT circuit uses an IC with a pair of comparators, the LM393N by National Semiconductor. The letters "LM" identify an IC from National Semiconductor, and "N" indicates a DIP package. A *comparator* is a differential amplifier with a very high voltage gain, typically about 200,000. Figure 11.18b shows the schematic symbol for a comparator. The triangle is the usual symbol for an amplifier. The differential inputs are labeled $+$ and $-$ so that you know which input is subtracted. Typically, one of the inputs is a voltage that changes, and we label it as V_i. The other is a fixed voltage that we call V_r, with the r standing for "reference." You can understand what the comparator does if you consider the output circuit for the comparator, which is just an npn transistor with its collector open-circuited (Figure 11.18c). People call this an *open-collector* output. We have two possibilities. If V_i is even slightly less than V_r, the gain of the comparator is so large that we will have a large base current in the output transistor. The input need only be a few microvolts below the reference voltage to completely turn the transistor on. The output transistor will present a low impedance to ground for an outside circuit. However, if the input voltage V_i is even slightly greater than V_r, the output transistor will turn off. We can reverse this if we apply V_i to the $-$ input and V_r to the $+$ input.

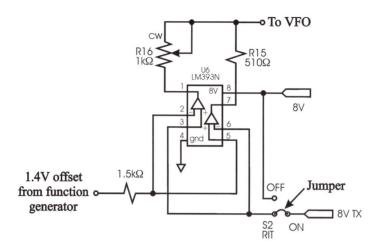

Figure 11.20. The RIT circuit. The original NorCal 40A circuit has a switch S2 to turn the RIT on and off. In our measurements we will not use the switch, but we will add a jumper so that the RIT is always on.

Now we can understand how the RIT circuit works (Figure 11.20). If the transmitter is on, 8V TX will be at 8V. This is greater than the reference voltage, 1.4 V, and it means that the left comparator will be off, and the right comparator will be on. R16 is disconnected, and R15 is effectively connected to ground, and the VFO will work in the same way it did in the last laboratory exercise.

However, for receiving, 8V TX will be below the 1.4-V reference. This disconnects R15, and shorts R16 to ground in its place. R16 is a 1-kΩ pot. Its resistance ranges from near zero to twice as large as R15. The pot setting adjusts the varactor voltage just the way the VFO Tune pot does. However, since the RIT pot varies over only 1 kΩ rather than the 10 kΩ that the VFO Tune pot covers, we get a finer frequency control. We will use the RIT for precise frequency adjustment of the receiver.

Now we are ready to build the RIT circuit. The first thing that you should do is to take the 510-Ω resistor lead that you connected to ground last time and move it to the empty R15 hole. Next pop the comparator IC into the U6 holes, checking the orientation carefully. Solder each of the pins for the comparator IC. Follow with the pot R16. Break off the locking tab to the left of the shaft with a pair of pliers, so that the tabs do not get in the way when we add the front cover.

There are several special connections (Figure 11.21). In the S2 outline, add a ground jumper wire across the two holes at the edge of the board. This is just another ground connection for probes. In addition, add a short jumper to bypass S2 in the two holes that are shown in the figure. Then solder one end of a 1.5-kΩ resistor to hole 2 of U2. This is our 1.4-V reference connection, and for this measurement it can be provided by an offset voltage from a function generator.

I. Now plug in the power supply and turn it on. The counter should be connected to the RX VFO loop. Also insert a shorting plug into the Key jack to turn on 8V TX. Use the big VFO Tune pot R17 to set the frequency close to 2,100 kHz. Record this

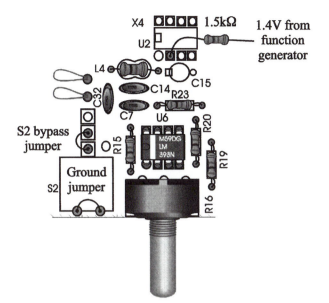

Figure 11.21. Jumpers and connections for the RIT circuit.

frequency. While the transmitter is on, the RIT should not work. You should verify that twiddling the RIT pot does not change the frequency appreciably. Now remove the shorting plug from the Key jack. This should turn off 8V TX and allow the RIT to work as a fine frequency control. Full counterclockwise rotation should give the minimum frequency, and full clockwise rotation should give the maximum. Record the RIT tuning range. Remove the 1.5-kΩ resistor from U2.

Mixers

Mixers shift signals from one frequency to another. In radios, mixers allow us to move a signal from the operating frequency that we use for transmission and reception to an audio frequency that we can hear. Mixers are fundamentally more complicated than amplifiers and oscillators because there are inputs and outputs at several frequencies. Because of the different frequencies that go in and out, mixers need filters to choose the frequencies that we want. Figure 12.1a shows the schematic symbol for a mixer. It is traditional to call the input the *RF* signal, for radio frequency. In addition, an oscillator signal, the *Local Oscillator*, or *LO*, is applied to the mixer. It is called the local oscillator because it is generated in the receiver itself, that is, locally. The output is called the *IF*, for *Intermediate Frequency*, and it is either at the sum of the RF and LO frequencies or at the difference. We characterize a mixer by a conversion gain G, which is defined the same way as for an amplifier, except that the input and output frequencies are different:

$$G = P/P_+, \tag{12.1}$$

where P is the IF output power and P_+ is the available power from the RF source.

Our receiver uses two mixers (Figure 12.1b). In the first mixer, the RF Mixer, the RF input mixes with the VFO to produce the IF. This IF is in turn the input for the second mixer, the Product Detector. There it mixes with the Beat-Frequency Oscillator, or BFO, to produce the audio output. This type of receiver, with an IF, is called a *superheterodyne* receiver, in contrast to the *direct-conversion* receiver, which has only a product detector. The superheterodyne receiver has the advantage that filters can reduce all of the spurious responses. We will measure the spurious responses in the exercises. The transmitter uses only one mixer (Figure 12.1c), where the VFO and the Transmit Oscillator mix to produce the output RF signal.

12.1 Gilbert Cell

The mixers in the NorCal 40A are called *Gilbert Cells*. This circuit was invented by Barrie Gilbert in 1967, and it is one of the most important circuits in

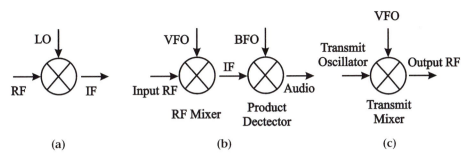

Figure 12.1. (a) Mixer schematic symbol and notation for the inputs and outputs. (b) The NorCal 40A receiver mixers. (c) The NorCal 40A Transmit Mixer.

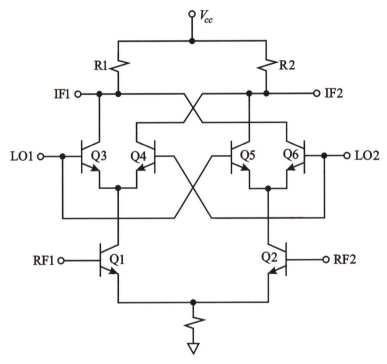

Figure 12.2. The Gilbert-cell mixer.

communications electronics. Figure 12.2 shows the mixer. It is complicated because there are some crossing wires. It may be easiest to start at the bottom, which is a long-tailed pair differential amplifier, with the RF signals as inputs. However, inserted between the collectors and the collector resistors are four cross-coupled transistors. These are driven by the local oscillator. To understand the effect of the local oscillator assume that the voltage at LO1 is large, so that Q3 and Q5 turn on. This connects Q1 to R1 and Q2 to R2, and we have a normal differential amplifier. Now consider what happens if the voltage at LO2 is large, so that Q4 and Q6 are turned on. This connects Q1 to R2 and Q2 to R1. We have a differential amplifier again, but the outputs are interchanged. This changes the sign of the outputs. The effect is that the output is alternately multiplied by $+1$ and -1 depending on the sign of the local oscillator.

The Gilbert Cell is an *active* mixer, which combines an amplifier with a mixer. We contrast active mixers with diode mixers that have loss. The gain from active mixers provides a big advantage because it reduces the gain that we need in the rest of the circuit. In addition, active mixers are tolerant of large reactances in the IF output lines. This is useful because IF crystal filters have large reactances at stop-band frequencies. The disadvantage of active mixers is that they can overload easily. Diode mixers, in contrast, are less likely to overload, but they require careful attention in the IF circuit and much more local-oscillator power than active mixers.

12.2 Mixer Mathematics

In a mixer, the LO drives transistor switches that steer the input to the output alternately in phase and out of phase. Mathematically this is like multiplying the input by a square wave. Notice that this is a linear network, in that we are always multiplying by $+1$ or -1. However, we say it is *time varying*, because the multiplier changes with time. Linear time-varying networks can shift the frequency of a signal, in contrast to linear *time-invariant* networks that leave the frequency unchanged. In practice, the switches do not change phase instantaneously, and so the input is really multiplied by a waveform that is somewhere between a square wave and a sine wave. But to keep the mathematics simple we will use a square wave. It is easy to sketch the output waveform (Figure 12.3) but not easy to perceive the different frequency components. We start with an input RF cosine signal, which we write as

$$V_{rf}(t) = V_{rf}\cos(\omega_{rf}t). \tag{12.2}$$

The Fourier components for the LO are given in Appendix B, Section 2:

$$V_{lo}(t) = \frac{4}{\pi}\left(\cos(\omega_{lo}t) - \frac{\cos(3\omega_{lo}t)}{3} + \frac{\cos(5\omega_{lo}t)}{5} - \cdots\right). \tag{12.3}$$

We multiply the two expressions

$$V(t) = V_{rf}(t) \cdot V_{lo}(t) \tag{12.4}$$

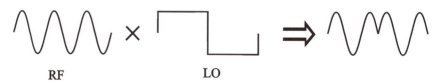

RF   LO

Figure 12.3. Mixing as the multiplication of an input RF cosine wave by a square-wave local oscillator.

and expand to get

$$
V(t) = \frac{2V_{rf}}{\pi}\left(\cos(\omega_{lo}t - \omega_{rf}t) - \frac{\cos(3\omega_{lo}t - \omega_{rf}t)}{3} + \frac{\cos(5\omega_{lo}t - \omega_{rf}t)}{5} - \cdots\right)
$$
$$
+ \frac{2V_{rf}}{\pi}\left(\cos(\omega_{lo}t + \omega_{rf}t) - \frac{\cos(3\omega_{lo}t + \omega_{rf}t)}{3} + \frac{\cos(5\omega_{lo}t + \omega_{rf}t)}{5} - \cdots\right).
$$

$$(12.5)$$

From these sums, we can identify the sum and difference frequency terms V_+ and V_-:

$$
V_+(t) = \frac{2V_{rf}}{\pi}\cos(\omega_+ t), \tag{12.6}
$$

$$
V_-(t) = \frac{2V_{rf}}{\pi}\cos(\omega_- t), \tag{12.7}
$$

where

$$
\omega_+ = \omega_{lo} + \omega_{rf}, \tag{12.8}
$$
$$
\omega_- = |\omega_{lo} - \omega_{rf}|. \tag{12.9}
$$

I have added absolute-value signs for the difference frequency, because ω_{lo} can be larger or smaller than ω_{rf}, and we work with positive frequencies. These are the mixer products for the fundamental frequency component of the local oscillator. Either the sum or the difference frequency could be used for the IF. We would need a filter to remove the one that we do not want. We can also pick out the sum and difference frequency terms for the third harmonic of the LO, V_{3+} and V_{3-}:

$$
V_{3+}(t) = -\frac{2V_{rf}}{3\pi}\cos(\omega_{3+}t), \tag{12.10}
$$

$$
V_{3-}(t) = -\frac{2V_{rf}}{3\pi}\cos(\omega_{3-}t), \tag{12.11}
$$

where

$$
\omega_{3+} = 3\omega_{lo} + \omega_{rf}, \tag{12.12}
$$
$$
\omega_{3-} = |3\omega_{lo} - \omega_{rf}|. \tag{12.13}
$$

The theory predicts that these components would be a factor of three below the fundamental components, but usually they would be less than this. We could go on to the higher harmonics, but life is short. This is a complicated set of components. Unfortunately, the real picture is even more complicated. In this analysis, there are no products with even-order harmonics of the LO, and there are no components at the RF or LO frequencies or their harmonics. However, in a real mixer, all these will be present to some extent.

12.3 Spurious Responses

We have seen that when an input sine wave mixes with a square-wave local oscillator, we generate many frequency components. In a receiver, we usually ask
this question in another way. An antenna receives a great many signals on different frequencies simultaneously. We ask, what frequencies in addition to the RF
frequency give an output at the IF frequency? Signals at these other frequencies
interfere with the signal that we want, and the outputs are called *spurious responses*,
or spurs for short. For example, consider the frequencies shown in Figure 12.4. In
the figure, the LO is below the RF, and the IF is difference frequency. These are
the relationships between the frequencies in the RF Mixer of the NorCal 40A, but
other choices are possible.

Now consider the difference of f_{lo} and f_{if}. This is called the *image frequency*,
because the RF signal and the image sit symmetrically around the IF frequency.
We write the image frequency f_i as

$$f_i = f_{if} - f_{lo}. \tag{12.14}$$

We can rearrange the formula to get

$$f_{if} = f_{lo} + f_i. \tag{12.15}$$

This formula means that if a signal at the image frequency is present at the mixer
input, it will produce a spurious output at the IF, just like an input at the RF frequency. We suppress this spurious response by adding a band-pass filter that passes
an input at the RF but blocks the image (Figure 12.4).

Signals can also cause a spurious response by mixing with the third harmonic
of the LO. Consider a signal at the frequency given by the difference

$$f_{3\downarrow} = 3f_{lo} - f_{if}. \tag{12.16}$$

This frequency is indicated on the left in Figure 12.5. We can write

$$f_{if} = 3f_{lo} - f_{3\downarrow}, \tag{12.17}$$

and this means that $f_{3\downarrow}$ will cause a spurious response. Similarly, we can look at

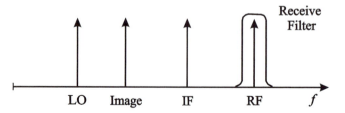

Figure 12.4. Mixer frequencies and the image frequency. A band-
pass filter centered on the RF frequency prevents a signal at the
image frequency from reaching the mixer.

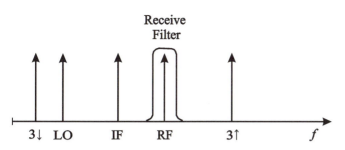

Figure 12.5. Spurious responses caused by mixing with the third harmonic of the LO.

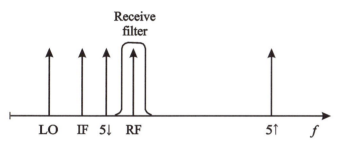

Figure 12.6. Spurious responses caused by mixing with the fifth harmonic of the LO.

the sum

$$f_{3\uparrow} = 3f_{lo} + f_{if}.$$ (12.18)

This frequency is on the right in Figure 12.5. It will also produce a spurious response. In the NorCal 40A, the spurs from the third harmonic of the VFO are well away from the RF frequency, and thus the band-pass filter does a good job of suppressing them. However, the lower frequency $f_{3\downarrow}$ is in the frequency band for AM radio stations, which are often close and powerful. It is easy to hear a spurious response in the NorCal 40A from them.

Finally, we consider the spurs from the fifth harmonic of the LO (Figure 12.6). We write

$$f_{5\downarrow} = 5f_{lo} - f_{if},$$ (12.19)

$$f_{5\uparrow} = 5f_{lo} + f_{if}.$$ (12.20)

In the NorCal 40A, $f_{5\uparrow}$ is high, and the band-pass filter does a good job of rejecting it. Unfortunately, $f_{5\downarrow}$ is close to f_{rf}, and it is hard for the band-pass filter to reject it. The frequency $f_{5\downarrow}$ is one of the largest spurious responses.

12.4 Broad-Band Receivers

The NorCal 40A VFO can tune only over a range of about 50 kHz. This is fine for the frequency range that it was intended for, but for many applications, much

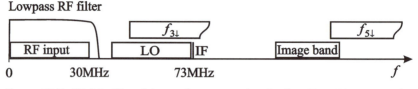

Figure 12.7. RF, LO, IF, and image frequencies for the first IF at 73 MHz in the Kenwood 850 transceiver.

broader bandwidths are needed. However, even assuming that the VFO could be redesigned for a broader bandwidth, the receiver would be limited by spurious responses. If we tried to increase the tuning range significantly, $f_{5\downarrow}$ would enter the pass band of the RF Filter. We can, however, make a receiver cover a much broader bandwidth without spurious responses if we use an IF frequency that is much higher than the RF frequency. At first glance, this does not seem like a great idea. A higher IF frequency takes us farther away from the audio frequencies that are our eventual goal. However, higher IF frequencies have the great advantage that the spurs are even higher in frequency. This means that we can block all of them with a low-pass filter. In practice, these receivers need to use more than one IF frequency, and this allows additional filter stages for better spur rejection.

As an example of a broad-band design, we consider the Kenwood 850, an extremely popular transceiver that covers the frequency range from 100 kHz to 30 MHz. It has three IFs, and so we call this a *triple-conversion* receiver. The first is at 73 MHz, the second at 8.83 MHz, and the third at 455 kHz. Figure 12.7 shows the RF input and LO range for the first IF. We can calculate the LO frequencies that are required by subtraction:

$$f_{lo} = f_{if} - f_{rf}. \tag{12.21}$$

The maximum LO frequency $f_{\overline{lo}}$ comes at the minimum RF frequency $f_{\underline{rf}}$. We write it as

$$f_{\overline{lo}} = f_{if} - f_{\underline{rf}} = 73\,\text{MHz} - 0.1\,\text{MHz} = 72.9\,\text{MHz}. \tag{12.22}$$

The minimum LO frequency $f_{\underline{lo}}$ comes at the maximum RF frequency $f_{\overline{rf}}$. We get

$$f_{\underline{lo}} = f_{if} - f_{\overline{rf}} = 73\,\text{MHz} - 30\,\text{MHz} = 43\,\text{MHz}. \tag{12.23}$$

We can calculate the corresponding maximum and minimum image frequencies as sums:

$$f_{\overline{i}} = f_{if} + f_{\overline{lo}} = 73\,\text{MHz} + 72.9\,\text{MHz} = 145.9\,\text{MHz}, \tag{12.24}$$

$$f_{\underline{i}} = f_{if} + f_{\underline{lo}} = 73\,\text{MHz} + 43\,\text{MHz} = 116\,\text{MHz}. \tag{12.25}$$

The image frequencies are far away from the input RF frequencies, and we can easily remove them with a low-pass filter.

The harmonic spurs $f_{3\downarrow}$ and $f_{5\downarrow}$ are also quite high. We write them as

$$f_{3\downarrow} = 3f_{lo} - f_{if}, \tag{12.26}$$
$$f_{5\downarrow} = 5f_{lo} - f_{if}. \tag{12.27}$$

The minimum spur frequencies come at the minimum LO frequencies:

$$\underline{f_{3\downarrow}} = 3\underline{f_{lo}} - f_{if} = 129\,\text{MHz} - 73\,\text{MHz} = 56\,\text{MHz}, \tag{12.28}$$
$$\underline{f_{5\downarrow}} = 5\underline{f_{lo}} - f_{if} = 215\,\text{MHz} - 73\,\text{MHz} = 142\,\text{MHz}. \tag{12.29}$$

These are well above the RF frequency band, and an input low-pass filter can remove them.

12.5 Key Clicks

Transmitter pulses have controlled rise and fall times to prevent clicking sounds on nearby frequencies. These are called *key clicks*, and they interfere with other operators. Key clicks point to an important problem in commercial communications. Companies want to have as many channels as possible in their frequency allocation, because they make more money when there are more channels. If the transmitter transitions are too fast, the power spreads in frequency, and the channel spacing must be large to avoid interference. We can understand this effect by using our mixer mathematics. We start with transmitter pulses that have zero rise and fall times (Figure 12.8). We will let the peak voltage be 2 V, and assume that the "on" and "off" time are the same. We let the load resistance be 1 Ω so that we can calculate the power easily. The power when the pulse is on is 2 W, and the average power is 1 W.

To find the frequency components, we multiply the carrier, $2\cos(\omega t)$, by the Fourier series for rectangular pulses with a maximum value of one and a minimum value of zero. We write the pulses as

$$V_r(t) = \frac{1}{2} + \frac{2}{\pi}\left(\cos(\omega_k t) - \frac{\cos(3\omega_k t)}{3} + \frac{\cos(5\omega_k t)}{5} - \cdots\right), \tag{12.30}$$

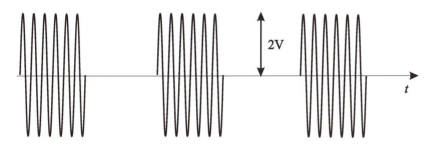

Figure 12.8. Transmitter pulses with zero rise and fall times.

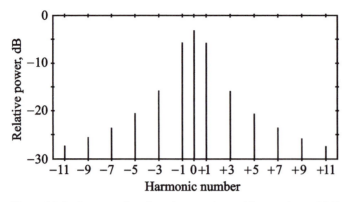

Figure 12.9. Spectrum for a keyed transmitter with zero rise and fall times, in dB relative the total power.

where $\omega_k = 2\pi f_k$ is the keying frequency. These coefficients are derived in Appendix B, Section 2. The product can be written as

$$V(t) = \cos(\omega t) + \frac{2}{\pi}\left(\cos(\omega t - \omega_k t) - \frac{\cos(\omega t - 3\omega_k t)}{3} + \frac{\cos(\omega t - 5\omega_k t)}{5} - \cdots\right)$$
$$+ \frac{2}{\pi}\left(\cos(\omega t + \omega_k t) - \frac{\cos(\omega t + 3\omega_k t)}{3} + \frac{\cos(\omega t + 5\omega_k t)}{5} - \cdots\right).$$

$$(12.31)$$

The largest frequency component is at the carrier frequency ω. The power in this component is 1/2. In addition there are sidebands around the carrier frequency that are caused by keying the transmitter on and off. These are shown in Figure 12.9. The horizontal axis is frequency, and the vertical axis is power in dB relative to the total power transmitted. This plot of frequency components is called a *spectrum*.

We get an interesting result if we add the power for each component. This is just the total power, which is one watt:

$$1 = \frac{1}{2} + 2 \cdot \frac{2}{\pi^2}\left(1 + \frac{1}{9} + \frac{1}{25} + \frac{1}{49} + \cdots\right).$$

$$(12.32)$$

The factor of 2 for the right term comes from the fact that we are summing two identical series, one for the upper sidebands and one for the lower sidebands. We can invert this formula to find an expression for the sum

$$\frac{\pi^2}{8} = 1 + \frac{1}{9} + \frac{1}{25} + \frac{1}{49} + \cdots = \sum_{n \, \text{odd}} \frac{1}{n^2}.$$

$$(12.33)$$

This is an elegant way to derive this formula.

For isolated frequency components such as transmitter harmonics, it makes sense to specify a maximum power level. For example, in the United States, the FCC requires that for transmitters like the NorCal 40A, with power levels of 5 W or below, the spurious components should be at least 30 dB below the carrier.

However, this kind of specification does not make much sense for keying side-bands, because a receiver may pick up many components at one time. The effect of any one component may be negligible, but the cumulative effect of all the components may be a problem. For keying sidebands, the corresponding question is: How wide does the channel need to be so that the power spilling out at frequencies above the channel is 30 dB below the transmitted power? Since the sidebands are symmetrical, the power below the channel would also be 30 dB down. We write the proportion of the sideband power at higher frequencies p as

$$p = \frac{2}{\pi^2} \left(\frac{1}{(n+1)^2} + \frac{1}{(n+3)^2} + \cdots \right), \tag{12.34}$$

where n is the number of the keying harmonic that marks the channel boundary. For large harmonic numbers, we can approximate this sum by an integral

$$p \approx \frac{2}{\pi^2} \cdot \frac{1}{2} \int_n^\infty \frac{dx}{x^2} = \frac{1}{n\pi^2}. \tag{12.35}$$

The factor of 1/2 before the integral takes into account the fact that only odd keying harmonics are generated. We can rewrite this as

$$n = \frac{1}{\pi^2 p}. \tag{12.36}$$

To find how large the bandwidth needs to be, we set $p = 0.001$ and get $n = 101$. This means that for 10-Hz keying, we would need a large channel that extends 1 kHz above and 1 kHz below the carrier frequency.

We can reduce the bandwidth dramatically if we increase the rise and fall times. The pulse shapes do not have simple mathematical forms, but we can approximate them by exponentials with a time constant τ (Figure 12.10a). To find the bandwidth that we would need for this waveform, we need to see how the frequency components are affected. One way that we can do this is to consider the frequency response of an RC network that produces these exponential waveforms (Figure 12.10b). This is the delay network we studied in Problem 3. We can

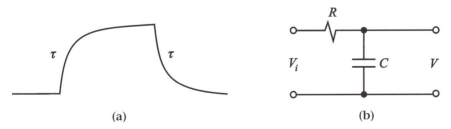

(a)   (b)

Figure 12.10. Pulse with exponential rise and fall (a). RC network with exponential waveforms (b).

relate the output voltage V to the input voltage V_i by the phasor relation

$$\frac{V}{V_i} = \frac{1}{1 + j\omega\tau},$$ (12.37)

where the time constant is $\tau = RC$. For the nth keying harmonic, we write

$$\frac{V}{V_i} = \frac{1}{1 + jn\omega_k\tau},$$ (12.38)

where n is the number of the keying harmonic. In our calculations, we are interested in the power rather than the voltage, so that we can consider

$$\left|\frac{V}{V_i}\right|^2 = \frac{1}{1 + (n\omega_k\tau)^2}.$$ (12.39)

This lets us modify the formula for the outside power p (Equation 12.34):

$$p = \frac{2}{\pi^2}\left(\frac{1}{(n+1)^2} \cdot \frac{1}{1 + (n+1)^2(\omega_k\tau)^2} + \frac{1}{(n+3)^2} \cdot \frac{1}{1 + (n+3)^2(\omega_k\tau)^2} + \cdots\right).$$ (12.40)

We can also approximate this sum by an integral, given by

$$p \approx \frac{1}{\pi^2}\int_n^\infty \frac{1}{x^2} \cdot \frac{1}{1 + (x\omega_k\tau)^2}\,dx.$$ (12.41)

We will assume that $(x\omega_k\tau)^2 \gg 1$ so that we can write

$$p \approx \frac{1}{\pi^2}\int_n^\infty \frac{dx}{x^4(\omega_k\tau)^2} = \frac{1}{3\pi^2 n^3(\omega_k\tau)^2},$$ (12.42)

or

$$n = (3p(\pi\omega_k\tau)^2)^{-1/3}.$$ (12.43)

For $p = 0.001$ we get

$$n \approx (f_k\tau)^{-2/3}.$$ (12.44)

Commercial transmitters often use a time constant of $\tau = 3$ ms. For $f_k = 10$ Hz, this gives

$$n = 10.$$ (12.45)

This reduces the required channel width dramatically, from 2 kHz to 200 Hz.

FURTHER READING

A good reference for practical details and measurements is "Mixers, modulators, and demodulators," by David Newkirk and Rick Karlquist. This is Chapter 15 in the *ARRL Handbook*, published by the American Radio Relay League. *Mastering Radio Frequency Circuits through Projects and Experiments* by Joseph Carr, published by McGraw-Hill, has a good chapter on the SA602AN mixer that is used in the NorCal 40A. A comprehensive advanced reference is *Microwave Mixers,* by Stephen Maas, published by Artech House.

PROBLEM 28 – RF MIXER

This is the first chance you get to use the SA602AN integrated circuit. This is a classic IC that was developed by the Signetics Corporation. Philips bought Signetics, and so it is now sold under the Philips name. The letters "SA" refer to a Philips product. The "A" suffix indicates that this is an improved version of an earlier part numbered 602. The SA602AN is used in many different communications systems, and it is very worthwhile to study the manufacturer's data sheets for this IC in Appendix D. It contains an amplifier, mixer, oscillator, and regulator, and it operates at frequencies as high as 500 MHz. The current consumption is quite low, only 2.4 mA, and this makes it quite suitable for battery-operated equipment. It costs only about $2. The NorCal 40A uses SA602ANs for the RF Mixer, the Transmit Mixer, and the Product Detector. The block diagram is shown in Figure 12.11. The supply voltage goes to pin 8 and should be between +4.5 and +8 V. There should be a bypass capacitor to ground nearby to prevent oscillations in the supply line. Pins 6 and 7 are the base and emitter for an oscillator transistor. Alternatively, you can connect an external oscillator to pin 6. An external oscillator should have a voltage of at least 200 mVpp. The RF inputs are pins 1 and 2. The LO and the RF signal mix in a Gilbert Cell, and the IF output appears at pins 4 and 5.

We can apply the input at either pin 1 or pin 2. The unused pin will need a bypass capacitor to ground. The input impedance at either pin is nominally 1.5 kΩ, but it actually varies considerably with frequency. Alternatively, the signal can be applied differentially between the two pins. The DC-bias voltage at the inputs is 1.4 V. The bias voltage is generated internally, so that we do not need any external bias circuitry. The circuits applied to the input pins should not draw any current, so that they do not upset

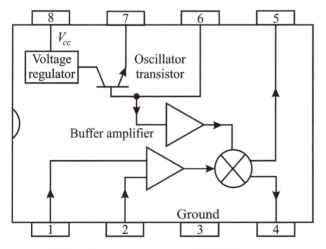

Figure 12.11. Block diagram for the SA602AN mixer.

the bias voltage. The NorCal 40A uses this voltage as the reference voltage for the RIT comparator.

We can take the IF output from either pin 4 or pin 5. Alternatively, since the two IF outputs are 180° out of phase, we can take the difference between the two pins to get an output that is twice as large. Aside from the larger voltage, this approach has other advantages. Often outside signals that couple into the lines will cause similar voltages on both lines, and this means that they do not affect the voltage difference. In addition, even-order spurious components tend to appear on both lines at similar levels, so that if we take the difference, these spurs will drop out. The output impedance for each pin is 1.5 kΩ, or 3.0 kΩ if we use a differential output. The DC voltage at the outputs is about 1.2 V below the supply voltage (6.8 V for an 8-V supply). To avoid upsetting the DC bias, the external input and output circuits should not draw any DC current.

In the RF Mixer, we do not use the internal oscillator but inject the VFO from the outside. We will use the internal oscillator as a fixed crystal oscillator for the Transmit Mixer and the Product Detector, however. First install the Receiver Mixer U1, together with the two bypass capacitors C5 and C8. Install the RF Gain pot R2. You may need to move the transformer T2 out of the way so that R2 and T2 do not touch each other. R2 serves as an attenuator at the input to keep the radio from overloading.

The RF Mixer output goes to the IF Filter, which removes the sum frequency response, and from there to the Product Detector U2. In the previous problem we applied the reference voltage for the RIT comparator to this pigtail. This time we will use the input of U2 to provide the reference. Install U2 and the bypass capacitor C15, being careful to orient both parts correctly. C15 is a very large capacitor (2.2 μF) to suppress coupling between the RIT comparator and the Product Detector.

A. Now we measure the conversion gain of the RF Mixer. Connect the power supply and turn it on. Connect the counter to the VFO loops and use the VFO Tune and RIT pots to adjust the VFO frequency to 2.1 MHz. The VFO Tune serves as a coarse adjustment and the RIT as a fine adjustment. It is important to leave the counter connected for the rest of this lab. The counter has 30 pF of capacitance, and the cable adds more. Removing the counter would pull the VFO frequency up enough to push a signal right out of the pass band of the IF Filter. Make sure that the RF Gain pot R2 is fully clockwise. This gives maximum gain. Connect the function generator to the Antenna jack J1. The amplitude setting should be for a 50-mVpp sine wave. Look over your work in Problem 14 to find the center frequency for your IF Filter. Use this frequency for your IF. Now calculate the correct frequency f_{rf} for the function generator, and set the function generator to this frequency. Be careful to get the frequency right, or else you will not likely see a response. Connect the 10:1 scope probe to pin 1 of U2 to see the IF. Because the IF is not at the same frequency as the function generator, you cannot use the sync output from the function generator to trigger the scope. You should set up the scope for internal triggering. Adjust the tuning capacitors for the RF Filter (C1 and C2) for maximum IF output. You should vary the frequency slightly to make sure that you are at the frequency of minimum loss for the IF Filter. Now find the conversion gain in dB for the RF Mixer, assuming that the input impedance of U2 is 1.5 kΩ.

B. For the remaining scope measurements, you will probably find it convenient to switch in the low-pass filter if one is available to reduce the noise on the screen. This will reduce the scope signal level, and you will need to take this reduction into account. We can control the signal level with the RF Gain pot R2. To see the range of attenuation that it offers, rotate it fully counterclockwise. How much attenuation in dB is provided by the pot? You will probably want to increase the function generator amplitude setting to its maximum value to make the IF output visible at all. The IF will be small, and it can be challenging to get a good trace on the scope. The mixer bias gives this signal a relatively large DC offset, and you will need to switch the scope coupling to AC to get rid of the DC component. Otherwise the trace sails off the scope screen. You may also see multiple traces, indicating that the triggering is not consistent. If you have trouble with the triggering, you may be able to eliminate the traces you do not want by adjusting the triggering level. After you make the attenuation measurement, you should return the function generator to its original amplitude setting, 50 mVpp. Restore the RF Gain pot to the clockwise position. Make sure that you do not forget to do this, or the remaining measurements will be off.

C. Next we look for spurious responses. There are many spurs, but for now we will measure two of the largest ones. Calculate the image frequency f_i for the VFO. Set the function generator to f_i and look for the IF response. You will need to set the function-generator amplitude to its maximum output value to see it. By how many dB is the image response suppressed compared to a signal at the RF frequency?

D. The other spur that we will look for is $f_{5\downarrow}$, from the fifth harmonic of the LO. This is tougher to find than the image. This spur is close to the normal receive frequency, and for this reason, the RF Filter does not reject it well. Calculate $f_{5\downarrow}$ and $f_{5\uparrow}$. Set the function generator to $f_{5\downarrow}$. This time, you need to be more careful with the function-generator level, because the output saturates if you set the amplitude too large. You might try an amplitude setting of 1 Vpp. It is important that the VFO frequency be precisely 2,100 kHz during this measurement, because we are mixing with the fifth harmonic. A 100-Hz error in the VFO frequency will cause a 500-Hz shift in f_{5-}, and this will push us clear out of the pass band of the IF Filter. It is easy to get this large a shift if something comes near the air-variable capacitor C50 or the big inductor L9. You should vary the function-generator frequency slightly to make sure that you are at the peak of the response. By how many dB is the $f_{5\downarrow}$ spur suppressed compared to a signal at the RF frequency?

E. Can you find the error in Figure 4 of the data sheet for the SA602AN in Appendix D?

PROBLEM 29 – PRODUCT DETECTOR

The Product Detector is the second mixer in the receiver (Figure 12.12). The input to the Product Detector is the IF signal, after it has passed through the IF Filter. The local oscillator for the Product Detector is called the Beat-Frequency Oscillator, or BFO. The output is at the difference frequency, which is an audio frequency. We use an audio

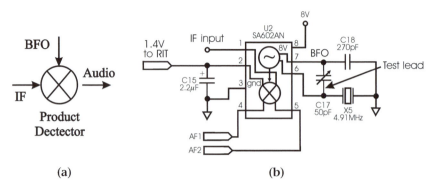

Figure 12.12. (a) Product Detector inputs and outputs, and (b) the schematic for the NorCal 40A Product Detector.

frequency of 620 Hz. This matches the resonant frequency of the tuned speakers that you made for Problem 17.

The BFO uses the internal oscillator circuit in the SA602AN (Figure 12.12b). It is a Clapp oscillator like the VFO, except that the resonator is a crystal instead of an LC circuit, and the transistor is a BJT rather than a JFET. The divider circuit consists of C17 and C18. C17 is a variable capacitor, and this allows the BFO to be set precisely. We will put the BFO 620 Hz above f_{if}.

A. Install C17, C18, and X5. Plug in the power supply and turn it on. Use short test leads to probe pin 6 to see if there is an oscillation. You may need to adjust C17 to see it, because the circuit often does not oscillate when C17 is in the low-capacitance part of its range. Then move the test leads from the scope to a counter. Adjust the capacitor C17 for the minimum oscillation frequency, and record this frequency.

B. For comparison, calculate the minimum oscillation frequency that we should be able to reach with the BFO, assuming that the range of the variable capacitor C17 is 7 pF to 70 pF. For the crystal X5 use the inductance L and capacitance C that you measured in Problem 14. Set the oscillation frequency 620 Hz above the center frequency of your IF Filter by adjusting C17. If you cannot reach this frequency, get as close as you can.

C. Measure the temperature coefficient for the BFO in the same way that you did for the VFO. Try to make sure that the BFO components and the thermometer both get plenty of hot air, so that their temperatures will be similar. The frequency shift should be small, and it is a good idea to measure the frequency to the nearest hertz.

Next we look for the output audio signal. Install a 3-kΩ resistor in the C19 holes. Leave room to hook probes on each end. This resistor will act as the load for the Product Detector. Set the function generator to a 7,010-kHz sine wave, with an amplitude of 50 mVpp. Put a 10:1 scope probe at one end of the resistor. You will need to set the scope channel for AC coupling, because the SA602AN has a large DC voltage on its outputs. Adjust the

VFO Tune pot and the RIT pot until you see an audio signal on the scope. You will need to slow down the trace to see this. The waveform may be difficult to interpret because you are seeing the raw mixer output with all the frequency components.

Next we set the RF Gain pot for an attenuation of 20 dB, so that we do not overload the Product Detector. Start with the RF Gain pot set to minimum attenuation (fully clockwise). Move the scope probe to pin 1 of the Product Detector, and readjust the trace speed so that you can look at the IF. Tune the VFO for maximum scope voltage. Then adjust R2 to reduce the voltage by a factor of ten.

We will use the multimeter for AC voltage measurements for the rest of this lab. The advantage of using the multimeter is that both inputs of the multimeter are isolated from ground, and you do not short out the output the way a scope ground lead would. In addition, a multimeter usually only measures the voltage for frequencies up to about 100 kHz. Thus the sum frequency at 10 MHz and the higher-order products are ignored. Make sure that the multimeter leads do not get near the VFO inductor or air-variable capacitor, or they will pull the VFO frequency.

D. Measure the gain of the receiver from the Antenna jack J1 through the Product Detector. The input is the available power P_+ from the function generator, and the output is the power P delivered to the 3-kΩ load. You will need to add 20 dB to this number as an allowance for the loss in the RF Gain pot. You should note that the multimeter measures rms voltage, not peak-to-peak voltage.

E. Next we look for the $f_{3\downarrow}$ spur for the RF Mixer. This is in the AM band, and it is the one that causes the most trouble in this receiver. Calculate the frequency of the $f_{3\downarrow}$ spur for $f_{rf} = 7,010$ kHz. Set the function generator to $f_{3\downarrow}$, and use the maximum voltage setting. There is no danger of saturating the output this time, and you need all the voltage you can muster to get through the RF Filter. Set the RF Gain pot fully clockwise for minimum attenuation. The voltage will be small. You should vary the function-generator frequency slightly to make sure that you are at the peak of the response. This requires patience, because a multimeter is slow. By how many dB is the $f_{3\downarrow}$ spur suppressed compared to a signal at the RF frequency?

F. Finally, we investigate the spur at f_{if}. In principle, this spur should not be there, but in practice there is some leakage through the RF Mixer. Set the function generator to f_{if}. Use as big a voltage as you can without saturating the output. The bigger the output, the more it will stand out from the noise, and the easier it will be to measure it accurately. You can check for saturation by making sure that the output is proportional to the input. Try cutting the input voltage in half, and verify that the output also drops by a factor of two. If it does not drop this much, the output is saturated, and you should start over at the lower input voltage. By how many dB is the f_{if} spur suppressed compared to a signal at the RF frequency? One thing that makes this spur different from the others is that it does not vary when you tune the radio, because it does not depend on the VFO frequency. You can go right from the bottom of the tuning range to the top without affecting the spurious response. Remove the 3-kΩ resistor that you installed as the load for the Product Detector.

PROBLEM 30 – TRANSMIT MIXER

The Transmit Mixer takes the signal from the VFO and mixes it with its internal oscillator, the Transmit Oscillator (Figure 12.13). Its output is a low-level signal that is amplified by the transmitter–amplifier chain. In the receiver mixers, we take differential outputs, but in the Transmit Mixer, we only use one of the outputs. If we do this, we cut the output voltage in half and the output resistance in half. The available power is proportional to the voltage squared and inversely proportional to the resistance. Consequently, our available power drops by a factor of two.

A. To start, you should install all of the components in Figure 12.13 except for C31. This leaves the VFO disconnected initially, which makes it easier to measure the Transmit-Oscillator frequency. You will also need to insert the shorting plug to turn on the transmitter. You should adjust C34 to set the Transmit-Oscillator frequency to the center frequency of your IF Filter. This aligns the transmitter and receiver frequencies and sets the Transmit Oscillator 620 Hz below the BFO. If you study the resonators for the two oscillators, you will see that the Transmit Oscillator has an additional inductor, L5, which is an 18-μH inductor. Calculate how much you would expect this inductor to lower the oscillation frequency.

Connect the Antenna jack J1 to the scope with a 50-Ω load. Now install the capacitor C31 (5 pF). This capacitor couples the TX VFO output to the Transmit Mixer. Adjust the filter capacitor C39 to peak the transmitter output. Now set the Drive Adjust pot R13 for an output of 30 Vpp. Connect the output to the counter, again with the 50-Ω load in parallel. Adjust the output frequency to 7,010 kHz with the VFO Tune pot. The RIT pot should have no effect on this frequency. You should repeat the scope adjustments of C39 and R13 as a final transmitter adjustment.

B. The coupling capacitor C31 has two additional functions. First, the VFO and the Transmit Mixer input have different DC voltages, and the capacitor ensures that they stay that way. Also, the TX VFO voltage is too large for the input of the Transmit Mixer, and the capacitor acts as a potential divider to reduce it. Calculate the voltage attenuation factor for the potential divider. Measure the TX VFO voltage. Use the TX VFO voltage and the voltage attenuation factor to calculate the input power to

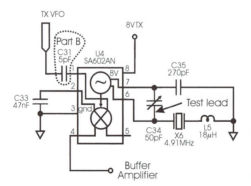

Figure 12.13. The Transmit Mixer and the Transmit Oscillator. C31 should not be installed until after the frequency is set in Part A.

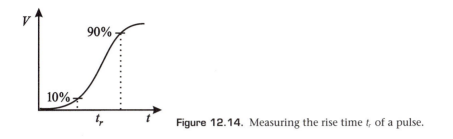

Figure 12.14. Measuring the rise time t_r of a pulse.

the Transmit Mixer. Calculate the gain in dB of the entire transmitter chain, starting from the Transmit-Mixer input and ending with the 30-Vpp output.

Now we look at the keying waveform. Replace the shorting plug with a keying relay cable, and drive it with a function generator set for a 10-Hz, 5-Vpp square wave. These pulse lengths are similar to those used in actual communications. You should use a sync cable to trigger the scope. Adjust the scope until you see the keying pulses on the screen. It is important that keyed transmitters not start or stop suddenly, because this causes interference. Operators on nearby frequencies hear annoying clicks. It is traditional to specify the rise and fall times of the keying envelope, which is the outline of the scope pattern. Previously we have characterized rise and fall times in terms of t_2, the time that it takes the trace to come a factor of two closer to its final value. The rise and fall waveforms for a transmitter are not exponentials, however, and so we will use a different definition. It is traditional to define the rise time of a pulse as the time it takes to go from 10% to 90% of the final value (Figure 12.14). The fall time is defined in a similar way. In the NorCal 40A, the rise time is determined by the Transmit-Oscillator build-up. The fall time is controlled by the Driver capacitor C56. In addition, the rise and fall times are influenced by the input–output characteristic of the Power Amplifier. It is not easy to calculate these times, and we will measure them instead.

C. Measure the rise and fall times of the keying envelope. It takes some thought to display the beginning and ending parts of the keying waveform with a time scale that allows accurate measurements. There is a substantial keying delay, and you may find that the leading edge of the pulse dances around the screen. Here are some things that you might consider to help make your measurements. You can use the trigger slope switch to determine whether the scope triggers at the beginning or the end of the pulse. You can also use the trigger level control to set where the sweep starts on the waveform. If you see several traces on the screen at the same time, try the holdoff control to control the delay between sweeps. This often helps in getting rid of traces that you do not want. When you finish the measurements, leave the holdoff control at the normal minimum position. Otherwise the delay reduces the number of sweeps and dims the trace.

D. The Transmit Mixer, in addition to producing the transmitter signal, has spurious frequency components that are of the form

$$f_{mn} = m f_{vfo} + n f_{to}, \tag{12.46}$$

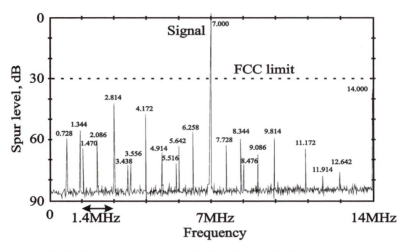

Figure 12.15. Frequency components for the NorCal 40A. The vertical axis shows the power on a dB scale, at 10 dB per division, and the horizontal axis shows the frequency, with a scale of 1.4 MHz per division. The different frequency components appear as lines on the plot, and the frequency is noted at each component.

where f_{to} is the frequency of the transmit oscillator, and m and n are integers. Figure 12.15 is a spectrum-analyzer plot that shows the power levels of the different frequency components from the NorCal 40A. Each of the components that is larger than 80 dB below the carrier has the frequency listed beside it. To make the plot, the spectrum analyzer measured the power at 14-kHz intervals from 0 to 14 MHz. Therefore, the frequencies are not precise, and you should allow for errors of up to 14 kHz. Find n and m for each of the lines. You should use the lowest values of n and m that give a frequency within 14 kHz of the line.

13

Audio Circuits

There are two audio circuits in the NorCal 40A, the Automatic Gain Control, or AGC, and the Audio Amplifier. The AGC is an attenuator with JFETs that act as variable resistors. The Audio Amplifier is the LM386N-1, made by National Semiconductor. This integrated circuit appears in many different audio systems, and it costs about a dollar. The "–1" indicates a supply voltage range from 4 to 12 V. A "–4" version is available that allows a supply voltage up to 18 V. For the LM386N-1, the maximum output power is about one watt. In standby, it draws about 4 mA, which is a reasonable level for running off batteries.

13.1 Audio Amplifier

Figure 13.1 shows the schematic for the LM386N-1. It is more complicated than the previous circuits that we have looked at, and because it is an integrated circuit, most points are not accessible for measurements. There are three stages of amplification. The input is a differential amplifier, but it is made with pnp transistors rather than npn transistors. This turns things upside down. Each input has a pair of stacked pnp transistors. The stack gives the effect of a single transistor with a current gain of β^2. The differential amplifier is followed by a common-emitter stage. The output is a Class-B emitter follower that is similar to the one we discussed in Section 10.6. The only difference is that the pnp transistor has been replaced by a combination of a pnp transistor and an npn transistor that acts like a single pnp transistor, but with the large current gain that is characteristic of an npn transistor. The feedback resistance R_f and the emitter resistance R_e determine the gain of the amplifier.

The collector loads for the differential amplifier comprise a pair of transistors wired together as shown in Figure 13.2a. This circuit is called a *current mirror*. Notice that the bases are connected together and the emitters are connected together. Thus the base–emitter voltage for each transistor is the same. In Section 9.4 we found that the collector current I_c and the base–emitter voltage V_b in an active transistor are related by the equation

$$I_c = I_{cs} \exp(V_b / V_t). \tag{13.1}$$

If V_b is the same for both transistors, then the collector current is also the same, provided that the saturation current I_{cs} is identical for each. By putting the two

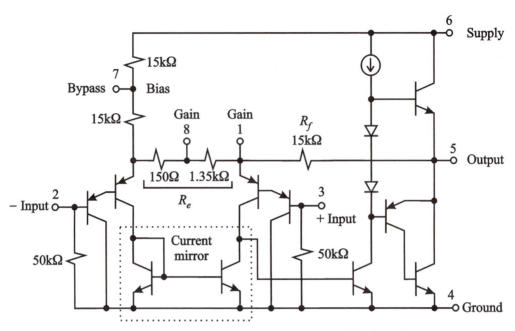

Figure 13.1. Schematic for the LM386N-1 Audio Power Amplifier, from the National Semiconductor data sheet in Appendix D.

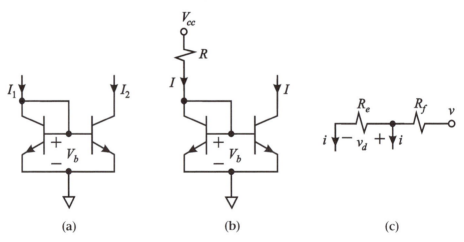

Figure 13.2. (a) Current mirror. (b) Making a current source with a current mirror. (c) Gain calculation for the Audio Amplifier. The two currents labeled i are forced to be equal by the current mirror.

transistors close together on the same piece of silicon, I_{cs} can be matched closely. Therefore

$$I_1 \approx I_2, \tag{13.2}$$

provided β is large, so that the base currents are much smaller than the collector currents.

One application of a current mirror is to make current sources (Figure 13.2b). We can set the current with a resistor to the supply. The current in the resistor is given by

$$I = (V_{cc} - V_b)/R. \tag{13.3}$$

The current in the other transistor is now forced to be I.

The current mirror forces the collector currents on each side of the differential amplifier to be the same. This property of the current mirror allows us to calculate the gain of the Audio Amplifier. A simplified circuit is given in Figure 13.2c. Here v is the AC output voltage. The currents labeled i at each end of the emitter resistance R_e are forced to be equal by the current mirror. The voltage across R_e is the differential input voltage v_d. We can understand this if we consider the base–emitter voltage drops in the pnp transistors on each side of the differential amplifier. Since the drops are the same on both sides, the difference in voltage between the two inputs is the same as the voltage across R_e.

The current that flows through the feedback resistor R_f is approximately $2i$. We can neglect the current that flows in the two 15-kΩ bias resistors, because these have a much higher impedance than the pnp emitters, which are basically forward-biased diodes. Consequently, the feedback current will flow in the emitters and not in the bias resistors. We write

$$2i \approx v/R_f. \tag{13.4}$$

Here we have assumed that the output voltage v is much larger than the input voltage v_d because of the gain of the amplifier. We can also write an expression for the current in R_e:

$$i = v_d/R_e. \tag{13.5}$$

If we eliminate the current between these two equations, we get the voltage gain

$$G_v = v/v_d = 2R_f/R_e. \tag{13.6}$$

One thing that is interesting about this expression is that the gain only depends on the feedback and emitter resistances, and not on the β of the transistors, which is quite variable. The gain is fixed even if β varies with the drive level, which ordinarily would cause distortion. In this amplifier, $R_e = 1.5$ kΩ and $R_f = 15$ kΩ, and so we expect a voltage gain of 20. We can reduce the effective value of R_e to 150 Ω by connecting a bypass capacitor between pins 1 and 8. This increases the gain to about 200. In addition, we can provide high-frequency roll-off by connecting a capacitor in parallel with R_f.

13.2 Op Amps

For the Audio Amplifier, we found that the gain is determined by the feedback resistance. This is negative feedback, and it reduces the gain. This is in contrast to the positive feedback that we used to make an oscillator. Negative feedback is one of the most important ideas in electrical engineering, and it is due to Harold Black of the Bell Telephone Laboratories. You might wonder why we would want

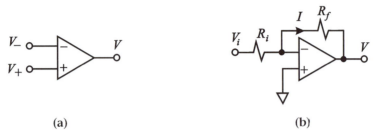

Figure 13.3. (a) Schematic symbol for an op amp, and (b) inverting amplifier circuit.

to reduce gain, but it turns out that it is relatively easy to build amplifiers with a lot of gain. It is more difficult to control the gain, reduce distortion, increase the input impedance, and reduce the output impedance. Negative feedback can help us with all of these.

This approach is taken to an extreme in operational amplifiers, or op amps, which are differential amplifiers with very high gain, like the LM393N comparator in the RIT circuit we built in Problem #27. Op amps typically have a voltage gain of 100,000 and they would not be used as an amplifier without negative feedback because the outputs would saturate. In addition, they have high input impedance and low output impedance. We will consider an example to see how the feedback works. Figure 13.3 shows the schematic symbol for an op amp. There are differential input terminals labeled − and +. In analyzing op amps, it is common to make two assumptions. Horowitz and Hill call these the Golden Rules.

Rule I. The differential gain is extremely high, and we will call it ∞. This means that if the output V is not at the ground or supply (as it would be in a comparator), then the difference between V_- and V_+ is infinitesimal. We write

$$V_- = V_+. \tag{13.7}$$

Rule II. The impedance of each input is extremely high, and we will call it ∞. This means that the inputs draw no current, and we can write

$$I_- = I_+ = 0. \tag{13.8}$$

We show an inverting amplifier in Figure 13.3b. If the + input is grounded, then $V_- = 0$ (Rule I). We call this a *virtual ground*. Moreover, since the − input draws no current, the current in both resistors is the same. We can therefore write two expressions for the current I:

$$I = V_i/R_i \tag{13.9}$$

and

$$I = -V/R_f. \tag{13.10}$$

If we eliminate the current between these two expressions, we get the voltage gain:

$$G_v = V/V_i = -R_f/R_i. \tag{13.11}$$

Notice the similarity to the gain calculation for the Audio Amplifier (Equation 13.6). The output voltage is proportional to the value of a feedback resistor. The input voltage is proportional to the value of the resistor in the input circuit.

13.3 JFETs as Variable Resistors

When we studied bipolar transistors, we considered two different regions of operation. If the collector voltage is more than a few tenths of a volt, the output impedance is large, and the output current is relatively independent of the load. In our circuit models, we used a current source for the output. We called this the active region, and we used it for our Class-A amplifier. However, we found that when the base current is large and the collector voltage is small, the effective output impedance is small. We used this "on" mode for the switches and the Class-C amplifier.

JFETs also have two modes of operation. Figure 13.4 shows a plot of drain current versus drain voltage with gate voltage as a parameter. In the active region, the current is relatively independent of the drain voltage. The current increases with gate voltage from zero at the cut-off voltage V_c to I_{dss} at $V_{gs} = 0$. In FETs, the active region is also called the saturation region, which is quite confusing, because it does not correspond to the saturation region for a BJT.

If $V_{ds} < V_{gs} - V_c$, the drain current is no longer independent of drain voltage. This is called the *linear* region because the current is approximately linear in the voltage. This means that the JFET acts like a conductance that is controlled by the gate voltage. The current is usually written approximately as

$$I_d = V_{ds}\left(\frac{2I_{dss}}{V_c^2}\right)(V_{gs} - V_c - V_{ds}/2). \tag{13.12}$$

This is different from BJTs where the current is quite nonlinear at low collector voltages. The JFET can also be used with negative drain voltages and currents.

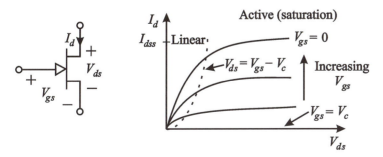

Figure 13.4. Drain current characteristics for a JFET.

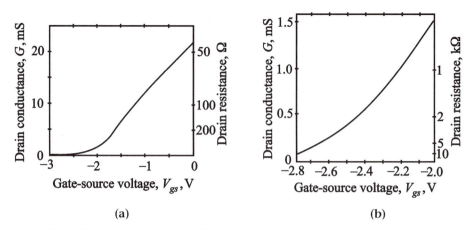

Figure 13.5. Drain conductance in the linear region for the J309 JFET as a function of the gate voltage (a), and the high-resistance region near cutoff (b).

This is like swapping the role of the source and drain. JFETs work well this way, although usually g_m drops somewhat and the Miller capacitance increases because the gate is usually placed closer to the source than the drain. This is quite different from BJTs, which usually work poorly backwards. We use JFETs in the linear region as a variable conductance in the Automatic Gain Control. To see how this works, let us calculate the conductance G as

$$G = \frac{I_d}{V_{ds}} = \left(\frac{2I_{dss}}{V_c^2}\right)(V_{gs} - V_c). \tag{13.13}$$

We have neglected the term $V_{ds}/2$, which is small in the linear region. This equation predicts a linear relation between the gate voltage and the conductance. This is only approximately true in practice. Figure 13.5a shows the measured conductance versus gate voltage for the J309 JFET that is used in the AGC. For gate voltages above -1.5 V, the relationship is relatively linear, covering a resistance range from 200 Ω to 50 Ω. Figure 13.5b shows the high-resistance region of the same plot. This covers the range where the resistance is greater than 500 Ω. The curvature is quite noticeable.

FURTHER READING

Horowitz and Hill cover JFET circuits in Chapter 3 and op amps in Chapter 4 in *The Art of Electronics*, published by Cambridge University Press.

PROBLEM 31 – AUDIO AMPLIFIER

We build the Audio Amplifier in three stages, so that we can see how different components determine the frequency response. The final result will be an amplifier with a band-pass characteristic that eliminates both high-frequency hiss and low-frequency rumble. In the first stage (Figure 13.6), we consider how the output circuit affects the gain. The output

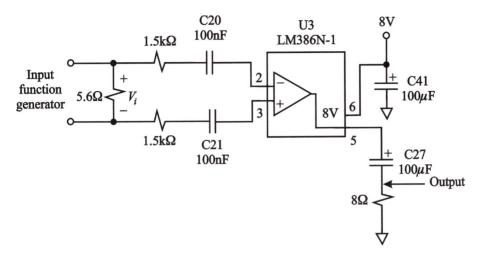

Figure 13.6. Measuring the frequency response of the output network of the Audio Amplifier.

voltage is taken from pin 5. The DC voltage there is half the supply voltage. Because we do not want DC current in the speaker we use a large electrolytic capacitor C27 (100 μF) to couple the output to the speaker. In the same way, we have coupling capacitors C20 and C21 (100 nF) at the input so as not to upset the input bias conditions. We will use an 8-Ω resistor in place of the speaker temporarily. The speaker impedance varies with frequency, and this would confuse the measurements.

Install the amplifier U3, the coupling capacitors C27, C20, and C21, and the supply bypass capacitor C41 (100 μF). For the load, solder an 8-Ω resistor across the outer holes of the R8 outline. Leave enough lead length that you can get probes on C20, C21, and the 8-Ω resistor. Connecting a function generator is complicated because the audio amplifier has a lot of gain, and even low input voltages can saturate the output. We use a potential-divider circuit to reduce the input voltage to prevent the output from saturating. Connect a 5.6-Ω resistor and a pair of 1.5-kΩ resistors as shown in Figure 13.7. The drain holes in the Q2 and Q3 outlines are convenient for this. Connect the function generator leads at the input, and connect the multimeter and scope probes at the output. Be careful when connecting the scope ground, because one can often cause oscillations by hooking up the scope leads backwards. Since the multimeter measures rms voltage, it is convenient to use rms amplitude settings on the function generator. This way, you do not have to worry about factors of $\sqrt{2}$ in the gain calculations.

A. Calculate how the input voltage V_i relates to the amplitude setting on the function generator, assuming that the input impedance of the amplifier is very high. You will need this ratio later to calculate the gain of the amplifier and the loss of the input network.

B. The output coupling capacitor gives a high-pass characteristic. Measure the voltage gain G_v at a frequency that is high enough that you get full gain. Also measure the 3-dB roll-off frequency f_l. Now calculate what you expect for G_v from Equation 13.6. Calculate the value you expect for f_l, assuming that the output impedance of the amplifier is small.

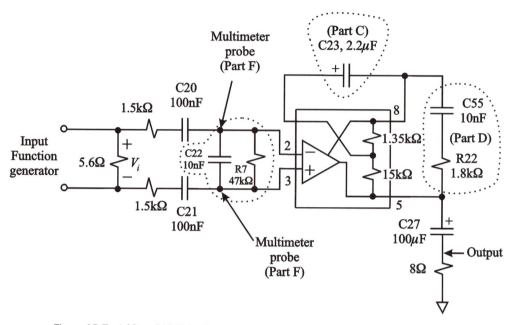

Figure 13.7. Adding C23 (2.2 μF) to increase gain (Part C). Adding C55 (10 nF) and R22 (1.8 kΩ) for a low-pass response (Part D). Adding the input network C22 (10 nF) and R7 (47 kΩ) for low- and high-frequency roll-off (Part F).

C. Now we increase the gain by bypassing the internal 1.35-kΩ emitter resistor. Install C23 (2.2 μF). C23 connects between pins 1 and 8 of the amplifier (Figure 13.7). Study Figure 13.1 to see how this affects the operation of the amplifier. You should work with a low input voltage now because the gain is large. Check frequently that the output is a clean sine wave and not saturated. The bypass capacitor only shunts the internal resistor at high frequencies, so this also gives a high-pass characteristic, but at a higher roll-off frequency than the output circuit. Measure G_v and f_l again. Now calculate the values that you would expect, neglecting the effect of the internal 1.35-kΩ resistor.

D. Next we add a low-pass response to the amplifier by installing the bypass network C55 (10 nF) and R22 (1.8 kΩ) for the internal 15-kΩ feedback resistor. These connect between pins 5 and 8 (Figure 13.7). You will need to study Figure 13.1 again to understand how it works. Combined with the high-pass response we have already measured, we get a band-pass characteristic. Make a plot of the gain in dB as a function of frequency from 100 Hz to 10,000 Hz, making sure that the input voltage is low enough that the output does not saturate. For the input power P_+, use the available power from the function-generator and resistor network. For the output power P, use the power that is delivered to the 8-Ω load. The plot will be compressed at the low-frequency end if you use a linear frequency scale. For this reason, for the frequency axis, you should plot $\log_{10} f$. This will give a more symmetric plot. This kind of plot is called a Bode plot, and it is convenient because the capacitive roll-offs show up as straight lines with a slope of 1. Bode plots are widely used for studying

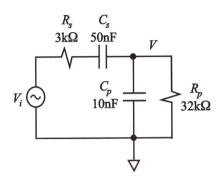

Figure 13.8. Simplified version of the input circuit.

amplifier response and checking stability against oscillation. Incidentally, a log plot should also guide you in choosing frequency increments. You should use much bigger increments at higher frequencies, or the plots will take a long time. One way to do this is to use equal increments of $\log_{10} f$, for example, the sequence, 2.0, 2.2, 2.4, ..., 3.6, 3.8, 4.0. This lets you make an excellent plot from 100 Hz to 10,000 Hz with only 11 data points.

E. Find the peak gain in dB and the corresponding frequency. What are the upper and lower 3-dB frequencies, f_u and f_l? Now calculate what you would expect for f_u. In this calculation, ignore the effect of R22. The role of R22 is to help prevent oscillations at high frequencies.

Now we turn to the input circuit. Install C22 (10 nF) and R7 (47 kΩ) (Figure 13.7). These provide additional roll-off at both low and high frequencies. Inside the amplifier, each input effectively has an internal 50-kΩ resistor in parallel with the pnp transistor pair. This adds a 100-kΩ resistor in parallel with R7. We can draw a simplified version of the input circuit (Figure 13.8). I have combined the input resistance of the amplifier, 100 kΩ, and R7 (47 kΩ) to make an equivalent resistance R_p given by

$$R_p = R7 \parallel 100\,\text{k}\Omega = 32\,\text{k}\Omega. \tag{13.14}$$

In addition, the coupling capacitors C20 and C21 are combined to make an equivalent capacitor C_s. The voltage V_i is the input voltage from Part A.

F. Find an expression for the loss of the network V_i/V. You should be able to put it in the form

$$\frac{V_i}{V} = L\left(1 + \frac{jf}{f_u} + \frac{f_l}{jf}\right), \tag{13.15}$$

where L is the loss in the pass band and f_u and f_l are the 3-dB roll-off frequencies. Find expressions for L, f_l, and f_u, and calculate the values for our circuit. Now measure them. It is convenient to do this with the multimeter probes attached to the amplifier side of the coupling capacitors C20 and C21. This is a differential input, and so you should not attach the scope ground there. When you finish the measurements, remove the 5.6-Ω and 1.5-kΩ resistors that you used to drive the Audio Amplifier. However, leave the 8-Ω load resistor in place, because we will need it for the next two exercises.

PROBLEM 32 – AUTOMATIC GAIN CONTROL

The AGC attenuates large signals to try to keep the audio output at reasonable levels for a wide range of inputs. The NorCal 40A uses a pair of JFETs as variable resistors in front of the Audio Amplifier. The AGC is a complicated circuit, with an attenuator, detector, and a keying section, and we will take it in several stages in this exercise and the next.

We start with the attenuator, or more precisely, dual attenuators (Figure 13.9). Install Q2 and Q3, R5, D5, D6, and R6. Leave enough room at the anode connection of D5 that you can attach a multimeter probe. Each JFET forms an attenuator for an audio amplifier input. R5 is a network of four identical 2.2-MΩ resistors in 2:1 potential dividers at the gates of the two JFETs. The input voltage for the attenuators is set by the AGC Threshold pot R6. The diodes prevent the gate voltage from rising above the JFET source voltage.

Once again, the function-generator voltage is too large, and it would cause the Audio Amplifier to saturate if we do not attenuate it first. Install a 300-kΩ resistor in one of the C19 holes, and attach the function-generator probe to the other end of the resistor. Either hole can be used. The function-generator ground should be attached to one of the ground loops on the side of the board.

Attach the oscilloscope and the multimeter to measure the output voltage of the Audio Amplifier as you did in the last problem. The function-generator setting should

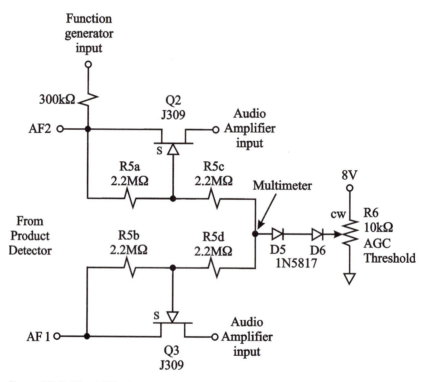

Figure 13.9. The AGC attenuators.

be for a 620-Hz sine wave. R6 should be set fully clockwise. This turns the attenuator off by setting the gate-source voltage to zero and leaves the JFETs with minimum resistance.

Adjust the amplitude setting of the function generator until the output voltage of the Audio Amplifier is 1 Vrms.

A. In the next set of measurements, you will need to move the multimeter lead that is not grounded back and forth between the 8-Ω load resistor and the anode of D5 (Figure 13.9), which is the control voltage for the attenuator. You will also need to switch back and forth between AC and DC voltage measurements. Make a plot with the audio output voltage on the y axis and the DC control voltage (the anode of D5) on the x axis. You should plot the audio voltage on a log scale because it varies over a wide range. This plot requires care and thought, because you will find that sometimes large changes in the control voltage make little change in the output, whereas at other times small changes in the control voltage cause large changes. You should try to keep a reasonable spacing between the points on the graph.

B. What is the maximum control voltage that you measure? What device supplies this voltage? Use this voltage and your plot to infer the cut-off voltage V_c for the JFET.

C. What is the minimum control voltage that you measure? D5 and D6 are Schottky diodes rather than pn diodes. Can you think of what could go wrong with the circuit if we used pn diodes instead?

D. Your plot should show that the range of control voltages that change the output is only a small part of the total range. For this reason we must set the AGC Threshold pot R6 at the beginning of the region where changes in the control voltage make a difference. Adjust the pot so that the audio output is reduced by 1 dB from its value at the full clockwise position. This is the setting that we use for the later measurements.

Now we can install the AGC capacitor C29 (10 μF) and the coupling capacitor C30 (2.2 μF) (Figure 13.10). C30 couples a portion of the output audio voltage to the AGC. When the audio voltage is sufficiently negative, it pulls down C29 through the rectifier diode D5. This reduces the input control voltage to the attenuator and increases attenuation. D6 prevents R6 from shunting D5, by turning off when the audio output voltage is negative.

E. The input network consists of the function generator, the 300-kΩ resistor, and a 1.5 kΩ resistor in the Product Detector. How does the open-circuit voltage for this network relate to the amplitude setting of the function generator? This open-circuit voltage should be the input voltage for the next plot. Plot the audio output voltage as the input voltage varies from 0.5 mVrms to 50 mVrms. Both the input and output voltages should be plotted on a log scale. You should find that the AGC has little effect at low voltages, but it should reduce the output at high voltages. Find the slope of the plot in the high-voltage region. Leave the 300-kΩ resistor in place; we will need it for the next exercise.

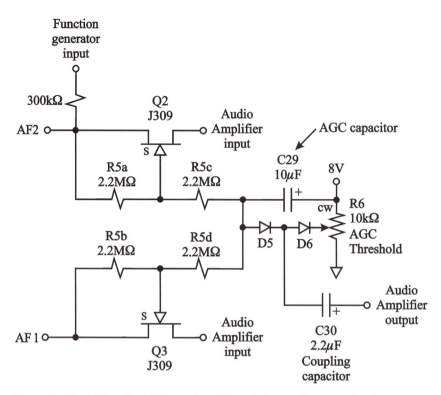

Figure 13.10. Adding the AGC capacitor C29 and the coupling capacitor C30.

PROBLEM 33 – ALIGNMENT

In this exercise we finish the transceiver, and do a checkout and alignment. We start with a measurement of AGC timing. We know from our transmitter measurements that the rise time for a pulse is in the range of 1 to 3 ms. The audio tones are at 620 Hz. Hence the rise time is only one audio cycle. The AGC should attenuate a strong signal on that time scale. People call this the *attack*. The attack for the NorCal 40A AGC is shown in Figure 13.11. The first cycle shows distortion at both the top and the bottom, indicating that the amplifier is overloaded. On the bottom half of the second cycle the amplitude is still large but the distortion is gone. On the top half of the second cycle the attack is complete, and the output is much reduced. The high volume at the beginning of an attack has an annoying pop.

When a signal ends, the AGC capacitor C29 *recovers* by discharging through the gate potential-divider network R5. Because the resistors and the capacitor are large, recovery is much slower than attack. This ensures that the AGC does not have to attack at each pulse, which would be quite irritating.

A. We can make a qualitative measurement of the recovery time. Set up the function generator to drive the AGC and audio amplifier through the 300-kΩ resistor as in the previous problem, and set up the oscilloscope to see the output of the Audio Amplifier. Use a snail's pace sweep speed of 0.5 s/division. Start with a 620-Hz,

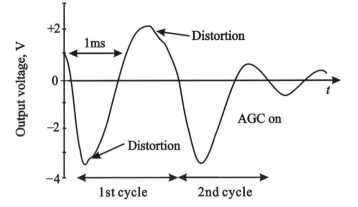

Figure 13.11. The attack for the NorCal 40A AGC. A 50-mV$_{rms}$ signal is applied suddenly at the input of one of the AGC JFETs. The plot shows the audio output. During the attack, the audio amplifier charges the AGC capacitor in the bottom half of each cycle.

0.1 Vrms setting on the function generator and find an appropriate voltage scale for the scope. You should see a vertical band sweep slowly across the screen. Now change the amplitude setting to 3 Vrms, but do not change the scope scale. This should give a much larger output voltage, and the band should shoot off the screen in both directions. Now change the amplitude setting back to 0.1 Vrms. Initially the scope voltage will be small, because the AGC attenuates the signal. However, eventually the AGC capacitor will discharge, and the scope voltage will increase. Measure the time it takes for the scope voltage to reach the final value. This is the recovery time. The measurement is a little tricky, and you may need to repeat it several times before you have confidence that you have the time right.

B. Now calculate approximately how long the recovery should take. You may assume that the AGC capacitor C29 will discharge 1 V during the recovery.

Before we add the final parts, you should take out the temporary resistors. Remove the 300-kΩ resistor that we used for the function-generator input and the 8-Ω resistor that we used for the audio output. Replace the 1-Ω resistor in the S1 holes with a bare wire jumper.

Now we can complete the assembly. The parts that we need to install are circled in Figure 13.12.

Install C19. This capacitor provides additional high-frequency roll-off.

Install the Audio Frequency (AF) Gain pot R8 (500 Ω). This pot allows us to adjust the output volume. It should be set fully clockwise. This gives the maximum output voltage for multimeter measurements.

Install the Speaker jack J4 and C26. The capacitor provides additional high-frequency audio roll-off for removing hiss.

The remaining parts are for muting while transmitting. Even with the Receiver Switch on, the transmitter gives a larger voltage in the receiver mixers than any signal that would be received by an antenna; therefore, the designer does not try to handle this with the AGC alone. The muting circuit includes the three diodes D1, D2, and D3 and

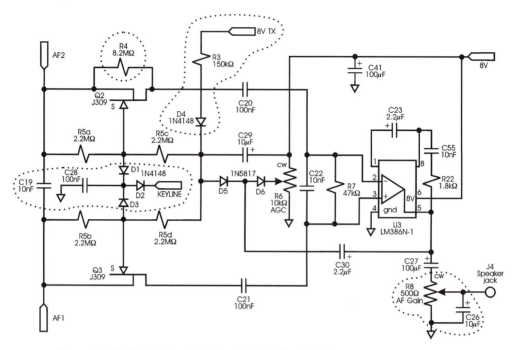

Figure 13.12. The final parts to install in the NorCal 40A.

the mute capacitor C28 (100 nF). We can understand what they do if we keep in mind what happens to KEYLINE during a pulse. At key down, KEYLINE goes low. This pulls the two gates of the JFETs all the way down to two diode drops above ground, cutting them off completely. In key up, KEYLINE rises, and this turns off D2. C28 charges through R5a and R5b. The gate voltages rise, the JFETs go into the linear region, and we are ready to receive again. The charging time for C28 has to be long enough to allow the transmitter to shut down completely. D1 and D3 ensure that the two gate circuits do not interact.

Install R3 (150 kΩ) and D4. This charges the AGC capacitor C29 during transmit so that the AGC recovers. The diode D4 prevents the AGC capacitor from discharging in reception when 8 V TX is low.

This leaves only one component, R4 (8.2 MΩ). R4 allows a little bit of the transmitter signal to leak around the JFET, so that you can hear your sending. This tone is called a *sidetone*. The volume is determined by the resistance. Some people find the sidetone with this resistor too loud and that a 15-MΩ resistor gives a better level.

Check the board to make sure that you have not forgotten to put in any components. You should mount the board in its box, add the knobs, and plug in your speaker.

Check that the AF Gain pot R8 is fully clockwise. This sets the volume to the maximum level.

Set the RF Gain pot R2 fully clockwise. You should not move this pot during the measurements. This position gives the maximum signal.

The RIT pot R16 should be centered.

The VFO Tune pot R17 should also be centered.

Plug in the power supply and turn it on. You should hear a hiss from the speaker. The function generator should be set for a sine wave at 20 mVrms. You will need an attenuator to reduce the level well below this. Set a frequency between 7,020 and 7,030 kHz. If you work in a lab where others also make measurements, you should set the kilohertz digit to the bench number. This ensures that the frequency at each bench differs by at least a kilohertz. Otherwise it can be difficult to tell the difference between your function generator and someone else's transmitter. Connect the attenuator to the Antenna jack J1 with a coaxial cable. It is a good idea to keep the cables separated so that signals do not couple around the attenuator.

A good place to start is with 80 dB of attenuation. Adjust the VFO Tune pot carefully to try to find the signal. If you cannot find it, reduce the attenuation to 50 dB or 30 dB. Once you find the signal, adjust the RIT to give maximum volume. Then take some time to enjoy the sound. There is almost no better feeling in electrical engineering than hearing the first tones out of a receiver you built, unless it is hearing someone answer your transmitter.

Now we tune the RF Filter. Hook the scope probes onto the speaker leads so that you can see the audio signal. Remember that one of the speaker leads is grounded. If you get the leads backwards you will short out the signal. Now tune the RF-Filter capacitors C1 and C2 for a maximum signal on the scope. They should be already well tuned, but this is a good time to check. Now we need to check the BFO frequency. Adjust the RIT pot for a maximum signal on the scope. This puts the signal through the center of your IF Filter. Hook up the counter to the output and check the audio frequency with a counter. Adjust the BFO capacitor C17 for an audio frequency of 620 Hz.

C.   Next we find the receiver gain. Attach the multimeter leads to the speaker so that you can measure the output AC voltage. There is only a limited range of input powers that we can use to measure the gain. If the attenuation is set too high, stray coupled voltages dominate the input. If the attenuation is set too low, the AGC will kick in, and that will also upset the measurement. One way you can check the stray voltage is to switch all the attenuation in, and see what voltage is left on the multimeter. You need to work at an output voltage that is much higher than this, as high as you can get without provoking the AGC. Start with 80 dB of attenuation and try switching in an additional 6 dB. This would be 86 dB in all. If the voltage drops by a factor that is close to 2, then you have a good place to work. If the signal does not change at all, you may be picking up an AM station or another student's transmitter. You may need to shift the function-generator frequency a kilohertz or so to avoid this interference. Once you find a 6-dB range where the response is linear, go ahead and calculate the power gain, assuming that the speaker has a resistance of 8 Ω. You should get around 100 dB. If it is less than 90 dB, you have a problem, and you will need to check the signal levels at different stages in the receiver, taking into account the fact that the RF Filter and IF Filter have a loss of about 5 dB each and the mixers have a gain of about 18 dB.

D.   One of the most important spurs is the response to the wrong sideband. This is the image of the Beat-Frequency Oscillator. The reason this is a problem is that if you hear a signal in the wrong sideband and transmit there, the other operator is unlikely to hear you. How much should you have to shift the frequency to hear a 620-Hz tone from the wrong sideband? Set the function generator to this frequency. You will probably not hear the signal, but if you decrease the attenuation, eventually you should hear it. Keep reducing the attenuation until the output voltage is the same as it was during the gain measurement. By how many dB does the IF Filter suppress the spurious sideband response?

The next two alignment steps are critical. The transmitter frequency needs to be adjusted so that it is the same as the signal being received. Otherwise another operator would not be able to hear the transmission. For this we need to set the sidetone frequency and find the center for the RIT knob. To start, connect the scope to the Antenna jack J1 with a 50-Ω load. Connect a switch to the Key jack J3 and turn the switch on to key the transmitter. You should peak the output by adjusting the Transmit Filter capacitor C39. Then set the Drive pot R13 to give 2.25 W output. You should also hear a sidetone at a reasonable volume level. Check the frequency of the sidetone with a counter. Set it to 620 Hz by adjusting the Transmit Oscillator capacitor C34.

Now we need to find the center position for the RIT pot R16. This is the position where the receive frequency matches the transmit frequency. Start by measuring the transmitter frequency. You should use the 50-Ω load in parallel when you connect the counter. You might need the counter's filter and attenuator because the transmitter voltage is large. Write the transmitter frequency down. Next we need to set the function generator to this same frequency. Often counters and function generators do not agree on the frequency, and so you should plug the function generator into the counter, and set it to match the transmitter. Now go back and connect the function generator to the radio through the attenuator. Adjust the RIT until the audio frequency is 620 Hz again. Use an indelible pen to put a center mark above the white dot on the knob, so that you can find this position again. This procedure is complicated enough that it is a good idea to check that the transmit and receive frequencies match when the RIT is set to the center mark.

Make marks with an indelible pen around the VFO Tune knob to indicate the frequencies 7,000, 7,010, 7,020, 7,030, and 7,040 kHz. This will make it convenient to set the transmitter and receiver to these frequencies in Problem 35.

14

Noise and Intermodulation

Fundamentally, a receiver is limited in sensitivity by noise that competes with the signal we want. A receiver is also limited in handling strong signals by its nonlinearities, which produce intermodulation products that block reception. *Noise* is a random voltage or current that is present whether a signal is there or not. We distinguish noise from *interference*, which is an unwanted signal coupled into the circuit, and from *fading*, which is a variation in the signal level, caused by interference between radio waves arriving by different paths. There are many different sources of noise. Several forms are caused by bias currents. In diodes, the random arrival times of electrons cause *shot* noise. Another current noise is $1/f$ noise, where power varies inversely with the frequency. This $1/f$ noise is found in contacts, and it is associated with energy states at interfaces called traps. It can often be reduced by improving the fabrication process. However, even in the absence of bias currents there is noise associated with resistors. It is called Johnson noise after John Johnson at the Bell Telephone Laboratories, who first measured it.

14.1 Noise

On an oscilloscope, noise makes a trace appear as a band that evokes the feeling of grass. We can write the noise as a function of time $V(t)$, but we would not be able to predict its value at a future time. We would expect the average over time to be zero, because the voltage will be positive at some times, and negative at others, and these periods cancel each other out. However, the time average of $V^2(t)$ is not zero, because $V^2(t)$ is positive. This means that noise has an rms value, V_{rms}, given by

$$V_{rms} = \sqrt{\frac{1}{\tau} \int_\tau V^2(t)\, dt,} \tag{14.1}$$

where τ is an averaging time. Noise also has an average power P_n, given by

$$P_n = V_{rms}^2 / R, \tag{14.2}$$

where R is the load resistance. We characterize a receiver's output by the *signal-to-*

noise ratio (SNR), given by

$$SNR = P/P_n, \tag{14.3}$$

where P is the output signal power. Different applications require very different signal-to-noise ratios, but a good way to compare receivers independently of the application is to ask how much input power is needed to give a 1:1 signal-to-noise ratio at the output. This is called the *minimum detectible signal*, or MDS. The MDS is actually quite an appropriate measure for the NorCal 40A, because a 1:1 SNR is about the lowest that can be used for receiving Morse code. We calculate the MDS by dividing the output noise by the gain G, and write

$$MDS = P_n/G. \tag{14.4}$$

Another approach is to measure the output noise when no signal is present. Then we find the MDS as the input signal that doubles the output power. In the exercises, we measure voltage instead of power, and so we will find the input signal that raises the output voltage by a factor of $\sqrt{2}$. For typical receivers, the MDS is small, much less than a femtowatt. The next unit prefix down is *atto*, for 10^{-18}. People do talk about attowatts and attofarads, but not often. It is more common to give powers in dB, using one milliwatt as the reference. For example, 10 aW is written as −140 dBm, where "m" denotes a reference power of a milliwatt.

Noise power does not appear at one frequency only; rather it is distributed over all frequencies. This means that we need to talk about noise power density at a particular frequency rather than noise power. We define the *noise power density N* as the noise power per unit bandwidth. The units are W/Hz. If N is constant with frequency, we can write

$$P_n = NB, \tag{14.5}$$

where B is the bandwidth. If N varies with frequency, we need to integrate N over the frequency range we are interested in to find P_n. A receiver's bandwidth is determined primarily by the bandwidths of the IF and audio filters, and a wide range of bandwidths are used in practice. Many receivers allow an operator to switch between narrow- and wide-bandwidth filters. Narrow filters are good for reducing noise, but wide filters make it easier to find signals.

The MDS that we defined depends on the bandwidth, because noise power is usually proportional to bandwidth. It is convenient to have a measure that does not depend on the bandwidth, because the bandwidth is determined for the most part by filters that make only a modest contribution to the receiver noise. Noise is associated primarily with mixers and amplifiers. The *noise-equivalent power*, or NEP, has the same relation to N that the MDS has to P_n. We write

$$NEP = N/G. \tag{14.6}$$

One way to think about the NEP is that it is the noise density we would need at the input to produce all of the noise that we observe at the output. People say it is the noise *referred* to the input.

14.2 Noise Phasors

We can use phasors for calculating how circuits affect noise just as we do for ordinary AC voltages. We write V_n for a noise phasor. However, noise phasors are different from ordinary phasors because we need to consider the bandwidth. We will use a bandwidth of 1 Hz to make it easy to relate the phasors to the noise power density N. In addition, the noise phasors are random variables, and we need probability theory to describe them. We will state our results in terms of expected values. If this terminology is not familiar to you, it is reasonable to think of it as an average. We indicate an expected value with an overline. For example, for a noise voltage V_n with a probability density function p, we write the expected value of $|V_n|^2$ as

$$\overline{|V_n|^2} = \int |V_n|^2 p \, dA, \tag{14.7}$$

where dA is an element of area in the complex V_n plane. We can relate this to the power density N by writing

$$N = \frac{\overline{|V_n|^2}}{2R} \quad \text{W/Hz}, \tag{14.8}$$

where R is the resistance. The units of V_n are a little strange, $\text{V}/\sqrt{\text{Hz}}$. We can take this as a reminder that the noise voltage increases as the square root of the bandwidth, in contrast to the noise power, which is proportional to the bandwidth.

We will use several arithmetic properties of expected values in our calculations. These follow from Equation 14.7. For a constant α, we write

$$\overline{\alpha |V_n|^2} = \alpha \overline{|V_n|^2}, \tag{14.9}$$

because we can bring a scalar multiple out of an integral. Now consider that we want to add two noise voltages V_1 and V_2. We can expand the expected value of $|V_1 + V_2|^2$ as

$$\overline{|V_1 + V_2|^2} = \overline{|V_1|^2} + \overline{|V_2|^2} + \overline{V_1 V_2^*} + \overline{V_1^* V_2}. \tag{14.10}$$

The last two terms are called *correlations*. Notice that the correlations are complex conjugates, so that the sum is real. The correlations indicate when part of each noise voltage comes from the same physical source. If the noise voltages come from two different sources, such as two different resistors, the sources are *independent*, and the correlation is zero:

$$\overline{V_1 V_2^*} = 0. \tag{14.11}$$

We can therefore write the power density of the sum, N, as

$$N = N_1 + N_2, \tag{14.12}$$

where N_1 and N_2 are the noise power densities for V_1 and V_2.

14.3 Nyquist's Formula

The formula for noise in resistors was first derived by Harry Nyquist, who worked at Bell Labs with Johnson. Nyquist used a statistical physics argument similar to the derivation of Planck's formula for blackbody radiation. In fact, you can think of Johnson noise as blackbody radiation in a circuit. First we need to understand why resistors have noise. You can make a resistor hot by applying a voltage or a current, and this means that the thermal energy associated with the vibrations of atoms couples to the voltages and currents. However, even if you do not apply a voltage, the thermal vibrations produce noise voltages and currents through this coupling. By this logic you would not expect a capacitor or inductor to produce noise, because they do not get hot when you apply a voltage or current. For these elements, the energy transfer and storage are electric and magnetic, and there is no coupling between thermal vibrations and voltages.

Nyquist used transmission-line theory to derive his formula, but it is easier for us to use an RLC circuit. We consider a resistor R at an absolute temperature T that is connected to a series resonant circuit (Figure 14.1a). The connecting wires couple the LC resonator to the vibrations of the atoms inside the resistor. Our calculation takes several steps. First we use circuit theory to find the capacitor voltage in terms of the resistor voltage. Then we integrate over frequency to find the energy stored in the capacitor. In Figure 14.1a, V_n is the resistor noise voltage phasor. We write the capacitor voltage V_c by a potential-divider formula:

$$V_c = \frac{1}{j\omega C} \frac{V_n}{R + j\omega L + 1/(j\omega C)} = \frac{V_n}{-\omega^2 LC + j\omega RC + 1}. \tag{14.13}$$

We can write the expected value of $|V_c|^2$ as

$$\overline{|V_c|^2} = \frac{\overline{|V_n|^2}}{|1 - \omega^2 LC + j\omega RC|^2}. \tag{14.14}$$

In thermal equilibrium, the energy that is stored in the inductor and capacitor is given by the Equipartition Theorem from classical thermodynamics, which specifies that the expected value of the energy associated with a resonance is the thermal energy kT. We last saw the thermal energy in Chapter 9 in connection with diode and transistor currents. Here k is Boltzmann's constant, 1.38×10^{-23} J/K. We can find the stored energy kT by multiplying by $C/2$ and integrating over frequency.

(a)   (b)

Figure 14.1. (a) Deriving the Nyquist noise-voltage formula from an RLC circuit. (b) Calculating the available noise power density from a resistor with a matched load.

This gives us

$$kT = \frac{C}{2} \int_0^\infty \overline{|V_c|^2} \, df = \frac{C}{2} \int_0^\infty \frac{\overline{|V_n|^2} \, df}{|1 - \omega^2 LC + j\omega RC|^2}. \tag{14.15}$$

Now we assume that the LC circuit has a very high Q. This gives the integrand a large peak at the resonant frequency that dominates the integral. We will let the Q be high enough that we can assume that $\overline{|V_n|^2}$ is constant over the frequency range that is important for the integral. Later on, we will see that $\overline{|V_n|^2}$ is independent of frequency anyway. The high-Q assumption lets us bring $\overline{|V_n|^2}$ out from under the integral sign. We can write

$$kT = \frac{C \, \overline{|V_n|^2}}{2} \int_0^\infty \frac{df}{|1 - \omega^2 LC + j\omega RC|^2}. \tag{14.16}$$

This integral looks difficult, but it is one of a family of integrals that can be attacked through the calculus of residues. This is an elegant technique in complex analysis that lets one turn truly awful looking integrals into simple expressions. This one is given as integral #3.1123 in *Table of Integrals, Series, and Products* by Gradshteyn and Ryzhik, published by Academic Press:

$$\int_0^\infty \frac{df}{|1 - \omega^2 LC + j\omega RC|^2} = \frac{1}{4RC}. \tag{14.17}$$

If we substitute for this integral in the previous equation, we get

$$kT = \frac{\overline{|V_n|^2}}{8R}, \tag{14.18}$$

which gives

$$\overline{|V_n|^2} = 8kTR. \tag{14.19}$$

This is the Nyquist noise formula. Notice that the noise voltage is independent of frequency. Because equipment for measuring noise invariably gives the rms voltage, it is more common to see this formula as

$$V_{rms} = \sqrt{4kTR} \;\; \text{V}/\sqrt{\text{Hz}}. \tag{14.20}$$

We can write an elegant alternative statement of the Nyquist noise formula if we consider the available noise power from a resistor. We can calculate this as the power dissipated in a matching load (Figure 14.1b). The load voltage is $V_n/2$, and the available power density N is given by

$$N = \frac{\overline{|V_n/2|^2}}{2R} = kT. \tag{14.21}$$

In words, the available power density from a resistor is kT, independent of the resistance. This is so convenient that people commonly use temperature as a measure of noise power density, even when it is not Johnson noise. We call this the *effective noise temperature T_e*, and we write it as

$$T_e = N/k. \tag{14.22}$$

People also define a noise temperature T_n for receivers, amplifiers, mixers, and attenuators by dividing the NEP by k to get

$$T_n = \frac{\mathrm{NEP}}{k} = \frac{N}{Gk}. \tag{14.23}$$

We have given the noise temperature a simple definition here, but there are complications. The formulas depend on whether a receiver amplifies one side-band or both. Furthermore, there are matching issues that we are neglecting, and at very high frequencies, there are quantum-mechanical corrections. We will not worry about these things, but you should realize that there is a lot more to this than we cover here.

A particularly interesting example is the noise temperature of an antenna. Antennas are not ordinarily made with resistors and thus they produce very little noise by themselves. However, they pick up natural radio waves. A plot of antenna noise for a wide range of frequencies is shown in Figure 14.2. At the operating frequency of the NorCal 40A, 7 MHz, the noise temperature is extremely high, millions of kelvins. This comes primarily from lightning in tropical thunderstorms. At frequencies from 30 MHz to 1 GHz, the temperature is lower but still

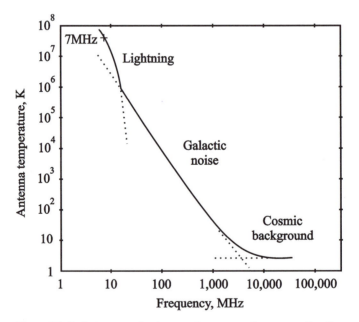

Figure 14.2. Antenna noise temperature versus frequency. For the frequency range from 30 MHz to 1 GHz, this is the noise temperature for a directive antenna pointed at the center of the Galaxy. For frequencies above 1 GHz, this is the noise temperature for an antenna at a high, dry site pointed straight up. This plot is adapted with permission from Figure 8.6 in an extremely interesting book, *Radio Astronomy*, 2nd edition, by John Kraus, published by Cygnus-Quasar. For radio astronomers, this noise *is* the signal. This book has an extensive discussion of astronomical radio sources and excellent coverage of receivers and antennas for radio astronomy.

large, and the dominant source is the black hole at the center of the Galaxy. At higher frequencies, the noise is quite small, and carefully designed antennas that point out to space receive the cosmic background radiation that is the dying embers of the primordial fireball. The cosmic background radiation has a noise temperature of 3 K.

A receiver designer tries to make the receiver noise lower than the antenna noise, so that the sensitivity is limited by the antenna rather than the receiver. It is a lot easier to do this at 7 MHz, where the antenna noise is enormous, than at 3 GHz, where the antenna noise is very low. However, the frequency range from 1 GHz to 10 GHz presents incredible opportunities for long-distance communication. For example, the *Voyager* spacecraft, now beyond the orbit of Pluto, speaks to us with a puny 10-W transmitter that has only slightly more power than the NorCal 40A.

14.4 Attenuator Noise

Now we find the noise from a resistive attenuator. We can use Nyquist's formula for this. We let L be the loss factor and N_a be the output noise power density (Figure 14.3a). Attenuators are commonly designed so that if they are terminated with a particular resistance R at the input, the resistance looking into the output port is also R. We will assume that this is the situation, and further assume that the resistor and attenuator have the same temperature T. We let the output available noise power density with a resistor R at the input be N' (Figure 14.3b). The noise N' includes noise from both the attenuator and the resistor. We can apply Equation 14.21 directly and write the available noise power density N' as

$$N' = kT. \tag{14.24}$$

This noise is in two parts. There is noise from the resistor that passes through the attenuator. This is given by kT/L. The rest is produced by the attenuator. We write

$$N' = kT/L + kT(1 - 1/L). \tag{14.25}$$

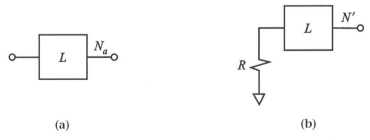

(a)   (b)

Figure 14.3. Attenuator with a loss factor L and an output noise density N_a (a). Attenuator with an input resistance R (b).

We can identify the right term as attenuator noise, and write

$$N_a = kT(1 - 1/L). \tag{14.26}$$

An attenuator with little loss has little noise. However, as the loss increases, the noise density approaches kT. We can use Equation 14.23 to write the attenuator noise temperature T_a as

$$T_a = N_a L/k = T(L - 1). \tag{14.27}$$

In this formula, we have multiplied by the loss L. This is equivalent to dividing by the gain in Equation 14.23, because loss is the reciprocal of the gain.

Although these formulas have been developed for attenuators, we can apply them to filters in the pass band if the loss in a filter is dominated by resistance in the inductors and capacitors.

14.5 Cascading Components

We calculate the total noise in receivers by adding the noise powers from the antenna and the different receiver stages. We can do this because the noise that comes from different parts of a receiver will usually be uncorrelated. There are exceptions, such as fluctuations in power-supply voltages that affect many components simultaneously. We also have to be careful to refer all the noise components to the same place in the system before we add them. We consider an amplifier chain with three amplifiers that are each characterized by a gain G_i, an output noise density N_i, and a noise temperature T_i (Figure 14.4). We include an antenna noise temperature T_a.

We write the output noise density N as

$$N = G_3 G_2 G_1 k T_a + G_3 G_2 N_1 + G_3 N_2 + N_3. \tag{14.28}$$

Notice that the noise from the antenna is amplified by the entire chain, but noise from the last amplifier appears directly. Usually this means that the noise in the early stages dominates the noise performance of a receiver. One way to see this in the NorCal 40A is to turn up the AGC. This attenuates the mixer and filter noise, so that only the noise from the Audio Amplifier is left. If you try this, the speaker sound simply goes away. We can rewrite this formula to give the receiver noise

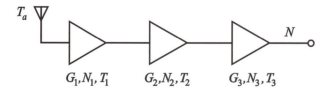

Figure 14.4. Finding the noise for an amplifier chain.

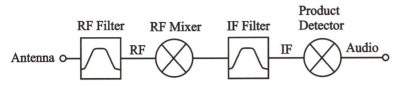

Figure 14.5. Input filters and mixers in the NorCal 40A receiver.

temperature T_r as

$$T_r = T_a + T_1 + T_2/G_1 + T_3/(G_1 G_2). \qquad (14.29)$$

The noise-temperature contributions of the later stages are reduced by the gain of the earlier stages.

There is an alternative to noise temperature called *noise figure* that is often quoted by manufacturers. The noise figure F is related to the noise temperature T_n by the formula

$$T_n/T_0 = F - 1, \qquad (14.30)$$

where T_0 is a reference temperature, usually 290 K. For example, Philips specifies that the SA602AN mixers in the NorCal 40A have a noise figure of 5 dB. This corresponds to a mixer noise temperature of 630 K.

As an example, let us predict the noise temperature of the NorCal 40A. We consider the first four elements of the receiver shown in Figure 14.5. We assume that each filter is at a physical temperature of 290 K and has a loss factor of $L = 3.2$ (5 dB). Assuming that the loss in the filter is due to resistance, we write the noise temperature of each filter using Equation 14.27:

$$T_f = 290(L - 1) = 630 \, \text{K}. \qquad (14.31)$$

For the mixers, we take numbers from the data sheets, which specify a gain of 18 dB ($G = 63$) and a noise figure of 5 dB ($T_m = 630$ K).

We use Equation 14.29 to write the noise temperature T_r of the receiver as

$$\begin{aligned} T_r &= T_f + T_m L + T_f L/G + T_m L^2/G \\ &= 630 + 2{,}020 + 30 + 100 = 2{,}780 \, \text{K}. \end{aligned} \qquad (14.32)$$

The terms represent the contribution, in order, of the RF Filter, the RF Mixer, the IF Filter, and the Product Detector. This corresponds to a noise figure of 10 dB, which is 4 dB less than the measured value. The largest component is the RF Mixer at 2,020 K. The IF Filter and the Product Detector contribute much less because their noise temperatures are divided by the large gain of the RF Mixer. Even though this noise temperature sounds high, antenna temperatures at 7 MHz are much higher than this, so that the receiver noise is usually not a problem.

14.6 Measuring Noise

We can measure the MDS of a receiver in several ways. The most direct is to measure the output noise power and the gain, and divide. Some care is needed to make sure the AGC is off when you measure the gain. Another approach is find the input signal power that gives an output that is twice the original output noise power. Here the challenge is to introduce a very small signal with a known power level. Many function generators do not provide small signals, and an adjustable attenuator is needed. Care must be taken to prevent the signal from leaking around the attenuator.

For measuring the receiver noise temperature T_r, we need a noise source with a known power density. Some function generators provide this feature. We can adjust the noise density until the output power doubles. If it is not convenient to vary a noise source continuously, we can use two different sources with known effective temperatures. If a receiver has good noise performance, one can use a resistor at two temperatures for this purpose. It is common to use room temperature and the temperature of liquid nitrogen, 77 K, because the resistor can be immersed in liquid nitrogen. We assume that two different source temperatures, T_1 and T_2, are available, and that we measure the output power in each case. We write the results as

$$P_1 = \alpha(T_r + T_1), \tag{14.33}$$
$$P_2 = \alpha(T_r + T_2), \tag{14.34}$$

where α is a proportionality constant. The quotient of P_1 and P_2 is called the Y factor. Since the Y factor is usually quoted as a number greater than one, we will assume $T_1 > T_2$ and write

$$Y = \frac{P_1}{P_2} = \frac{T_r + T_1}{T_r + T_2}. \tag{14.35}$$

Now we can solve for the receiver noise temperature T_r to get

$$T_r = \frac{T_1 - YT_2}{Y - 1}. \tag{14.36}$$

14.7 Intermodulation

Earlier we studied spurious responses that result from signals that give an output at the same audio frequency as the signal we want, even though they are at the wrong frequency. These responses are suppressed by filters. There is a different spurious component that is generated when there is more than one strong signal at the input. This is illustrated in Figure 14.6, where two input signals at closely spaced frequencies f_1 and f_2 are present. A nonlinear response in a mixer or amplifier produces signals at harmonic combinations of f_1 and f_2. These frequencies are

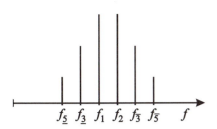

Figure 14.6. Intermodulation products that are close to the input frequencies f_1 and f_2.

called *intermodulation products*. Many intermodulation products are at quite different frequencies from the input signals, and this means that the RF Filter can block the inputs. However, there are four products that are quite close in frequency to the input signals, and this means that the RF Filter cannot stop them. These are the third-order products

$$f_{\underline{3}} = 2f_1 - f_2, \tag{14.37}$$
$$f_{\overline{3}} = 2f_2 - f_1 \tag{14.38}$$

and the fifth-order products

$$f_{\underline{5}} = 3f_1 - 2f_2, \tag{14.39}$$
$$f_{\overline{5}} = 3f_2 - 2f_1. \tag{14.40}$$

These are also shown in Figure 14.6. If the receiver is tuned to these frequencies, we may hear an interfering tone.

Now let us consider how the intermodulation products come about. We represent the response by a Taylor series

$$V = G_v V_i + G_2 V_i^2 + G_3 V_i^3 + G_4 V_i^4 + G_5 V_i^5 + \cdots, \tag{14.41}$$

where G_v is the voltage gain, and the other coefficients show the non linear behavior. Assume that the input voltage V_i contains two frequency components at f_1 and f_2:

$$V_i = V_1 \cos(2\pi f_1 t) + V_2 \cos(2\pi f_2 t). \tag{14.42}$$

The higher-order terms, $V_i^2, V_i^3, \ldots$, generate intermodulation products. At low input levels these components are below the receiver noise. However, at higher levels, the intermodulation products increase rapidly, producing spurious tones.

The most important intermodulation products are the third- and fifth-order ones, because the second- and fourth-order products are relatively far away. We will calculate third-order coefficients. For this, we need to find a large number of cosine products. To make things simpler, we assume that $V_1 = V_2 = V$ and $f_1 < f_2$. We can interpret the product as being made up of a sum frequency and a difference frequency. We write

$$V_1 \cos(2\pi f_1 t) \cdot V_2 \cos(2\pi f_2 t) = \frac{V^2}{2}[\cos(2\pi f_2 t + 2\pi f_1 t) + \cos(2\pi f_2 t - 2\pi f_1 t)].$$
(14.43)

When we expand the product

$$V_i^3 = (V \cos(2\pi f_1 t) + V \cos(2\pi f_2 t))^3$$
(14.44)

we get all the possible sum and difference combinations of three frequencies chosen from f_1 and f_2. There is a common coefficient of $V^3/4$. The sum and difference frequencies often repeat, and our job is to count the number of repetitions, rather like dice and card combinations. You should work through each of these carefully so that you will be ready to find the fifth-order products in Problem 35. We consider the frequencies in two groups. First are the sum frequencies. There are four of these: $3f_1$, $2f_1 + f_2$, $2f_2 + f_1$, and $3f_2$. The third harmonics $3f_1$ and $3f_2$ appear once, and the mixed sums $2f_1 + f_2$ and $2f_2 + f_1$ appear three times each. Now consider the difference frequencies. There also four of these: $2f_1 - f_2$, f_1, f_2, and $2f_2 - f_1$. It may be hard to see the original frequencies f_1 and f_2 as arising from a difference, but we can write

$$f_1 = f_1 + f_2 - f_2$$
(14.45)

and

$$f_1 = f_1 + f_1 - f_1.$$
(14.46)

The differences $2f_1 - f_2$ and $2f_2 - f_1$ each appear three times. The original frequencies f_1 and f_2 appear nine times each. The count for each frequency is shown in Figure 14.7 on top of each line. The presence of the original frequencies indicates that intermodulation will modulate the original signals in addition to producing products at other frequencies.

Notice that the coefficients for the sum frequencies form a line from Pascal's triangle (Figure 14.8). The coefficients in the difference-frequency group are also derived from the same line of the triangle except that they are multiplied by a factor of three, which is itself a coefficient from the same line.

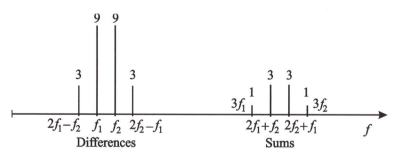

Figure 14.7. Third-order coefficients for the intermodulation products for $V_i = V[\cos(2\pi f_1 t) + \cos(2\pi f_2 t)]$. There is a common factor of $V^3/4$ for each coefficient.

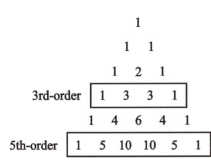

Figure 14.8. Pascal's triangle. The numbers in each row are obtained by adding the pair of numbers above. The coefficients for third- and fifth-order products are boxed.

14.8 Dynamic Range

To see how intermodulation products affect reception, we plot output power versus input power for an intermodulation product and an ordinary signal (Figure 14.9). The input for the signal is a single carrier, whereas the inputs for the intermodulation product comprise two carriers of equal power. It is traditional to consider the input power to be the power of one of the input carriers, rather than the total. On a log scale, the slope gives the order of product. A signal with linear gain has a 1:1 slope, whereas intermodulation products have steeper slopes. In theory, third-order intermodulation products have a slope of three, and fifth-order products have a slope of five. In practice, the situation is more complicated, because both third- and fifth-order products appear at f_3 and $f_{\tilde{3}}$. Often the slope becomes steeper at higher power levels, because the fifth-order products overtake the

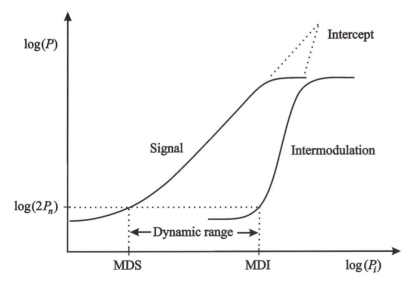

Figure 14.9. Finding the dynamic range for a receiver. Plot of the output signal and intermodulation product P versus the input power P_i on log scales. The outputs saturate at high levels because of the AGC. People often extrapolate the linear portion of the curves until they intersect. Manufacturers often quote the input or output powers associated with the intercept as a measure of the quality of the amplifier or mixer.

third-order products. Other amplifiers and mixers are not adequately described by a small number of Taylor-series terms, and the slope of the intermodulation products may be lower than three. Computer programs that simulate intermodulation products often assume a slope of three. It is therefore a good idea to be cautious in interpreting the simulations and to measure the products yourself.

The MDS is the input signal power that gives an output SNR of 1:1. We can identify it as the signal that gives a total output power, signal plus noise, of $2P_n$, where P_n is output noise. In the same way, we can identify the *minimum detectible intermodulation input*. This is labeled MDI on the plot. This is the input power level that gives a total output, tone plus noise, of $2P_n$. The difference between the MDS and the MDI is called the *dynamic range*. We write it as

$$\text{Dynamic range} = \text{MDI} - \text{MDS}. \tag{14.47}$$

The dynamic range is invariably quoted in dB. It is a measure of the range of useful signals for the receiver. It is the difference between signals that are just strong enough to be heard and signals that are just strong enough to cause interfering intermodulation products. Good receivers have a dynamic range of 100 dB. If the noise power increases, the two curves approach each other, and the dynamic range decreases. For example, in the NorCal 40A, antenna noise is usually 30 dB above receiver noise, and this reduces the dynamic range considerably. Assuming that the slope of the signal is 1:1, and that the slope of the intermodulation product is 3:1, the MDS will increase by 30 dB, while the MDI only increases by 10 dB. Thus the dynamic range drops by 20 dB. We can get some of this dynamic range back by adding an attenuator. For example, if we add 15 dB of attenuation, the MDS drops by 15 dB, but the MDI falls by only 5 dB, giving us a 10-dB improvement in dynamic range. The NorCal 40A includes an RF Gain pot to help improve the dynamic range.

A more fundamental solution to the intermodulation problem is to use a different mixer. In the SA602AN, dynamic range is limited by the exponential relationship between base voltage and collector current in the bipolar transistors of a long-tailed pair. This causes intermodulation products when the input signal levels approach the thermal voltage, 25 mV. Diode and FET mixers have better intermodulation performance, but they make the receiver more complex.

FURTHER READING

A good introduction to probability and random variables is given in *Probability, Random Variables, and Random Signal Principles*, by Peyton Peebles, published by McGraw-Hill. Residue calculus is covered in *Theory of Functions of a Complex Variable*, by A. I. Markushevich, published by Chelsea Publishing Company.

PROBLEM 34 – RECEIVER RESPONSE

First we make a plot of the audio frequency response of the receiver. This response is affected by both the IF Filter and the Audio Amplifier. Make the connections shown in

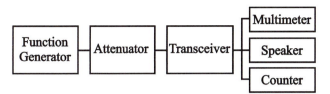

Figure 14.10. Connections for measuring the audio frequency response.

Figure 14.10, and set the function generator and attenuator for an input power of 100 fW at a frequency between 7,020 and 7,030 kHz. If you work in a lab where others also make measurements, you should set the kilohertz digit to the bench number to avoid interference. Tune the receiver so that the output audio frequency is 620 Hz. The output audio voltage should be near 100 mVrms. If it is appreciably lower, check the receiver components, particularly the filter capacitors C1 and C2.

A.   Readjust the attenuator and the function generator to give an audio output voltage of 100 mVrms. Now plot the audio voltage on a log scale as the audio frequency varies from zero to 1,200 Hz. Find the 3-dB bandwidth. This plot is a good check of the BFO setting. If the peak of the plot is outside of the frequency range from 600 Hz to 650 Hz, you should readjust the BFO.

For this part of the lab, you will need to work with a partner. We measure the MDS using weak input signals from another transceiver. Function generators often have a very limited power range, and it may be difficult to isolate a function generator from the receiver at low power levels. To get lower input signal levels, we use another NorCal 40A as the signal source, and run it off a battery to keep signals from coupling back through the wall outlets.

B.   Select one of the two transceivers to be the transmitter. Set the VFO Tune pot to mid range, and plug in a battery. You will probably find it convenient to plug a switch into the Key jack to turn the transmitter on and off. We need to reduce the power of the transmitter. What is the peak-to-peak voltage needed to deliver −40 dBm to a 50-Ω load? To get this voltage, first set the Drive pot R13 to minimum gain. Then mistune the Transmit Filter capacitor C39 to reduce the output power to −40 dBm. Make sure you have the 50-Ω load in parallel with the scope input, or else the settings will be off.

The other transceiver will act as the receiver. Make the connections shown in Figure 14.11, with the key switch off. The RF Gain control on the receiver should be fully clockwise for minimum attenuation. Measure the output noise voltage when the transmitter is off with a multimeter. Set the attenuator to give an input signal of −100 dBm,

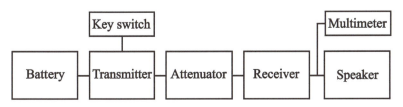

Figure 14.11. Measuring the receiver response.

and tune the receiver for maximum audio output. Once the receiver is tuned, it is a good idea to put the mouth of the speaker face down on the table to preserve the ears of others.

Now check the signal level at -150 dBm. This is well below the MDS, and you should not hear any tone in the noise. The multimeter reading should be the same as it was without any signal. Larger voltages indicate that the signal is leaking through somewhere. Sometimes the cable between the transmitter and the attenuator is the source of this leakage, and you may be able to reduce it by making a direct connection between the attenuator and the transmitter. Also try moving the transmitter and battery far away from the receiver.

C. Plot the output voltage on a log scale as the input power varies from -150 dBm to -50 dBm.

D. What is the MDS? This would be the signal that gives an output power of $2P_n$, or a multimeter reading of $\sqrt{2}V_{rms}$, where V_{rms} is the rms output noise voltage.

E. What is the weakest input signal that you can still hear? What is the signal-to-noise ratio at this level?

F. A function generator that can produce noise is useful for finding the NEP. For example, the HP33120A produces noise that is spread over a bandwidth of 10 MHz. With a function-generator setting of -30 dBm and an attenuator setting of 60 dB, what is the input noise power density to the receiver? Find the NEP as the input noise density that gives an output of $2P_n$. Divide the MDS by the NEP to find the bandwidth. People call this the *noise bandwidth*, because it is usually close to but not the same as the 3-dB bandwidth.

G. Now connect your receiver to an antenna. Tune to a part of the band where you do not hear a signal. What is the output voltage? Antenna noise at 7 MHz is primarily due to lightning, and this gives it a boom and crash sound that is different from the steady roar of receiver noise.

H. Use the plot you made in Part C to find the MDS for antenna noise.

I. What is the antenna noise temperature?

You should retune the transmitter to full power. Make a note of your MDS for receiver and antenna noise to use in the next problem.

PROBLEM 35 – INTERMODULATION

This problem is best done in groups of three, because two transmitters are needed. If function generators are used instead, you need to be careful, because function generators can also produce intermodulation products. Unfortunately these products are at the same frequencies we are interested in, and so they interfere with an intermodulation measurement.

A. Find the coefficients and frequencies for $[\cos(2\pi f_1 t) + \cos(2\pi f_2 t)]^5$. You may assume that f_2 is larger than f_1.

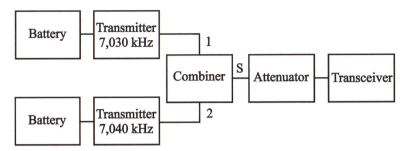

Figure 14.12. Measuring intermodulation products.

Choose two of your three transceivers to be transmitters and one to be the receiver. Set one transmitter for 7,030 kHz, and set the other for 7,040 kHz. Set the output power to a 50-Ω load for each to 2 μW. The two transmitters should be connected to a power combiner as shown in Figure 14.12. The connector labeled "S" is the sum port. A power combiner is different from an ordinary BNC tee. With a tee, power couples from one transmitter to the other, causing intermodulation in each transmitter. The power combiner isolates each transmitter from the other to prevent this. Power combiners have a combining loss of 3 dB, because half the power is dissipated in a resistor inside the combiner. This means that the power into the attenuator at each frequency is 1 μW. The isolation is not perfect, but the power coupled between transmitters is usually more than 20 dB below the power that goes to the sum port. This makes the input to the other transmitter less than 10 nW, which is small enough to prevent transmitter intermodulation products.

B. To hear the tone at f_3, set the attenuator so that the input power for each component is −40 dBm, and tune the receiver for a signal near 7,020 kHz. Now vary the input power with the attenuator. Plot the output audio voltage on a log scale versus the input power in dBm. Use a wide enough input range that the output extends from the noise floor up to 200 mVrms.

C. Now find the tone at f_5, and make a plot for it on the same graph.

D. From the graph and your measurement of the receiver MDS in the previous problem, find the dynamic range of your receiver.

E. From your measurement of the antenna MDS in the previous problem, find the antenna-limited dynamic range.

PROBLEM 36 − DEMONSTRATION

Present the transceiver that you built for inspection. The construction should be complete, and the solder connections should be neat.

A. Find a weak signal in the frequency range from 7,000 to 7,040 kHz. The receiver filters, the VFO, and the BFO will need to be properly adjusted to receive the signal.

B. Transmit a signal with at least 2 W of power within 200 Hz of the received signal. The sidetone should match the tone of the received signal.

15

Antennas and Propagation

So far, we have made measurements without saying anything about antennas, or about how power gets from the transmitter to the receiver. In our measurements, a 50-Ω load has taken the place of the antenna. However, instead of dissipating the power as heat, an antenna radiates power as electromagnetic waves. One thing that makes antennas interesting is that they necessarily involve both the voltages and currents that we study in circuits and the electric and magnetic fields that make up radio waves. This gives antennas a special place in the history of physics. They were the crucial components that Hertz developed in the 1880s to demonstrate that Maxwell's equations for electricity and magnetism are correct. In the 1960s, a special parabolic antenna allowed Arno Penzias and Robert Wilson at Bell Telephone Laboratories to discover the cosmic background radiation. That measurement earned them a Nobel Prize, and it gave an entirely new interpretation to the history of the universe.

An antenna is characterized by its impedance and pattern, which is a plot of where the power goes for a transmitting antenna. Traditionally, antennas have been analyzed as transmitters, and most antenna engineers think entirely in terms of transmitting antennas. If we know how an antenna transmits, we can use the reciprocity theorem to figure out how the antenna works in reception. The physical description of transmission and reception are actually quite different, and the physics of receiving antennas is in many ways as interesting as for transmitting antennas. In particular, a dipole can be analyzed in a simple way as a receiving antenna, and we will follow this approach. We need a few results from electromagnetic theory to start.

15.1 Radio Waves

Radio waves are predicted by Maxwell's equations. These are complicated, but we will consider the special case of plane waves in free space, where Maxwell's equations are dual to the transmission-line formulas. We write them in curl form with phasors, so that $\partial/\partial t$ becomes $j\omega$:

$$\nabla \times \mathbf{E} = -j\omega\mu_0\mathbf{H}, \tag{15.1}$$

$$\nabla \times \mathbf{H} = j\omega\epsilon_0\mathbf{E}. \tag{15.2}$$

The first equation is Faraday's law, and the second is Ampère's law. In the equations, **E** is the electric field, with units of V/m, and **H** is the magnetic field, with units of A/m. We use bold face to indicate a vector. μ_0 is the *permeability* and ϵ_0 is the *permittivity* (ϵ is the Greek letter *epsilon*). We add the zero subscript to indicate that these are for free space. μ_0 and ϵ_0 are given by

$$\mu_0 = 4\pi \times 10^{-7}\,\text{H/m} \approx 1.26\,\mu\text{H/m} \tag{15.3}$$

and

$$\epsilon_0 \approx 1/(36\pi)\,\text{nF/m} \approx 8.85\,\text{pF/m}. \tag{15.4}$$

In addition to numerical forms, I have given forms with factors of π because they sometimes simplify expressions. The π form for μ_0 is exact.

In a *plane wave*, we assume variation in only one direction, along the z axis, and we let the field vary as $\exp(j\omega t - j\beta z)$. This allows us to rewrite Maxwell's equations in algebraic form as

$$\beta\hat{\mathbf{z}} \times \mathbf{E} = \omega\mu_0\mathbf{H}, \tag{15.5}$$
$$\beta\hat{\mathbf{z}} \times \mathbf{H} = -\omega\epsilon_0\mathbf{E}, \tag{15.6}$$

where $\hat{\mathbf{z}}$ is the unit vector in the z direction, and $\times$ denotes a cross product. A cross product of two vectors is orthogonal to both vectors; thus these equations tell us that **E** and **H** are both orthogonal to $\hat{\mathbf{z}}$. In other words, **E** and **H** have only x and y components. For us it is sufficient to consider *linear polarization*, where the fields lie along one axis. Let us orient the x axis so that it is aligned with the electric field, so that there is an E_x component only. Equation 15.5 can then be written as

$$\beta E_x\hat{\mathbf{y}} = \omega\mu_0\mathbf{H}. \tag{15.7}$$

This means that **H** must be in the y direction (Figure 15.1a). We can rewrite Equations 15.5 and 15.6 as

$$\beta E_x = \omega\mu_0 H_y, \tag{15.8}$$
$$\beta H_y = \omega\epsilon_0 E_x. \tag{15.9}$$

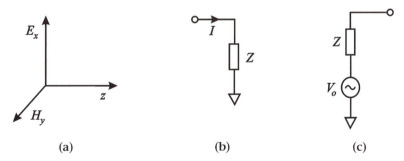

(a)   **(b)**   **(c)**

Figure 15.1. (a) Field directions for radio waves. For a linearly polarized wave propagating in the z direction, we can align the x axis with the electric field and the y axis with the magnetic field. (b) Current and impedance for a transmitting antenna. (c) Thevenin equivalent for a receiving antenna.

These are duals to the general transmission-line equations 4.41 and 4.42. The ratio of E_x and H_y is called the *wave impedance*, and it is written as η_0. This corresponds to the characteristic impedance of a transmission line. It is given by

$$\eta_0 = \frac{E_x}{H_y} = \sqrt{\frac{\mu_0}{\epsilon_0}} \approx 120\,\pi\,\Omega \approx 377\,\Omega. \tag{15.10}$$

We write the phase constant β as

$$\beta = \omega\sqrt{\mu_0\epsilon_0} \tag{15.11}$$

and the velocity as

$$c = 1/\sqrt{\mu_0\epsilon_0} = 299{,}792{,}458\,\text{m/s} \approx 3.00 \times 10^8\,\text{m/s}, \tag{15.12}$$

where c is the traditional symbol for the speed of light. We have exact expressions for both c and μ_0, and therefore we can work backwards to find ϵ_0 to any precision that we want. In transmission lines, the product of voltage and current gives power. The corresponding quantity for radio waves is power density, and we write the average power density in the z direction, S, in W/m^2 as

$$S = \text{Re}\left(\frac{E_x H_y^*}{2}\right) = \frac{|E_x|^2}{2\eta_0}. \tag{15.13}$$

15.2 Impedance

In a circuit, a transmitting antenna is a load, and we can characterize it by an impedance Z (Figure 15.1b). We can write the power delivered to an antenna in terms of the real part R and the current I. We write this power as P_t, with the t standing for "transmitter":

$$P_t = R|I|^2/2. \tag{15.14}$$

We distinguish between power that is absorbed by the antenna materials and power that is actually radiated. To do this we split the antenna resistance into two parts, a *radiation resistance* R_r and a *loss resistance* R_l. We can write the antenna resistance R as the sum

$$R = R_r + R_l, \tag{15.15}$$

and we write the radiation efficiency as

$$\eta = R_r/R. \tag{15.16}$$

In a circuit, a receiving antenna is a source, and we can write a Thevenin equivalent for it (Figure 15.1c). We find the Thevenin source impedance as the look-back impedance with the voltage source V_o turned off. This is just how we find the transmitting impedance. This means that *an antenna's impedance is the same for transmitting and receiving.*

15.3 Directions and Solid Angles

In calculations, we specify directions by a spherical coordinate system with angles θ and ϕ (Figure 15.2a). In this system, θ is the angle from the θ axis, and ϕ is the angle of rotation around the θ axis. This is particularly convenient for antennas with rotational symmetry, where we only need a plot as a function of θ. It is common in measurements to make plots as a function of the *elevation* and *azimuth* angles (Figure 15.2b). This is useful when the antenna has a particular orientation to the earth. The elevation angle is the angle above the horizon, and the azimuth angle is the relative bearing. These are actually quite similar to θ and ϕ, if we point the theta axis straight up. This direction is called the *zenith*. It is the angle from the zenith that actually corresponds to θ, and ϕ becomes the negative of the azimuth.

Antenna formulas are often conveniently written in terms of *solid angles*. We define a solid angle Ω for a surface S and a reference point P(Figure 15.3a). It is the area of the projection of the surface on a sphere with unit radius centered at P. The units are *steradians*. The solid angle for a closed surface that includes P is the

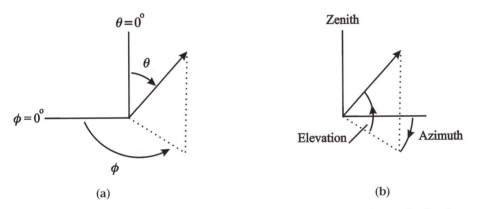

Figure 15.2. Specifying direction with a spherical coordinate system (a) and azimuth–elevation system (b).

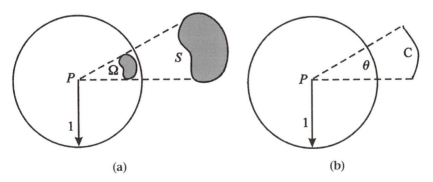

Figure 15.3. Defining a solid angle Ω as the area of the projection of a surface S onto a unit sphere (a). Defining a plane angle as the projection of a curve C onto a unit circle (b).

area of a unit sphere, which is 4π. This definition may seem strange, and it may help to make an analogous definition for ordinary plane angles. We can define an angle θ for a curve C and a reference point P. It is the length of the projection of the curve on a unit circle centered at P (Figure 15.3b). The angle for a closed curve that includes P is the circumference of the unit circle, 2π.

15.4 Transmitting Antennas

We define the *gain G* for a transmitting antenna as the ratio of the power density S to the maximum power density S_r from a reference antenna. We can write

$$G(\theta, \phi) = \frac{S(\theta, \phi)}{S_r}. \tag{15.17}$$

Power density has units of watts per square meter, but gain itself has no units. Gain is used only for transmitting, and we have different measures for receiving. For theoretical work, it is usually most convenient to define a reference *isotropic* antenna that is lossless and radiates uniformly in all directions. We can write the power density for an isotropic antenna S_i in terms of P_t as

$$S_i = \frac{P_t}{4\pi r^2}, \tag{15.18}$$

where P_t is the transmitter power and r is the distance from the antenna. We can then write the gain G as

$$G = \frac{S}{S_i} = \frac{4\pi r^2 S}{P_t}. \tag{15.19}$$

The gain is usually quoted in dB, and the letter "i" is added if it is necessary to specify that the reference antenna is isotropic. We could say that the gain of the isotropic antenna itself is 0 dBi in all directions. Gain is a function of angle, but it is common to say *the* gain for the maximum gain. In measurements, the reference antenna is often a half-wave dipole. In this case, we would write dBd to show that the reference antenna is a dipole. The gain of a dipole in free space is about 2 dBi, but reflections from the ground can raise or lower this as much as 6 dB.

In calculations it is common to assume that an antenna is lossless. Often the losses are small, and this is a reasonable approximation. For a lossless antenna, the integral of the gain over all angles has a particularly simple form:

$$\oint G(\theta, \phi) \, d\Omega = \oint \frac{4\pi r^2 S(\theta, \phi)}{P_t} \, d\Omega = \frac{4\pi}{P_t} \oint r^2 S(\theta, \phi) \, d\Omega. \tag{15.20}$$

A circle is added to the integral sign to indicate that the integral is over all angles. Because an area at a distance r scales by r^2 when we project it onto a unit sphere, we can think of $r^2 S$ as the power density per unit solid angle. We write this as

$$S_\Omega(\theta, \phi) = r^2 S(\theta, \phi), \tag{15.21}$$

where S_Ω is transmitter power density per steradian. We get

$$\int G(\theta, \phi)\, d\Omega = \frac{4\pi}{P_t} \oint S_\Omega(\theta, \phi)\, d\Omega. \tag{15.22}$$

The integral of S_Ω over all angles is just P_t. This gives us

$$\oint G\, d\Omega = 4\pi. \tag{15.23}$$

For a lossless antenna, the integral of the gain over all angles is 4π.

15.5 Receiving Antennas

We characterize a receiving antenna by an *effective length* and an *effective area*. The effective length is defined in terms of the Thevenin equivalent circuit for the receiving antenna (Figure 15.1c). We write the effective length h in terms of the Thevenin voltage V_o as

$$V_o = hE, \tag{15.24}$$

where h is the effective length and E is the incident electric field. The effective length is useful for characterizing short wire antennas, because the peak value is about half the physical length. As an example, consider the effective length of a pair of wires with total length l (Figure 15.4a). This antenna is called a *dipole*. We assume that the effective length is much less than the wavelength. We let θ be the angle from the dipole axis. The voltage at the terminals is just the potential difference between the two wires. This can be determined by symmetry if the potential of one wire is not affected by the presence of the other wire. The voltage of each wire is the potential at the midpoint, so that V_o is given by

$$V_o = (l/2)\, E \sin \theta. \tag{15.25}$$

One important point is that the antenna responds only to the component of the electric field along the wire. The cross-polarized component does not cause

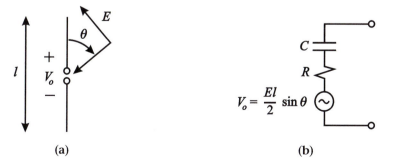

(a)   (b)

Figure 15.4. Calculating the effective length of a short dipole (a), and the Thevenin equivalent circuit (b).

a voltage. It is true in general that antennas with one pair of output terminals respond only to a single polarization. In our calculations, we will assume that the antenna is oriented to match the polarization of the incident waves. We can write the effective length as

$$h = (l/2)\sin\theta. \tag{15.26}$$

Figure 15.4b shows the Thevenin circuit. In addition to the radiation resistance R, there is also a capacitance C between the wires. We will consider these later.

The effective area A is defined in terms of the available power P_r from the antenna terminals. The r is for "receiver." We write

$$A(\theta, \phi) = \frac{P_r}{S(\theta, \phi)}, \tag{15.27}$$

where S is the incident power density. The effective area is often useful for characterizing large reflector antennas, because the peak effective area turns out to be about half the physical area. In addition, we will see that, by reciprocity, the effective area relates in a simple way to the gain:

$$A(\theta, \phi) = \frac{\lambda^2}{4\pi} G(\theta, \phi), \tag{15.28}$$

where λ is the wavelength. This formula indicates that the angle patterns for receiving and transmitting are the same. It also means that we can use the same antenna for transmitting and receiving in a communications system, and we can measure antenna patterns in either transmission or reception.

We can relate effective area to effective length through the Thevenin equivalent circuit. We write the available power as

$$P_r = \frac{|V_0|^2}{8R} = \frac{|hE|^2}{8R}, \tag{15.29}$$

where R is the antenna resistance. For an incident plane wave we can substitute for E from Equation 15.13 to get

$$P_r = \frac{|h|^2 S \eta_0}{4R}. \tag{15.30}$$

Now, if we compare this formula to the definition of effective area (Equation 15.27), we get

$$A = \frac{|h|^2 \eta_0}{4R}. \tag{15.31}$$

15.6   Friis Formula

We can use the gain and effective area to calculate signal levels in a communications system. We consider two antennas separated by a distance r (Figure 15.5). The transmitting antenna has a gain G with transmitter power P_t, and the receiving

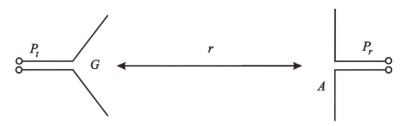

Figure 15.5. Deriving the Friis transmission formula to calculate signal levels transmitted between two antennas.

antenna has an effective area A with available power P_r. We write the far-field power density S as

$$S = \frac{P_t G}{4\pi r^2}. \tag{15.32}$$

The receiver power P_r is given by

$$P_r = SA = \frac{P_t G A}{4\pi r^2}. \tag{15.33}$$

This formula is due to Harald Friis of the Bell Laboratories. In most communications systems, we have additional loss because of ground reflections and absorption and refraction in the atmosphere.

Signals in the HF range from 3 MHz to 30 MHz can propagate long distances, because they reflect off the ionosphere. It is interesting to apply the Friis formula to communication with the NorCal 40A. Antenna gains vary widely, but reasonable values to consider at 7 MHz are $G = 1$ and $A = 150$ m². For a distance of 2,000 km and a transmitter power of 2 W, we can write

$$P_r = \frac{P_t G A}{4\pi r^2} = 6\,\text{pW}. \tag{15.34}$$

In practice, the loss would likely be somewhat greater than this because of losses in the ionosphere. Figure 15.6 shows typical power levels for a pair of NorCal 40As that are communicating with each other at this distance. The levels span a wide power range, over twelve orders of magnitude. We start in the transmitter, where the VFO output is a microwatt. The signal goes through four successive stages of amplification. We pick up power steadily and transmit two watts from the antenna. At a distance of 2,000 km, we could expect to receive a few picowatts, just above nighttime antenna noise. In the receiver we alternate filters with loss and mixers with gain. The signal level stays low until we reach the audio amplifier, where the power is still below a nanowatt. The audio amplifier gives a huge gain to bring us above a milliwatt, which is a comfortable level for the speaker.

Signals in the VHF range above 30 MHz usually are not reflected by the ionosphere. In this case, the actual range will often not be limited by the Friis formula, but by the curvature of the earth. It helps to have the antenna as high as possible. Figure 15.7a shows the geometry for calculating this line-of-sight limitation, and we can use right-triangle formulas for calculating the relation between the

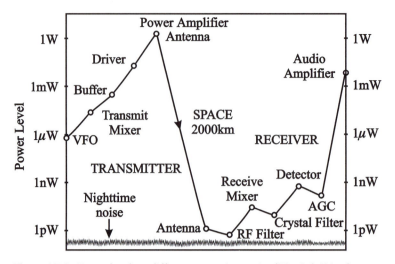

Figure 15.6. Power levels at different stages in a pair of NorCal 40As that are communicating over a distance of 2,000 km.

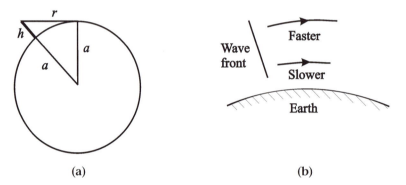

(a)   (b)

Figure 15.7. Line-of-sight path limit on a curved earth (a). Velocity variation in the atmosphere (b).

antenna height h and the range r to another antenna on the ground. We write

$$(a + h)^2 = a^2 + r^2, \tag{15.35}$$

where a is the radius of the earth, 6,370 km. We can expand and rewrite the formula approximately as

$$r \approx \sqrt{2ah}. \tag{15.36}$$

This formula is easy to apply, but the ranges it predicts are too short. The reason is that the velocity varies in the atmosphere, making the wave follow a curved path (Figure 15.7b). The waves bend toward the denser part of the atmosphere where the velocity is lower, and this gives us longer ranges than we might have expected. The effect varies from time to time and from place to place, but a reasonable approximation is to use an effective earth radius a_e that is 4/3 the actual radius, or 8,500 km. In addition, if the receiving antenna is also elevated, the range increases. We can write the range formula as

$$r = \sqrt{2a_e h_1} + \sqrt{2a_e h_2},$$ (15.37)

where h_1 and h_2 are the antenna heights.

15.7 Antenna Theorem

We found that for a lossless antenna, the integral over all solid angles of the gain was equal to 4π. This formula came directly from the definition of gain. The effective area for lossless antennas satisfies a similar relation, written as

$$\oint A \, d\Omega = \lambda^2.$$ (15.38)

This formula is called the *antenna theorem*. It is a beautiful result and not at all obvious. In fact, we might think that we could increase the integral by making the antenna larger. However, this is not so, because the received energy has to be coupled to the output transmission line. Our proof is due to Bernard Oliver of the Hewlett-Packard Laboratories. It requires Nyquist's formula for noise and formulas for blackbody radiation. We consider a lossless antenna in an absorbing cavity at the same temperature T (Figure 15.8). The antenna is connected to a matched load, which is also at temperature T. We consider the power flow between the antenna and the cavity. When the load and the cavity are at the same temperature, they are in thermal equilibrium, which is to say that the temperatures are not changing. This means that the noise power from the load radiated by the antenna must be the same as the power received by the antenna from the cavity, or else the temperatures would change.

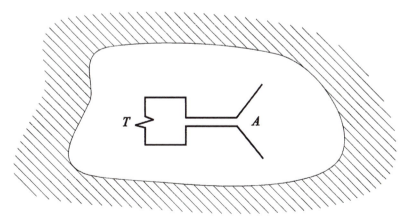

Figure 15.8. Antenna in a radiating cavity. Both the matched load and the cavity are at the same temperature T, and so the system is in thermal equilibrium.

The radiated power from the antenna is Johnson noise from the load, and the power density is given by Nyquist's formula as kT. This power must be the same as the power received by the antenna. To figure out the received power, we need to know the power radiated from the enclosure. This is given by the black-body radiation law. It is calculated in a manner similar to that for the Nyquist formula we derived in Chapter 14. We use the equipartition theorem to associate the energy kT with each of the resonant modes of a cavity. The quantum-mechanical version is called Planck's formula, and it applies when the photon energy is comparable to the thermal energy. We use the blackbody radiation formula from classical thermodynamics called the Rayleigh–Jeans law, which applies when the photon energy is much smaller than the thermal energy. Counting the resonant modes is difficult, and we will only quote the result here. It is written in terms of E, the energy per unit frequency per unit volume in a single polarization:

$$E = \frac{4\pi kT}{c\lambda^2}. \tag{15.39}$$

This is not a good final form for us, because we can only calculate the solid-angle integral if we know the incident power density per unit solid angle. This quantity is called the brightness B, and its units are W Hz^{-1} m^{-2} steradian^{-1}. We calculate the brightness from the energy density by multiplying by the velocity c, and dividing by the solid angle of a sphere, 4π, to take into account the fact that the power comes from all directions. This gives the brightness as

$$B = kT/\lambda^2. \tag{15.40}$$

We can write

$$kT = \oint BA \, d\Omega = \oint (kT/\lambda^2) A \, d\Omega. \tag{15.41}$$

We can express the received power as an integral such as this because the radiation that comes from different angles effectively comes from different sources and is uncorrelated. This simplifies to

$$\oint A \, d\Omega = \lambda^2, \tag{15.42}$$

which proves the theorem.

15.8 Reciprocity

The reciprocity principle allows us to predict the effect of interchanging the input and the output. For filters, it means that the loss in one direction is the same as the loss in the other. For antennas, reciprocity allows us to relate gain and effective area. However, we cannot just set them equal, because they have different units.

We will see that they are related by the formula

$$G/A = 4\pi/\lambda^2. \tag{15.43}$$

To start, we state, without proof, a form of the reciprocity theorem.

The positions of an ideal voltmeter and an ideal current source can be interchanged in a circuit without changing the voltmeter reading.

Reciprocity is a subtle idea, and there are a long list of conditions that must hold for a circuit to be reciprocal. The network must be linear and time invariant. In addition there can be no bias currents or fields. The conditions for reciprocity usually hold for the antennas themselves. However, the earth's magnetic field sometimes causes the ionosphere to show nonreciprocal behavior. In practice, it is fairly common for the signal-to-noise ratios at the two ends of a radio link to be quite different, even though the transmitter powers are the same. However, this is usually due to different noise and interference levels at the two receivers. This is not itself a violation of reciprocity. In the case we are considering here, where we are just interested in antenna properties, we can assume that the space between the antennas is reciprocal and use the Friis transmission formulas.

Consider the circuit shown in Figure 15.9. The transmitting antenna is driven by a current source I. We let the open-circuit voltage of the receiving antenna be V. We can write the transmitter power P_t as

$$P_t = |I|^2 R_1/2. \tag{15.44}$$

We can write the receiver power P_r using the Friis formula as

$$P_r = \frac{P_t G_1 A_2}{4\pi r^2}. \tag{15.45}$$

Notice that this is available power, rather than delivered power. The delivered power is actually zero, because the output is open-circuited. We can rewrite P_r in terms of the open-circuit voltage V as

$$P_r = \frac{|V|^2}{8R_2}. \tag{15.46}$$

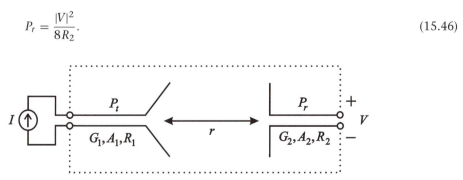

Figure 15.9. Deriving the reciprocal relation between transmitting and receiving antennas. Each antenna is characterized by a gain G, effective area A, and resistance R.

This lets us write the relation

$$\frac{|V|^2}{8R_2} = \frac{|I|^2 R_1 G_1 A_2}{8\pi r^2}. \tag{15.47}$$

Let us regroup to isolate the gain and effective area. We can write

$$G_1 A_2 = \frac{|V|^2 \pi r^2}{|I|^2 R_1 R_2}. \tag{15.48}$$

An interesting thing happens if we move the current generator to the second antenna. By reciprocity, the voltage at the first antenna must be V, and in fact, the entire right side of the equation does not change. However, on the left side, we swap the indices 1 and 2. Thus

$$G_1 A_2 = G_2 A_1 \tag{15.49}$$

or, as quotients

$$G_1/A_1 = G_2/A_2. \tag{15.50}$$

Notice that the left side depends only on antenna 1, and the right side depends only on antenna 2. At this point, we have not specified anything about the antennas or how they are pointed. This means that the ratio of gain to effective area is a universal constant, independent of the antenna orientation or type. It does not depend on whether the antenna is lossless or not.

To find the constant, we consider a particular class of antennas, lossless antennas, and their solid-angle integrals. We saw that from the definition of gain,

$$\oint G \, d\Omega = 4\pi, \tag{15.51}$$

and from the antenna theorem,

$$\oint A \, d\Omega = \lambda^2. \tag{15.52}$$

If the ratio G/A is a constant, independent of angle, then the ratio of the two solid-angle integrals will give us the same constant. We get

$$G/A = 4\pi/\lambda^2. \tag{15.53}$$

15.9 Dipoles

We can use the antenna theorem to find the radiation resistance R_r of a short, lossless dipole. We substitute for the effective area in terms of the effective length from Equation 15.31 and write

$$\lambda^2 = \oint A \, d\Omega = \oint \frac{|h|^2 \eta_0}{4R_r} \, d\Omega. \tag{15.54}$$

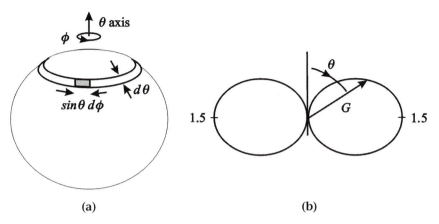

Figure 15.10. (a) Converting a solid-angle integral to the spherical coordinates θ and ϕ. We show a rectangular solid-angle element $d\Omega$ on the surface of a unit sphere. The side lengths can be written as $\sin\theta\,d\phi$ and $d\theta$. (b) Polar plot of the gain pattern for a short dipole antenna.

Substituting for h from Equation 15.26, we write R_r as

$$R_r = \frac{\eta_0 l^2}{16\lambda^2} \oint \sin^2\theta\,d\Omega. \tag{15.55}$$

For the solid-angle integration element $d\Omega$, we can substitute in terms of θ and ϕ (Figure 15.10a):

$$d\Omega = \sin\theta\,d\phi\,d\theta. \tag{15.56}$$

This gives

$$R_r = \frac{\eta_0 l^2}{16\lambda^2} \int_0^{2\pi} d\phi \int_0^{\pi} \sin^2\theta \sin\theta\,d\theta. \tag{15.57}$$

Because there is no ϕ variation, we get a factor of 2π when we integrate over ϕ:

$$R_r = \frac{\eta_0 \pi l^2}{8\lambda^2} \int_0^{\pi} \sin^2\theta \sin\theta\,d\theta. \tag{15.58}$$

The θ integral is best attacked by making the substitution $x = -\cos\theta$. This makes the integrand algebraic. The integration element dx is given by

$$dx = \sin\theta\,d\theta. \tag{15.59}$$

We can rewrite the integral as

$$R_r = \frac{\eta_0 \pi l^2}{8\lambda^2} \int_{-1}^{+1} (1 - x^2)\,dx. \tag{15.60}$$

The integral has the value 4/3, so that the radiation resistance becomes

$$R_r = \eta_0 (\pi/6)(l/\lambda)^2. \tag{15.61}$$

If we substitute 120π for η_0, we get

$$R_r = 20\pi^2(l/\lambda)^2. \tag{15.62}$$

The radiation resistance is proportional to the length squared. We can substitute back into Equation 15.31 to find the effective area as

$$A = \frac{3\lambda^2}{8\pi}\sin^2\theta. \tag{15.63}$$

We can find the gain by multiplying by $4\pi/\lambda^2$ to get

$$G = 1.5\sin^2\theta. \tag{15.64}$$

This pattern is shown as a polar plot in Figure 15.10b. The maximum gain, 1.5, is at $\theta = 90°$.

Dipoles have a resonance when each arm is a quarter wavelength long, or when the total length is a half wavelength. From a transmission-line point of view, this is what you might expect. We could consider the dipole as an opened-out transmission line, and we would expect a series resonance when the length of the transmission line is a quarter wavelength. Actually the resonant length for practical dipoles is a little shorter than this, about 0.48λ, once the effect of the wire thickness and end supports are taken into account. In practice, the resonant length will also be affected by the presence of the ground; hence dipoles often have to be adjusted for resonance after they are put up. We can estimate the resonant resistance for a dipole by setting $l = \lambda/2$ in Equation 15.62. We get $R_r = 49\,\Omega$. This is a fair estimate of the actual value. A more detailed theory gives $73\,\Omega$, but effects such as the presence of the ground, wire thickness, and insulation conspire to lower it to nearer $50\,\Omega$ in practice. This is a convenient resistance for connecting 50-Ω coaxial cable. Usually, however, we would not connect a coaxial cable directly to a dipole. The problem is that currents run down the outside of the shield and affect both the impedance and the pattern. A transformer called a *balun* is used to isolate the outside of the shield from the antenna.

15.10 Whip Antennas

A common variation on the dipole is a single vertical wire above a ground plane (Figure 15.11a). This is called a *monopole*, or informally, a *whip* or *stub*. The calculations are similar to those for a dipole, except for some factors of two. The effective length doubles because the reflection off the ground plane doubles the component of the electric field along the wire. This effect by itself would raise the radiation resistance by a factor of 4. The pattern above the ground plane is the same as that for the dipole, but no power is received for angles below the ground plane. The θ integral in Equation 15.58 becomes an integral from 0 to $\pi/2$ rather than from 0 to π. This reduces the radiation resistance by a factor of 2. The net

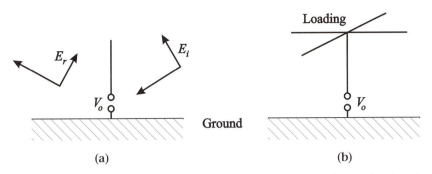

Figure 15.11. (a) Monopole above a ground plane, showing incident and reflected waves. The normal component of the reflected electric field is the same as that of the incident field, and this means that the resulting parallel to the dipole is twice the incident field. (b) Loading a whip antenna.

effect is to double the radiation resistance:

$$R_r = 40\pi^2(l/\lambda)^2. \tag{15.65}$$

A monopole is resonant when the length is a quarter of a wavelength. At low frequencies, it may not be convenient to make the whip this long. For example, the NorCal 40A operates in the 40-meter band. A quarter wavelength is 10 meters, which is too long for a car antenna. However, if we add additional wires at the top, the potential of the antenna can approach the potential at the top rather than half way up (Figure 15.11b). This is called *loading*. Loading doubles the effective length and quadruples the radiation resistance. We can write the radiation resistance for a loaded whip as

$$R_r = 160\pi^2(l/\lambda)^2. \tag{15.66}$$

We can also load the ends of dipoles. In Hertz's original experiments, the dipoles had spheres mounted on each end that had the effect of loading.

The dipoles and monopoles we have discussed so far have low gain. In 1926, Shintaro Uda, a professor at Tohoku University in Sendai, Japan, discovered an ingenious way to increase the gain. He added additional wires in front and behind the dipole (Figure 15.12a). These wires are called *parasitic elements*. They are not directly connected. However, currents are excited on the wires, and these currents reradiate energy with a phase that is determined by the length of the element. One parasitic wire is made shorter than the resonant length, and it acts as a *director*, strengthening the radiated beam in the forward direction. The other element is made somewhat longer than the resonant length, and it acts as a *reflector*, reducing the radiation in the back direction. With the correct dimensions, the gain can be increased substantially. These antennas are usually called Yagi antennas, after Hidetsugu Yagi, who was Uda's laboratory director at Tohoku, and who reported Uda's work in the United States. You may also hear them called Yagi–Uda antennas. The mechanical construction of Yagi antennas is quite simple, and this has made them popular around the world for communications and television.

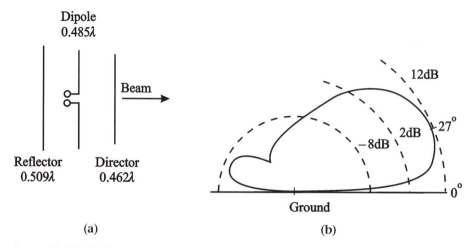

Figure 15.12. (a) Yagi antenna. (b) Calculated gain as a function of the elevation angle at 7 MHz for this antenna when it is 20 m above the ground. The array elements are 6 meters apart and have a diameter of 16 mm. This calculation was done with the *EZNEC* program.

The behavior of a Yagi antenna is complicated enough that you need a computer program to predict the patterns. Figure 15.12b shows the elevation pattern for the Yagi antenna in Figure 15.12a. The peak gain is 12 dB. The radiation behind the antenna is suppressed by more than 20 dB. The peak in the pattern is at an elevation of 27° for a 7-MHz antenna. The peak angle is determined by the height above the ground. It is the angle at which the reflected wave from the ground is in phase with the direct wave. This pattern is suitable for long-distance communication by reflection of radio waves off the ionosphere.

15.11 Ionosphere

Early investigators believed that the range of radio transmission would be completely limited by the curvature of the earth. However, radio waves can reflect off ionized electrons in the upper atmosphere. In fact, radio waves can bounce repeatedly, making worldwide radio communications possible. In the upper atmosphere, ultraviolet radiation and X rays from the sun strip electrons from atoms and molecules. This part of the atmosphere is called the *ionosphere*. Figure 15.13 shows a typical daytime electron-density plot. The peak electron density occurs at an altitude of 300 km. The electron density falls off at both higher and lower altitudes. At higher altitudes there are not many atoms and molecules to ionize. At lower altitudes, ultraviolet radiation and X rays become less intense, because they are absorbed in propagating through the atmosphere. The peaks on the plot are given the letter names D, E, F1, and F2. The more energetic ionizing radiation penetrates deeper into the ionosphere, so that X rays are primarily responsible for the D and E layers, and ultraviolet radiation is primarily responsible for the F layers.

The D layer is at an altitude of 70 km. The maximum ionization level occurs at noon each day and is about 10^{10} electrons per cubic meter. The D layer disappears

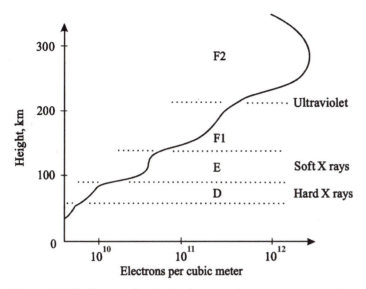

Figure 15.13. Electron density in the ionosphere on a summer day. The layers were named by Sir Edward Appleton in the 1920s. Historically, the first layer that was discovered was called the *electric* layer, or E layer for short. Then new layers were found below and above the E layer, and so these became the D and F layers. Adapted with permission from Figure 1.1 in *Ionospheric Radio*, by Kenneth Davies, published by Peter Peregrinus.

at sundown and reappears at sunrise. The D layer is responsible for most of the absorption in the ionosphere, and it prevents frequencies below 10 MHz from being used for long-range communications in the daytime.

The E layer is at 120 km. The maximum ionization, also at midday, is about 10^{11} electrons per cubic meter. Frequencies up to 15 MHz can be reflected 1,500 km in a single hop. Like the D layer, the E layer disappears at night. In addition, there are patches of ionization at the altitude of the E layer that are called *sporadic E*. The ionization in these patches can be so intense that frequencies up to 100 MHz can be reflected. The patches are about 100 km across, and they do not appear to be caused by solar radiation. One cause of sporadic E is friction at shear layers between fast moving air masses. This is the atmospheric equivalent of making sparks by scraping your shoes on the carpet. It occurs irregularly in summer months. Meteors also cause sporadic E, and this has been used for data transmission from remote weather stations.

The F layers have the highest ionization levels, and they are the main layers used for worldwide broadcasting and communications. Frequencies up to 50 MHz may be reflected 3,000 km in a single hop. The F1 layer disappears in the winter and at night, and the single layer that remains is just called the F layer. The typical peak ionization is about 10^{12} in the day at an altitude of about 300 km. At night the layer may drop 50 km in altitude and a factor of ten in ionization, and frequencies up to 15 MHz may be used. Unlike the E and D layers, whose ionization is determined by the position of the sun, many factors affect the F layer. Electron recombination is quite slow at these high altitudes, and the electron densities are strongly affected by diffusion from lower layers and long-distance drift. Because of this, there is still an F layer in the polar regions in the winter, even though there is no sun for months.

Solar activity has a strong effect on the F layer, and measurements of solar activity are as important for long-distance broadcasting and communications as weather reports are for farmers. The National Institute of Standards and Technology (NIST) broadcasts microwave solar power densities every hour from the standard-time stations WWV and WWVH. This microwave flux is measured every day at noon at 2.8 GHz at the Algonquin Radio Observatory in Algonquin Park, Ontario, Canada. It varies from a minimum of 65×10^{-22} $Wm^{-2}Hz^{-1}$ to a maximum of around 200×10^{-22} $Wm^{-2}Hz^{-1}$. In the broadcast, the exponent and units are dropped, and the announcer will just say 65 for the minimum value and 200 for the maximum. The 2.8-GHz radiation does not ionize the upper atmosphere. However, it is easy to measure, and surprisingly, it correlates well with the ultraviolet and X-ray fluxes, which are much more difficult to measure because they are absorbed in the ionosphere. Solar flux levels tend to repeat each solar rotation period (27 earth days), indicating that the areas on the sun that are responsible for the radiation last for quite a while. Solar flux also correlates with the number of sunspots (Figure 15.14). Sunspots are cool (at least by solar standards) regions with temperatures of 3,000 K or so, compared with 6,000 K, which is typical of the rest of the sun's surface. It might seem surprising that the number of sunspots correlates strongly with solar activity. However, sunspots have enormous magnetic fields, in the range of 0.4 teslas, which is similar to the magnetic field of a large laboratory magnet, and these magnetic fields have a strong effect on the solar plasma. Sunspots have a dramatic 11-year cycle (Figure 15.15). Near the peak of the cycle, frequencies as high as 50 MHz can be used for long-distance

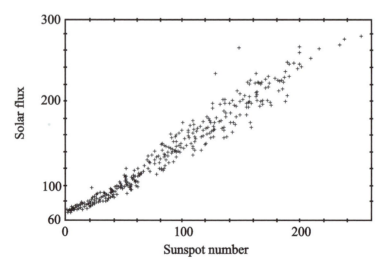

Figure 15.14. Monthly solar flux averages, compared with monthly averages of the sunspot number. Reprinted with permission from Chapter 23 of the *ARRL Antenna Book*, published by the American Radio Relay League.

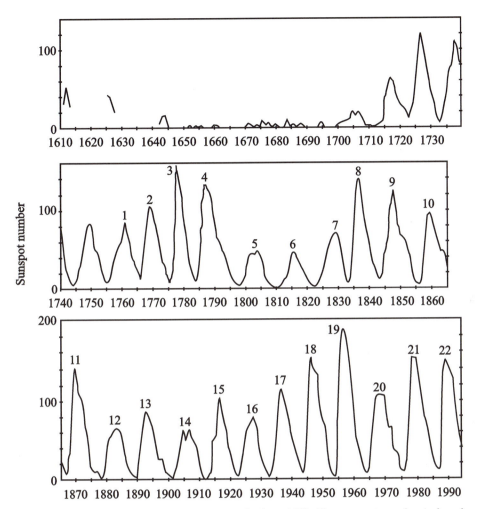

Figure 15.15. Annual sunspot averages going back to 1610. The sunspot number is found by counting the number of individual sunspots and the number of groups of sunspots. The sunspot number is traditionally calculated as the sum of the number of individual sunspots and ten times the number of sunspot groups. Near the minimum, the sunspot number is close to zero, and near the maximum, it can be above 200. The sunspot number is a crude way to measure solar activity, but measurements are available for almost 400 years. The extended quiet period during the 1600s is called the *Maunder minimum*. The cycles are numbered consecutively, starting with the cycle that began in 1755. Reprinted with permission from Figure 2.3 in *Ionospheric Radio*, by Kenneth Davies, published by Peter Peregrinus.

ionospheric communications. Near the sunspot minimum, frequencies higher than 15 MHz can be used only sporadically.

15.12 Radio Waves in the Ionosphere

To investigate how radio waves propagate in the ionosphere, we use Newton's second law to find an expression for the conductivity of the ionized electrons.

Newton's second law says that the applied force is equal to the rate of change of momentum. The force on a single electron due to an electric field is $q\mathbf{E}$, where q is the electronic charge, 1.6×10^{-19} C. Neglect the effects of magnetic fields and collisions, and we can write

$$q\mathbf{E} = m\frac{d\mathbf{v}}{dt}, \tag{15.67}$$

where m is the electron mass, 9.1×10^{-31} kg, and $\mathbf{v}$ is the velocity. We can rewrite this formula in phasor notation, replacing the time derivative by $j\omega$:

$$q\mathbf{E} = j\omega m\mathbf{v}. \tag{15.68}$$

We can solve for the velocity as

$$\mathbf{v} = \frac{q}{j\omega m}\mathbf{E}. \tag{15.69}$$

We relate the velocity to the current density $\mathbf{J}$ by multiplying by the charge density Nq, where N is the number of electrons per cubic meter. This gives us

$$\mathbf{J} = Nq\mathbf{v} = \frac{Nq^2}{j\omega m}\mathbf{E}. \tag{15.70}$$

The current lags the electric field because of the inertia of the electrons. Now we can see how this affects radio-wave propagation by including the current density $\mathbf{J}$ in Ampère's law. We write

$$\nabla \times \mathbf{H} = \mathbf{J} + j\omega\epsilon_0\mathbf{E} \tag{15.71}$$

and substitute for $\mathbf{J}$ to get

$$\nabla \times \mathbf{H} = \frac{Nq^2}{j\omega m}\mathbf{E} + j\omega\epsilon_0\mathbf{E}. \tag{15.72}$$

We can rewrite this equation in the form

$$\nabla \times \mathbf{H} = j\omega\epsilon\mathbf{E}, \tag{15.73}$$

where the effective permittivity ϵ is given by

$$\epsilon = \epsilon_0 - \frac{Nq^2}{\omega^2 m}. \tag{15.74}$$

We can then rewrite the phase constant β using Equation 15.11 as

$$\beta = \beta_0\sqrt{1 - \frac{Nq^2}{\epsilon_0\omega^2 m}}, \tag{15.75}$$

where β_0 is the phase constant for free space. The *refractive index n* is given by

$$n = \frac{\beta}{\beta_0} = \sqrt{1 - \frac{Nq^2}{\epsilon_0\omega^2 m}}. \tag{15.76}$$

The refractive index in the ionosphere is less than 1, which seems strange, because it means that the wavelength in the ionosphere is longer than in free space. Notice that the expression contains the electron mass in the denominator, indicating that the effect comes from the inertia of the ionized electrons.

15.13 Critical Frequency

To simply Equation 15.76, we define a *critical frequency* f_c given by

$$f_c = \frac{1}{2\pi} \sqrt{\frac{Nq^2}{\epsilon_0 m}}.$$ (15.77)

If we substitute for the constants, we find

$$f_c \approx 9.0\sqrt{N}.$$ (15.78)

We can rewrite Equation 15.76 as

$$n = \sqrt{1 - \left(\frac{f_c}{f}\right)^2}.$$ (15.79)

At frequencies below the critical frequency, the propagation constant becomes imaginary, and the wave attenuates. The wave is said to be *evanescent*. Physically, the power is reflected.

The critical frequency can be measured by transmitting pulses from a radar *sounder* straight up, and listening for reflections. Typically the pulses are 30 μs long and have a peak power in the range of 1 kW to 10 kW. The delay in the reflection tells us the height of the layer. Plots of the reflection delay versus frequency are called *ionograms*. Figure 15.16 shows a summer ionogram, with the E, F1, and F2 layers are clearly visible. The E layer is quite sharp, at a height of 100 km, with a critical frequency of 3.8 MHz. The F1 layer is at 200 km, with a critical frequency of 4.9 MHz. The F2 layer is at 370 km, with a critical frequency of 6.7 MHz.

Figure 15.17 shows how solar activity affects f_c. The E-layer critical frequency increases from 3 MHz to 4 MHz as the sunspot number moves from zero to 200. For the F1 layer, f_c increases from 4 MHz to 5 MHz. At higher sunspot numbers, the F1 layer is not always distinct. The most dramatic change is in the F2 layer, where f_c doubles from 5.5 MHz to 11 MHz.

Figure 15.18 shows how the critical frequencies depend on the time of day. For the F2 layer, f_c is relatively constant in the summer, but it peaks sharply in the winter in the early afternoon. Surprisingly, the highest critical frequencies are in the winter, even though the sun is at a lower angle than in the summer. This is a complicated effect involving collisions with molecules in the upper ionosphere. The upper ionosphere is primarily atomic oxygen. However, the density of molecular oxygen increases in the summer, and this reduces the electron recombination time, causing the ionization level to drop.

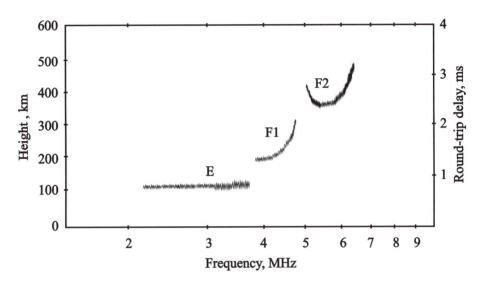

Figure 15.16. Summer daytime ionogram for Boulder, Colorado. The local time is 1:30 pm. The right axis gives the round-trip delay, and the left axis gives the apparent height. Reflection from the ionosphere causes additional delay. This causes the apparent height to be larger than the actual height, particularly near a layer boundary. Redrawn with permission from Figure 4.4a in *Ionospheric Radio*, by Kenneth Davies, published by Peter Peregrinus.

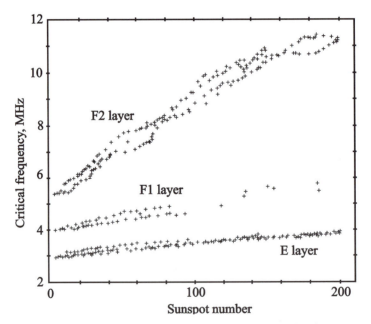

Figure 15.17. Critical frequencies for Washington, DC, for the summer noon F2 layer (upper data points), F1 layer (middle data points), and E layer (lower data points), plotted against the sunspot number over two sunspot cycles. Reprinted with permission from Figure 5.13 in *Ionospheric Radio*, by Kenneth Davies, published by Peter Peregrinus.

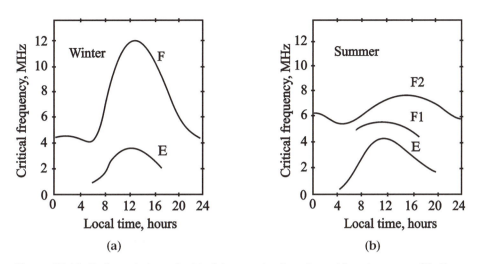

Figure 15.18. Daily variation of critical frequencies for winter (a) and summer (b). From Figure 9.2 in *Radio Wave Propagation*, by Lucien Boithias, published by McGraw-Hill.

15.14 Maximum Usable Frequency

You may have studied Snell's law in optics, which relates the propagation angles for a wave incident on a surface between layers with two refractive indexes (Figure 15.19). It is written as

$$n_i \sin \theta_i = n_t \sin \theta_t, \tag{15.80}$$

where n is the refractive index and θ is the angle from the normal. The subscript i is for the *incident* wave and the subscript t is for the *transmitted* wave. This bending of waves is called *refraction*. We can state Snell's law in a more general way:

$$n \sin \theta \text{ is a constant.} \tag{15.81}$$

This form also applies if there are more layers or if the refractive index varies continuously. There is also an interesting possibility that occurs if n_i is larger than n_t. We can have

$$n_i \sin \theta_i > n_t. \tag{15.82}$$

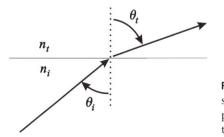

Figure 15.19. Snell's law in optics for transmission through an interface. The arrows show the propagation directions for the incident wave and the transmitted, or refracted, wave.

This makes it impossible to satisfy the Snell's law formula. We define a *critical incident angle* θ_c given by

$$\sin \theta_c = n_t/n_i, \tag{15.83}$$

where $\theta_t = 90°$. At larger incident angles, the wave is completely reflected. This allows us to make optical fiber waveguides by enclosing one dielectric by another with a lower refractive index. The waves bounce back and forth along the guide without loss as long as the incident angle is larger than the critical angle.

Because the refractive index of the ionosphere is less than 1, we can have a critical angle when radio waves are incident from the lower atmosphere. If we take the refractive index of the lower atmosphere to be 1, we can write the critical angle θ_c as

$$\sin \theta_c = n, \tag{15.84}$$

where n is the refractive index of the ionosphere. We can relate the critical angle to the critical frequency by substituting from Equation 15.79:

$$\sin \theta_c = \sqrt{1 - \left(\frac{f_c}{f}\right)^2}, \tag{15.85}$$

or

$$\cos \theta_c = f_c/f. \tag{15.86}$$

We can interpret this formula in two different ways. For a fixed frequency, we can think of it as determining the minimum incident angle that is completely reflected. For a fixed incident angle, we can calculate the maximum frequency that is completely reflected. This is called the *maximum usable frequency* f_m, and we write it as

$$f_m = f_c/\cos \theta_i. \tag{15.87}$$

We should realize that because of the curvature of the earth, the angle of incidence on the ionosphere is not the same as the launch angle on earth. The geometry is shown in Figure 15.20. We can use the sine law to relate θ_i and the launch angle θ:

$$\sin \theta = (1 + h/a) \sin \theta_i, \tag{15.88}$$

where a is the radius of the earth, 6,370 km, and h is height of the ionosphere. We can write the range r in terms of these angles as

$$r = 2a\phi = 2a(\theta - \theta_i). \tag{15.89}$$

We can use these formulas to relate the maximum usable frequency f_m and the range r to the launch angle θ when the F layer is at a height of 300 km; see Figure 15.21. The corresponding elevation angle is shown at the bottom. This plot shows that waves can be reflected at much higher frequencies than the critical frequency. The largest value of f_m, $3.4f_c$, and the greatest range, 3,840 km, are for low

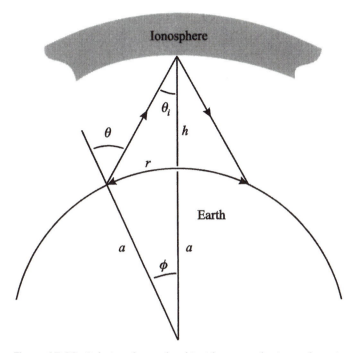

Figure 15.20. Relating the angle of incidence on the ionosphere θ_i to the launch angle θ on earth.

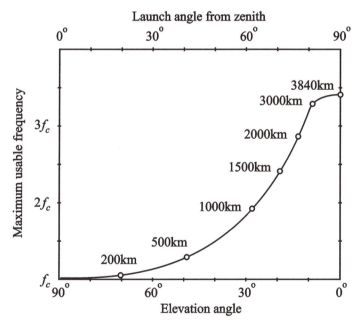

Figure 15.21. Maximum usable frequency f_m versus elevation angle for an F-layer height of 300 km. This curve is calculated from Equations 15.87 through 15.89. The range for a single reflection is marked on the curve. In practice, radio waves bounce repeatedly, and much greater ranges are possible.

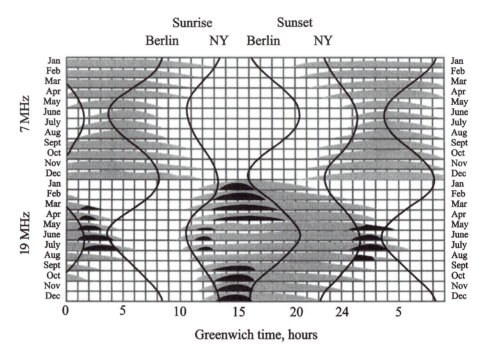

Figure 15.22. Received signal levels between Berlin and New York, measured over an entire year. The signal strength is indicated by the height of the vertical lines. This is Figure 45.4 in *Radio-Wave Propagation and the Ionosphere,* Volume 2, by Y. L. Al'pert, published by the Consultants Bureau, a division of the Plenum Publishing Company.

elevation angles. A peculiar feature of ionospheric radio communications is that when the frequency is greater than f_c, there is a minimum range. As the range decreases, the incident angle on the ionosphere becomes less than the critical angle, and the wave passes through the ionosphere without reflection. This minimum range is called the *skip* distance. The skip distance can also be determined from the figure. For example, if we operate at $2f_c$, the maximum elevation angle is 25°, and the skip distance is 1,100 km. To communicate at closer distances, we would need to reduce the operating frequency.

Figure 15.22 shows a plot of signal reception reports between Berlin and New York over an entire year for 7 MHz and 19 MHz. The distance is 6,000 km, so that two or three reflections from the F layer are needed. The two frequencies show many different effects. The 7-MHz signal is greatly absorbed by the D layer. This means that it has poor propagation during the day. However, the frequency is low enough that it is still reflected throughout the night, even as the electron concentration of the F layer drops. In contrast, 19 MHz is high enough that the signals are not significantly absorbed by the D layer. However, at night, the critical frequency drops and 19-MHz signals fade away. These patterns hold well when there is daylight all along the path, or darkness all along the path. The patterns follow the changes in the hours of sunrise and sunset. Notice also that 19-MHz propagation continues for several hours after sunset in Berlin. This is because the electron concentration in the F layer decays slowly after the sun sets.

FURTHER READING

The book *Antennas and Radio Wave Propagation,* by Robert Collin, published by McGraw-Hill, gives excellent coverage of both antennas and propagation. The *ARRL Antenna Book*, published by the American Radio Relay League, is good for information on building antennas. For studying wire antennas, simulation software is essential. My favorite program is called *EZNEC*. It is available from Roy Lewallen, P.O. Box 6658, Beaverton, OR 97007, w7el@teleport.com. A good reference for the detailed derivation of the blackbody radiation formula is given in *Thermal Physics*, by Charles Kittel, published by W. H. Freeman. Hertz's experiments were extraordinary, and his own descriptions of his experiments make good reading even today. See *Electric Waves*, by Heinrich Hertz, published by Dover. In addition, John Bryant's short book, *Heinrich Hertz, the Beginning of Microwaves*, published by the Institute of Electrical and Electronics Engineers, gives an excellent description of Hertz's experiments. For a proof of the reciprocity theorem, see Desoer and Kuh's *Basic Circuit Theory*, published by McGraw-Hill. The story of the discovery of the cosmic background radiation is told by Steven Weinberg in the *The First Three Minutes,* published by Bantam Books. Shintaro Uda has written an interesting book about the development of the Yagi–Uda antenna, *Short Wave Projector, Historical Records of My Studies in Early Days*, which has been privately published in English.

The standard reference on propagation in the ionosphere is *Ionospheric Radio*, by Kenneth Davies, published by Peter Peregrinus. You can learn more about early radio experiments in *200 Meters and Down* by Clinton DeSoto, published by the American Radio Relay League. Computer predictions of radio-wave propagation can be extremely helpful. I use *Miniprop Plus*, by Sheldon Shallon, published by W6EL Software, 11058 Queensland Street, Los Angeles, CA 90034–3029. The book, *Morse Code, the Essential Language,* by Peter Carron, published by the American Radio Relay League, has more information about the history and use of the code. For a discussion of Snell's law from the perspective of the principle of least time, see *The Feynman Lectures on Physics*, Volume 1, Chapter 26.

PROBLEM 37 – ANTENNAS

A. Use the relation between gain and effective area to rewrite the Friis transmission formula in terms of gain only. Consider a UHF voice radio system for communicating between airplanes. Assume that whatever frequency is finally chosen, the quarter-wave stub antennas will have a gain of 2. Find the maximum possible line-of-sight range between two airplanes at an altitude of 10 km. The required receiver power P_r is –90 dBm. Find the minimum transmitter power P_t required for successful transmission at this range for 100 MHz, 300 MHz, and 1 GHz.

Let us consider a whip antenna for using the NorCal 40A in a car (Figure 15.23). Tuning and matching are done by a tapped inductor, which is a coil with a third wire connected at one of the intermediate turns. This is a transformer where the primary is a part of the secondary. People call this an *autotransformer*, and it acts like an ordinary transformer,

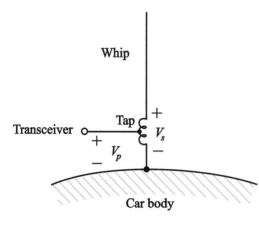

Figure 15.23. Whip antenna with a tapped inductor for using the Nor-Cal 40A in a car.

except that there is no DC isolation. The whip capacitance C is given approximately by

$$C \approx l \epsilon_0, \tag{15.90}$$

where l is the length.

B. Find the inductance required to resonate a 3-m whip. Assuming that the Q of the coil is 200, find the turns ratio required to give the transceiver a 50-Ω load. What is the radiation efficiency?

C. Repeat these calculations for a whip with capacitive end loading, assuming that the capacitance doubles.

PROBLEM 38 – PROPAGATION

The HF band from 3 MHz to 30 MHz is used by thousands of international broadcasters, ships, and amateurs. Many different types of transmissions can be heard, including AM, FM, single sideband (SSB), frequency-shift-keying (FSK), and the oldest of all – Morse Code.

Ionospheric reflection and absorption depend strongly on the frequency, latitude, solar activity level, the time of day, and the season of the year. Signal levels can change quickly, particularly at sunrise and sunset. In addition, on rare occasions, ionospheric storms shut down the HF bands for days. For this reason, some users have switched to microwave satellite links, which are not usually affected by ionospheric disturbances. Others, like amateurs and listeners to international broadcasters, find that the ionospheric "weather" makes the HF bands interesting.

One way to investigate propagation is to listen to *beacon* stations. These are stations that transmit throughout the day and night on a standard frequency. We can tune to these frequencies to find out how radio waves are propagating. Figure15.24 gives reception reports for several beacon stations, taken at hourly intervals over a 24-hour period. The first four stations all transmit at the same frequency, 14.1 MHz, but from different locations. These stations transmit Morse code call signs at regular intervals in a fixed sequence. In addition, the stations transmit a sequence of four tones at power levels of 100 W, 10 W,

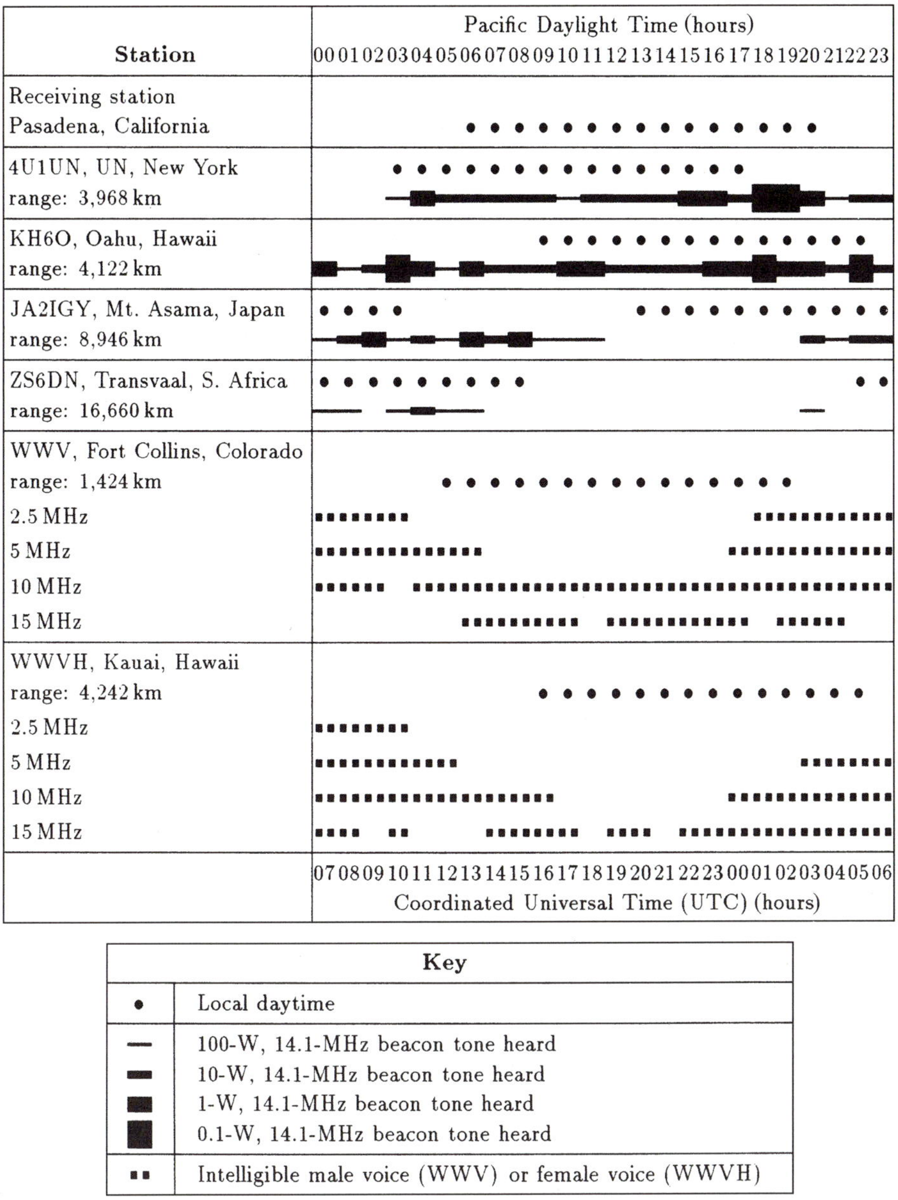

Station	Pacific Daylight Time (hours) 00 01 02 03 04 05 06 07 08 09 10 11 12 13 14 15 16 17 18 19 20 21 22 23
Receiving station Pasadena, California	
4U1UN, UN, New York range: 3,968 km	
KH6O, Oahu, Hawaii range: 4,122 km	
JA2IGY, Mt. Asama, Japan range: 8,946 km	
ZS6DN, Transvaal, S. Africa range: 16,660 km	
WWV, Fort Collins, Colorado range: 1,424 km 2.5 MHz 5 MHz 10 MHz 15 MHz	
WWVH, Kauai, Hawaii range: 4,242 km 2.5 MHz 5 MHz 10 MHz 15 MHz	
	07 08 09 10 11 12 13 14 15 16 17 18 19 20 21 22 23 00 01 02 03 04 05 06 Coordinated Universal Time (UTC) (hours)

Key	
•	Local daytime
—	100-W, 14.1-MHz beacon tone heard
▬	10-W, 14.1-MHz beacon tone heard
■	1-W, 14.1-MHz beacon tone heard
▉	0.1-W, 14.1-MHz beacon tone heard
▪▪	Intelligible male voice (WWV) or female voice (WWVH)

Figure 15.24. Hourly reception reports for June 26, 1993, for Pasadena, California. Courtesy of Kate Rutledge.

1 W, and 0.1 W. By listening to the tones, one can get a feeling for the propagation conditions.

In addition to the beacon stations, there are *time* stations that broadcast the time on particular frequencies. In the United States, the time stations are WWV in Fort Collins, Colorado, and WWVH in Kauai, Hawaii. They use AM. WWV transmits continuously on 2.5 MHz (2.5 kW), 5 MHz (10 kW), 10 MHz (10 kW), 15 MHz (10 kW), and 20 MHz (2.5 kW). WWVH transmits continuously on 2.5 MHz (5 kW), 5 MHz (10 kW), 10 MHz

(10 kW), and 15 MHz (10 kW). Every minute, the Coordinated Universal Time (UTC) is given. This is the same as Greenwich Mean Time, or you may hear it called "Zulu" in voice communications, or "Z" in Morse code. About 15 seconds before the minute, WWVH transmits a woman's voice beginning "At the tone, ..." About 7 seconds before the minute, WWV transmits a man's voice with the same message. This lets us know which stations are being received. The time stations also give ionospheric forecasts and navigation-system status reports. Figure 15.24 also gives reception reports for WWV and WWVH.

Discuss the signal reports in Figure 15.24. You should consider the range and whether the path was in daylight or darkness. Compare sunrise and sunset, and high and low frequencies.

PROBLEM 39 - LISTENING

Your transceivers are for communication by Morse Code. Samuel Morse was the inventor of the telegraph. Morse put a great deal of thought into developing a code for telegraph messages. The first code was based on a code dictionary, with numbers representing words. This is quite slow to encode and decode. Then Morse and his assistant, Alfred Vail, hit upon the idea of using different length pulses and spaces to represent letters. Initially, inking machines with electromagnets were constructed to show these patterns on paper, but operators found that with practice they could interpret the clicking sounds of the instruments by ear, and that they could copy down the message directly. After this, devices called *sounders* were made that were designed for reception by ear. Sometimes you hear sounders in Western movies. They give a distinctive *tick* and *tock* sound at the beginning and end of the pulses. A large network of telegraph lines was built around the world to send messages rapidly, and this was the dominant form of long-distance electrical communication for 100 years.

Morse and Vail incorporated several important ideas in their code (Figure 15.25). To make the sending as efficient as possible, they used shorter patterns for the common letters, like E, T, A, O, I, and N. To find out which letters were the most common, a visit was made to a newspaper printer and the letters in the type cases were counted. In this way, they made the code about 25% faster than it would be if the letters were assigned patterns randomly. Several lengths of pulses are used. You can see these if you compare the pulse lengths for E, T, L, and 0. In addition, they used two different lengths of spaces between the pulses that you can see by comparing I and O. Yet longer spaces separated letters and words.

In Europe, an Austrian, Frederick Gerke, developed a variation of the Morse code that was adopted there (Figure 15.26). Many of the letters are the same in both codes. However, Gerke simplified the code by using only one space length and only two pulse lengths ("dits" and "dahs"). Gerke's code was easier to learn than Morse's, but it was somewhat slower. One interesting character is the letter "O," which is three dahs in Gerke's code. This is much longer than Morse's O, which is a pair of dits. Morse made O short because O is the fourth most common letter in English. However, in German, O is an uncommon letter, ranking only 16th, and for this reason it was made long.

Both the American code and the European code were used when most messages were sent by land telegraph lines. However, when people began sending messages by radio,

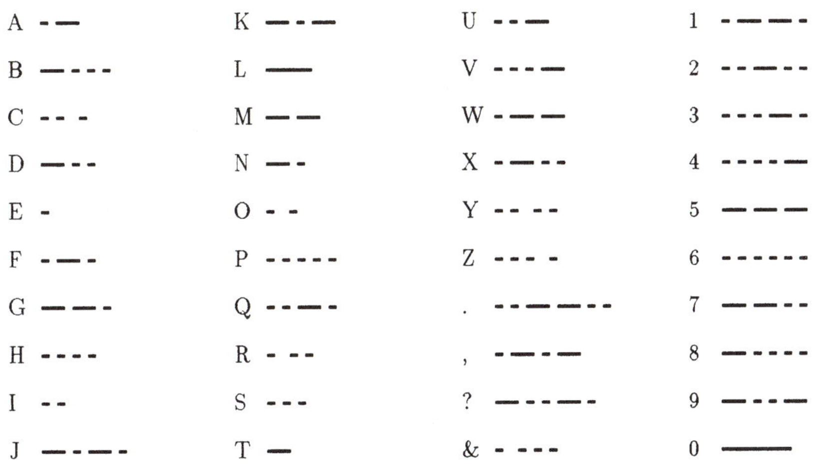

A ·—	K —·—	U ··—	1 ·———·
B —···	L ——	V ···—	2 ··———··
C ·· ·	M ——	W ·——	3 ···——·
D —··	N —·	X ·—··	4 ————·
E ·	O · ·	Y ·· ··	5 ———
F ·—·	P ·····	Z ··· ·	6 ······
G ——·	Q ··—·	. ··——··	7 ——··
H ····	R · ··	, ·—·—	8 —····
I ··	S ···	? ——··—·	9 ——·—
J —·—·	T —	& · ···	0 ——

Figure 15.25. The American Morse Code developed by Samuel Morse and Alfred Vail. The length of the lines represents the length of the pulses, and the gap between the lines represents the spacing between the pulses. This code is not in common use today, having been replaced by the International Morse Code.

A ·—	K —·—	U ··—	1 ·————
B —···	L ·—··	V ···—	2 ··———
C —·—·	M ——	W ·——	3 ···——
D —··	N —·	X —··—	4 ····—
E ·	O ———	Y —·——	5 ·····
F ··—·	P ·——·	Z ——··	6 —····
G ——·	Q ——·—	. ·—·—·—	7 ——···
H ····	R ·—·	, ——··——	8 ———··
I ··	S ···	? ··——··	9 ————·
J ·———	T —	= —···—	0 —————

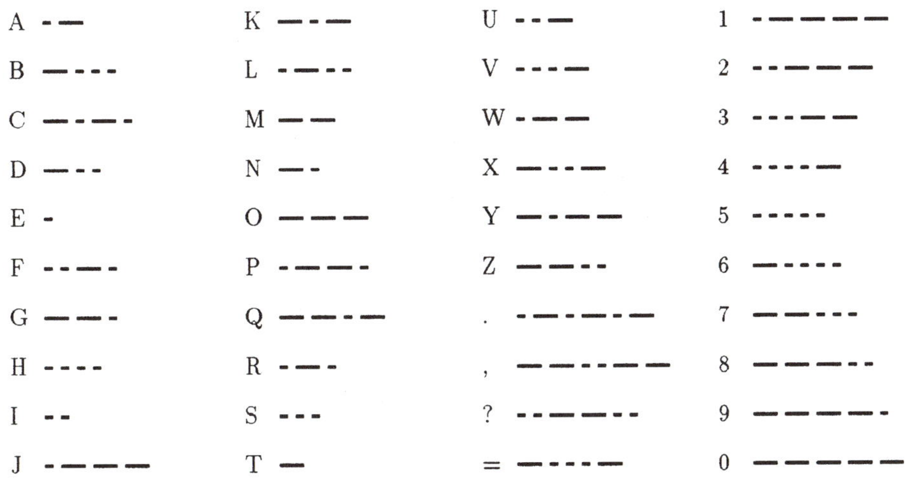

Figure 15.26. The International Morse Code developed by Gerke.

it became important to have a single code, and Gerke's simpler code won out. Known today as the International Morse code, it is universal for languages that use Roman letters. Similar codes have been developed for other writing systems, and the Japanese code is often heard on the air.

Today, Morse code is rarely used in commercial or military communications, where codes intended for automatic decoding are used instead. However, Morse's idea of representing common characters by shorter patterns is used in data compression algorithms, and the idea of conveying information by varying pulse and space lengths is very much alive in radio control systems. In some ways the code is an obsolete form of communications, but in other ways it is no more obsolete than sailing or horseback riding are for transportation. It requires lower power levels, less bandwidth, and simpler equipment than voice communications, and there are a large number of standard abbreviations that allow people who do not share a common language to communicate.

Use your radio to take down a Morse-code two-way conversation off the air. Amateurs call this a QSO. Include an interpretation. For each station, you should try to get the call sign, name, location, signal report, output power, and antenna type. You can do this with a decoding machine tuned to 600 Hz. Plug the audio output from your transceiver into the jack of the decoder. You will probably want to connect your speaker in parallel, because the speakers in decoders are often quite poor. Adjusting the radio takes some care. If the RF gain is set too high, atmospheric noise will cause false letters on a decoder. Usually these will appear as Es. Set the RF gain pot as high as you can without triggering Es. Now look for a signal. Tune carefully through the band with the VFO Tune pot, looking for strong signals. When you find a station, set the frequency precisely by adjusting the RIT pot to make the tune light on the decoder as bright as you can. You should start to see the copy appear. It will be difficult to receive if the signal is weak or if the sending is sloppy. Machines are not as good as people are at copying Morse code! Sometimes you will find a station that is using a radio teletype code called the Baudot code. The Baudot code uses FSK, with two tones 170-Hz apart, and you will hear a continuous warbling. Most decoders also have a setting for the Baudot code. You should try to tune on each of the tones, and see if one gives good copy.

Each station is identified by a call sign that is assigned by the communications authority in that country. An example would be KE6AGH. The first letter "K" identifies the station as American. The number "6" identifies the call district within a country, California in this case. Figure 15.27 shows the call districts in the United States, and Table 15.1 gives the call-sign prefixes for many countries.

Watch carefully to see when one operator quits sending and the other begins. You will usually need to adjust the RIT slightly, unless the two operators are on exactly the

Figure 15.27. United States call districts.

Table 15.1. Some Call-Sign Prefixes and Their Countries.

A, K, N, W United States	HL South Korea	SU Egypt
BV Taiwan	HP Panama	SV Greece
BY China	HR Honduras	TA Turkey
CE Chile	HS Thailand	TG Guatemala
CM–CO Cuba	HT Nicaragua	TI Costa Rica
CP Bolivia	I Italy	U Russia
CT Portugal	JA–JS Japan	VE Canada
CX Uruguay	LA Norway	VK Australia
DA–DL Germany	LU Argentina	VU India
DU Philippines	OA Peru	XE, XF Mexico
EA Spain	OE Austria	YB–YD Indonesia
EI Ireland	OH Finland	YS El Salvador
F France	OK Czech Republic	YV Venezuela
G United Kingdom	OM Slovak Republic	ZL New Zealand
HA Hungary	ON Belgium	ZP Paraguay
HB Switzerland	OZ Denmark	ZS South Africa
HC Ecuador	PA Netherlands	4X Israel
HH Haiti	PP–PY Brazil	7L, 7M Japan
HI Dominican Republic	SM Sweden	9M Malaysia
HK Colombia	SP Poland	9V Singapore

same frequency. There are several ways for an operator to turn over to the other. If an operator has a question, often the question will be sent, followed by BK. This is equivalent to "over" in a radio voice conversation. The other operator will start with BK. It is also common to first indicate the end of a message by sending an A and R run together ($\overline{AR}$). This sign is a run-together $\overline{FN}$ (for finished) in American Morse. Then the two call signs would be sent, followed by K, or $\overline{KN}$. For the last message in a QSO, an operator will send $\overline{SK}$. This is $\overline{30}$ in American Morse, and it was traditionally sent by telegraphers before a 30-minute lunch break.

Interpreting your copy takes some work, because it is quite different from spoken English. Table 15.2 gives a glossary, and Figure 15.28 gives a typical QSO. These forms are used world wide, and much of it dates back to American land-line telegraphy. For example, the abbreviation for "and" is like the & character in American Morse, and the abbreviation for a laugh is from the American Morse "ho." Many are just abbreviations to save time. Often they help the operator avoid sending an "O," which is quite long. "73" for "best regards" is ubiquitous. It derives from nineteenth-century land-line number codes. "73" has a wonderful sound in International Morse Code, and this may be why it is so common. The articles "a," "an," and "the" appear only rarely, and "is" may be omitted. "=" is typically used instead of a period or comma. You may see a series of them as a pause. It is a good idea to listen for a CQ, which is a call to any station. The CQ is sent before the QSO begins, and if you hear it, you can copy the entire QSO. The first exchange gives a signal report, name, and location. These are typically repeated to help if reception is poor. The first number in the signal report is the *readability*. The readability varies from 1 to 5, and indicates how easy it is to copy the signal. "5" indicates a signal that is easy to copy, and 2 indicates a signal that is quite difficult to copy. The second

Table 15.2. Morse-Code Vocabulary.

ABT about	QRL This frequency is busy.
AGN again	(Asks others to find another frequency)
ANT antenna	QRL? Is this frequency busy?
BK over ("go ahead")	(A check to avoid interfering with others)
BN been	QRM troubling interference
BURO bureau (an agency for exchanging	QRN troubling atmospheric noise
QSL cards between countries)	QRP low power (5 watts or less)
C degrees Celsius	QRS slow down
CONDX ionospheric conditions	QRT quit
CQ call to any station	QRU I have nothing more to say.
CUAGN see you again	QSB troubling fading
CUD could	QSL confirm the contact by a card
CW Morse code	QSO a contact
DE KD6PFK This is station KD6PFK.	QST a broadcast to amateurs
DP dipole	QTH PASADENA, CA gives location
DR dear (before a name, as "DR Kate")	R are, or received, OK ("roger")
DX foreign contacts	RIG transceiver
E E E error	RPT repeat, or report
EL, ELE antenna element	RST 599 a signal report
ES and	SIG signal
FB good ("fine business")	SKED a "scheduled" appointment
FER for	SUM some
FM from	SRI sorry
GA good afternoon	T zero (much shorter than "0")
GB good bye	TEMP temperature
GE good evening	TKS, TNX thanks
GL good luck	TT that
GM good morning	TU thank you
GN good night	U you
GUD good	UR your
HI, HEE telegraphic laugh	VEE a common antenna
HPE hope	VFB very FB
HR here	VY very
HV have	W watt or watts
HW? How do you copy?	WID with
INFO information	WL well, or will
K over ("go ahead")	WUD would
M meters	WX weather
/M from a vehicle ("mobile")	XYL wife ("ex YL")
N nine (much shorter than "9")	YL woman ("young lady,"
NW now	often sent after a name,
OM form of address for men ("old man")	to avoid being called an OM)
OP name ("operator")	YRS years
PSE please	73 best regards
PWR transmitter power	88 best regards (to a woman)

CQ CQ CQ DE KN6EK KN6EK KN6EK K

KN6EK DE JA0FCC +

JA0FCC DE KN6EK GE OM ES PLEASED TO MEET U –

UR RST 599 5NN – QTH PASADENA, CA PASADENA, CA –

NAME DAVE DAVE – HW? + JA0FCC DE KN6EK !

RR KN6EK DE JA0FCC GM DAVE SAN –

TNX FB RPT FM PASADENA, CA – UR RST 449 44N –

OP GAKU GAKU – QTH NAGANO NAGANO – HW? + KN6EK DE JA0FCC !

RR JA0FCC DE KN6EK ES TNX RPT FM NAGANO –

RIG HR IS HOMEMADE – RUNNING 2W – ANT 2 EL YAGI AT 25M –

WX CLOUDY ES TEMP 15C – HW? + JA0FCC DE KN6EK !

RR KN6EK DE JA0FCC ALL COPY DAVE SAN –

RIG HR KENWOOD 850 – PWR 100W – ANT HORIZONTAL DP AT 10M –

WX HR RAIN ES TEMP 10C – HW? + KN6EK DE JA0FCC !

RR JA0FCC DE KN6EK TNX INFO GAKU OM –

AGE 46 YRS – HV 3 CHILDREN – JOB TEACHER –

WHAT IS UR JOB GAKU SAN? BK

BK BUDDHIST PRIEST – I LIVE IN A TEMPLE – QSL VIA BURO ? BK

BK MY QSL VIA BURO SURE – TNX FB QSO GAKU SAN –

HPE CUAGN 73 # JA0FCC DE KN6EK !

RR KN6EK DE JA0FCC – TNX FER 1ST QSO DAVE SAN –

73 73 # KN6EK DE JA0FCC E S E

TU E E

Figure 15.28. Typical Morse code on the air. Work through the meaning of this QSO before you try to copy one. In the text, Microcraft decoder conventions are used: – for =, + for $\overline{AR}$, ! for $\overline{KN}$, and # for $\overline{SK}$.

number is the *strength*. The strength varies from 1 to 9, with 9 indicating a very strong signal, and 1 indicating a very weak signal. The third number indicates the tone quality of the signal. Nowadays, this is almost always a nine. The endings "E S E" and "E E" are traditional.

Equipment and Parts

Kent Potter

To do the exercises, you need the NorCal 40A transceiver kit, available from Wilderness Radio, tel. 650.494.3806, http://www.fix.net/jparker/wild.html, P.O. Box 734, Los Altos, CA 94023-0734. The NorCal 40A has its own home page on the World Wide Web at http://www.fix.net/~jparker/norcal.html. The kit includes all the parts, a metal box with silk-screen lettering, and excellent instructions.

We have included a list of the equipment and parts that are used in each exercise and a vendor list. For reference, we include vendors for the components in the NorCal 40A, even though it is cheaper and more convenient to buy the kit. Vendors and parts change from time to time. There are additional considerations for a class. Extra components need to be purchased to allow for failures, and it is a good idea to talk to Bob Dyer at Wilderness Radio directly to plan. Power supplies and batteries *must* be short-circuit protected, both for safety and cost. For the batteries, we add fuses and switches and wrap everything in heat-shrink tubing.

A.1 EQUIPMENT

The measuring equipment that is required is similar to what would be available in a university electronics laboratory, or on the electronics bench of a radio amateur. Here is a list of the equipment we use:

Tektronix 2215A oscilloscope with a 10:1 probe. (Any 50-MHz scope should do.)
Hewlett-Packard 33120A synthesized 15-MHz function generator.
Hewlett-Packard 3478A multimeter. (The Fluke 87 is a good substitute.)
Tenma 72-860 sound-level meter (35–130 dB, available from Newark).
Tripp-Lite PR-3A power supply (3-A supply, available from Marvac).
Fluke 1900A counter. (Any 10-MHz counter with 10-Hz resolution would do.)
Kay 860 attenuator (a 120-dB attenuator).

A.2 VENDORS

Amidon Corporation
240 Briggs Avenue
Costa Mesa, CA 92626
tel. 714.850.4660

Hamilton/Hallmark
tel. 800.841.5197

Kay Elemetrics Corporation
2 Bridgewater Lane
Lincoln Park, NJ 07035
tel. 201.628.6200

Mouser Electronics
tel. 800.346.6873
http://www.mouser.com

Tronser, Incorporated
2763 Route 20 East
Cazenovia, NY 13035
tel. 800.379.2444

Digi-Key
701 Brooks Avenue, South,
P.O. Box 677
Thief River Falls, MN 56701-0677
tel. 800.344.4539
http://www.digikey.com

Jameco
1355 Shoreway Rd.
Belmont, CA 94002-4100
tel. 800.831.4242

Microcraft Corporation
Box 513Q
Thiensville, WI 53092
tel. 414.241.8144

Newark Electronics
tel. 800.463.9275

A.3 PARTS

Problem 2 – Sources

12 V, 0.8 amp-hour battery, Yuasa NP0.8-12, Newark 87F636
510 Ω Resistor (4), Digikey 510QBK-ND
24 AWG solid, tinned, insulated hook-up wire, red and black
3M #306 solderless breadboard, Newark 92F1460
banana patch cords, red and black
extra fuses for batteries

We add a 2.1 mm power jack to the breadboard for supply connections.

Problem 3 – Capacitors

10 nF, 25 V ceramic disk capacitor, Digikey P4300A-ND
300 kΩ resistor, Digikey 300KQBK-ND
BNC to mini-hook adapter (2)
BNC tee
30 inch BNC cable (2)
18 inch BNC cable (2)
10:1 oscilloscope probe

One short BNC cable should be used to connect the function generator trigger
output to the external trigger (sync) input of the oscilloscope. With a BNC tee

at the channel 1 input, use the second short BNC cable and tee to connect the function generator first to the oscilloscope and then to the circuit.

Problem 4 – Diode Detectors

1N4148 diode, Digikey 1N4148CT-ND
3 kΩ resistor, Digikey 3KQBK-ND
10 nF, 25 V ceramic disk capacitor, Digikey P4300A-ND
breadboard
BNC to mini-hook adapter (2)
BNC tee
30 inch BNC cable (2)
18 inch BNC cable (2, 1 sync)

Connect the function generator and oscilloscope as in Problem 3. Connect the function generator output and channel 1 input to the input of the detector. The detector output goes to channel 2.

Problem 5 – Inductors

1 mH inductor, Mouser 43LS103
1N4148 diode, Digikey 1N4148CT-ND
PN2222 npn transistor, Newark
2 kΩ resistor (2), Digikey 2KQBK-ND
breadboard
BNC to mini-hook adapter (2)
BNC tee (2)
BNC 50 Ω load
30 inch BNC cable (2)
18 inch BNC cable (2, 1 sync)
10:1 oscilloscope probe
power supply

The oscilloscope and function generator are connected as in Problem 3, excepting the addition of a tee and 50 Ω load at the channel 2 oscilloscope input.

Problem 6 – Diode Snubbers

As in Problem 5.

Problem 8 – Series Resonance

NorCal 40A printed circuit board
C1, 8–50 pF, 250 VDC ceramic trimmer capacitor, Mouser 24AA024
L1, 15 μH inductor, Mouser 43LS155
BNC to mini-hook adapter (2)

BNC tee
BNC 50 Ω load
30 inch BNC cable (2)
18 inch BNC cable (sync)

It is convenient to hold the circuit board in a vise when installing parts and making some measurements. A suitable unit is the Panavise Type 301, with the Type 308 base, both available from Newark.

Problem 9 – Parallel Resonance

C37, 5 pF, 50 VDC ceramic disk capacitor, Mouser 140-CD50S2-500J
C38, 100 pF, 100 VDC ceramic disk capacitor, Digikey 1313PH-ND
C39, 8–50 pF, 250 VDC ceramic trimmer capacitor, Mouser 24AA024
T37-2 powdered iron toroidal core, Amidon
#28 AWG enameled wire
lighters and sandpaper for stripping wire
BNC to mini-hook adapter
30 inch BNC cable
18 inch BNC cable (sync)
10:1 oscilloscope probe

The enamel should be removed from the ends of the inductor winding by burning and then sanding the wire. Tinning the wires and checking for a shiny, even coating of solder will help avoid poor connections.

Problem 10 – Coaxial Cable

box for measuring voltage and current
BNC tee
BNC union
BNC 50 Ω termination
10 meter BNC cable
18 inch BNC cable (2, 1 sync)
antenna

For the velocity measurement, use the short BNC cables to connect the function generator trigger and output to the oscilloscope, with a tee at the channel 1 input. For the antenna, we use a Lakeview 40-meter Hamstick antenna (tel. 864.226.6990, http://www.hamstick.com). It should be mounted outside the building. Though small, it provides plenty of noise and signal for the lab measurements. We use a ZFSC-10 10:1 power splitter/combiner to connect each bench to an antenna. It is available from the Mini-Circuits Corporation, P.O. Box 350166, Brooklyn, New York 11235-0003 (tel. 718.934.4500, http://www.minicircuits.com). The 10-way splitter allows simultaneous use of the antenna by ten students. BNC cables are run from the splitter to each bench. BNC receptacles with 50 Ω resistors are mounted to each bench to terminate unused antenna lines. When measuring the antenna line length, remember the 10 dB of loss for each pass through the splitter. For the impedance measurement, an adapter is required to allow the oscilloscope to measure voltage and current. This box is described in Figure 4.16. A 1 Ω, quarter-watt resistor is connected across the channel 2 input, in series with the center conductor of the cable coming from the function generator. A transformer is used to avoid the common ground problem. It consists of seven turns of bifilar-wound #30 wire on a 73-mix ferrite bead. The #30 wires should be twisted together before they are wound on the bead. The voltage is measured at the output connector on channel 1.

Problem 12 – Resonance

BNC tee
10 meter BNC cable
18 inch BNC cable (2, 1 sync)
connections as for the first part of Problem 8

Problem 13 – Harmonic Filter

C45, C47, 330 pF, 50 VDC ceramic disk capacitor, Digikey P4030A
C46, 820 pF, 50 VDC ceramic disk capacitor, Digikey P4035A
J1, BNC pc-board-mount jack, Mouser 177-3138
T37-2 powdered-iron toroidal core, Amidon
#26 AWG enamelled wire
BNC to mini-hook adapter
BNC tee
BNC 50 Ω termination
30 inch BNC cable (2)
18 inch BNC cable (sync)

Problem 14 – IF Filter

C9-C13, 270 pF, 50 VDC ceramic disk capacitor, Digikey P4029A-ND
C14, 47 pF, 50 VDC ceramic disk capacitor, Digikey P4452A
L4, 18 μH inductor, Mouser 43LS185
X1-X4, 4.91520 MHz crystal, 20 pF, HC-49 case, Digikey CTX050-ND
plastic crystal spacer (4), Bivar BI-CI-192-028-SR, Electronic Hardware Ltd.
150 Ω resistor, Digikey 150QBK-ND
200 Ω resistor, Digikey 200QBK-ND
#22 AWG tinned wire
BNC to mini-hook adapter (2)
BNC tee

BNC barrel
18 inch BNC cable (2, 1 sync)

The function generator should be connected directly to the 150 Ω input resistor, and the output to channel 1 should use only the BNC mini-hook adapter and BNC barrel – no cable.

Problem 15 – Driver Transformer

R14, 100 Ω resistor, Digikey 100QBK-ND
FT37-43 ferrite toroidal core, Amidon
#26 AWG enamelled wire
200 Ω resistor, Digikey 200QBK-ND
1 kΩ resistor, Digikey 1KQBK-ND
BNC to mini-hook adapter (2)
BNC tee
BNC 50 Ω termination
30 inch BNC cable (2)
18 inch BNC cable (sync)

Problem 16 – Tuned Transformers

C2, 8–50 pF, 250 VDC ceramic trimmer capacitor, Mouser 24AA024
C4, 5 pF, 50 VDC ceramic disk capacitor, Mouser 140-CD50S2-500J
C6, 47 pF, 50 VDC ceramic disk capacitor, Digikey P4452A
FT37-61 ferrite toroidal core (2), Amidon
#22 AWG tinned wire
#26 AWG enamelled wire
#28 AWG enamelled wire
750 Ω resistor, Digikey 750QBK-ND
1.5 kΩ resistor (2), Digikey 1.5KQBK-ND
2.2 kΩ resistor, Digikey 2.2KQBK-ND
BNC to mini-hook adapter
30 inch BNC cable
18 inch BNC cable (sync)
10:1 oscilloscope probe

Problem 17 – Tuned Speaker

2.25 inch round speaker, Jameco 10840
2.5 inch (inside diameter) cardboard mailing tube, 16 cm long
#24 AWG 2-conductor speaker wire, 20 cm long
3.5 mm stereo phone plug, Mouser 17PP004

cork strip
foam blocks (2)
sound level meter
BNC to mini-hook adapter
BNC tee
BNC–banana adapter (to connect multimeter to BNC cable)
BNC-to-3.5 mm stereo phone jack adapter
30 inch BNC cable
18 inch BNC cable
glue gun

Use the short BNC cable to connect the function generator to a BNC tee at the multimeter and then the long cable to connect to the speaker. A BNC-to-3.5 mm stereo phone jack adapter can be fabricated to connect to the speaker. Alternatively, a BNC to mini-hook adapter can be used to clip to the speaker contacts. Hot-melt glue guns are available from craft and hardware stores – most any kind will work. A word of warning: Some do get hot enough to burn fingers, with the gun or the molten glue. Soft foam blocks, approximately 7 by 10 by 10 cm support the speaker and sound level meter. Foam is available from upholstery and craft shops, and they can often cut it to size. A band saw will work well for foam and cardboard tubes if the material is fed slowly. The cork strips are cut from eighth-inch Portuguese cork sheet, 1 cm wide and long enough to fit snugly inside the mailing tube with the ends butted together. A good source for the cork is ABC School Supply, 2990 E. Blue Star Anaheim, CA 92806, tel. 800.498.2990.

Problem 18 – Acoustic Standing-Wave Ratio

Problem 17 equipment
speaker tube extender
phenolic tube with scale
foam block

The extender tube for the speaker is made of the same mailing tube, 30 cm long. Cork and a slightly larger tube can be used to fabricate a slip-on joint. The phenolic tube is sized to slip over the microphone of the sound level meter (one-half inch for the Tenma meter). It is 30 cm long with a paper centimeter scale attached. An additional foam block is needed to support the extension tube.

Problem 19 – Receiver Switch

C3, 47 nF, 25 VDC ceramic disk capacitor, Digikey P4307A-ND
Q1, 2N4124 npn transistor, Digikey 2N4124-ND
R1, 1.8 kΩ resistor, Digikey 1.8KQBK-ND
BNC to mini-hook adapter (2)

BNC 50 Ω termination
BNC tee
30 inch BNC cable (2)
18 inch BNC cable (sync)
banana to mini-hook test leads (2, to power supply)

Problem 20 – Transmitter Switch

C42, 10 μF, 25 VDC electrolytic capacitor, Mouser 140-XRL25V10
C43, C57, 47 nF, 25 VDC ceramic disk capacitor, Digikey P4307A-ND
D7, 1N5817 Schottky diode, Newark
D11, 1N4148/1N914A switching diode, Newark
J2, 2.1 mm coaxial power jack, Mouser 16PJ031
J3, 3.5 mm stereo phone jack, Mouser 161-3500
Q4, 2N3906 pnp transistor, Digikey 2N3906-ND
R9, 47 kΩ resistor, Digikey 47KQBK-ND
R24, 150 kΩ resistor, Digikey 150KQBK-ND
U5, 78L08 voltage regulator, 8 V, 150 mA, Mouser 333-ML78L08A
1 Ω resistor, Digikey 1QBK-ND
30 inch BNC cable
18 inch BNC cable (sync)
10:1 oscilloscope probe
keying relay cable

The relay is a Magnecraft W171DIP-7, Newark part number 47F1142. A BNC jack is connected to the coil, center conductor to pin 2 (note that this is polarity sensitive because of the included diode). The normally open contacts are connected to a mono 3.5 mm phone plug using #24 speaker wire or similar.

Problem 21 – Driver Amplifier

C56, 10 μF, 25 VDC electrolytic capacitor, Mouser 140-XRL25V10
D10, 1N5817 Schottky diode, Newark
Q6, PN2222A npn transistor, Newark
R11, 510 Ω resistor, Digikey 510QBK-ND
R12, 20 Ω resistor, Digikey 20QBK-ND
R13, 500 Ω trimmer potentiometer, Digikey 36C53-ND
BNC to mini-hook adapter (2)
BNC 50 Ω termination
BNC tee
30 inch BNC cable (2)
18 inch BNC cable (sync)
banana to mini-hook test leads (2, to multimeter)
shorting plug, mono 3.5 mm phone plug

Note that only one side of R11 is installed for this lab. The shorting plug is a 3.5 mm mono (two-contact) phone plug with the tabs soldered together. Adding a length of wire or tubing can make it easier to tell a shorting plug from an new plug.

Problem 22 – Emitter Degeneration

As in Problem 21.
10:1 oscilloscope probe

Problem 23 – Buffer Amplifier

C36, 47 nF, 25 VDC ceramic disk capacitor, Digikey P4307A-ND
Q5, J309 JFET, Newark
R10, 510 Ω resistor, Digikey 510QBK-ND
1.5 kΩ resistor, Digikey 1.5KQBK-ND
BNC to mini-hook adapter
30 inch BNC cable
18 inch BNC cable (sync)
10:1 oscilloscope probe
banana to mini-hook test leads (2, to multimeter)
shorting plug

The 1.5 kΩ resistor will be used again in Problem 25.

Problem 24 – Power Amplifier

C44, 47 nF, 25 VDC ceramic disk capacitor, Digikey P4307A-ND
D12, 1N4753A 36 V, 1 W zener diode, 583-1N4753A
Q7, 2N3553 npn transistor, Hamilton/Hallmark
heat sink, Mouser 532-578305B00
plastic transistor spacer, Bivar BI-515-020, Electronic Hardware Ltd.
RFC1, 18 μH inductor, Mouser 43LS185
BNC to mini-hook adapter
BNC 50 Ω termination
BNC tee
30 inch BNC cable (2)
18 inch BNC cable (sync)
10:1 oscilloscope probe
banana to mini-hook test leads (2, to multimeter)

Problem 25 – Thermal Modeling

C48, 10 nF, 25 V ceramic disk capacitor, Digikey P4300A-ND
BNC to mini-hook adapter
BNC 50 Ω termination

BNC tee

30 inch BNC cable (2)

18 inch BNC cable (sync)

banana to mini-hook test leads (2, to multimeter)

keying relay cable

shorting plug

thermometer

heat sink compound

The 1.5 kΩ resistor should remain from Problem 23. A -10 to 110°C thermometer is necessary. VWR Scientific (tel. 800.932.500) catalog number 61067-913 is suitable. Heat sink compound is available from Digikey, part number CT40-5-ND.

Problem 26 – VFO

D8, MVAM108 varactor diode, Hamilton/Hallmark

D9, 1N4148/1N914A switching diode, Newark

C7, 10 nF, 25 V ceramic disk capacitor, Digikey P4300A-ND

C32, 150 pF, 50 VDC ceramic disk capacitor, Digikey P4026A

C49, 47 pF, 50 VDC ceramic disk capacitor, Digikey P4452A

C50, 2–25 pF air-variable capacitor, Tronser 10-1108-25023-000

C51, 390 pF, 50 VDC polystyrene capacitor, Mouser 23PS139

C52, C53, 1200 pF, 50 VDC polystyrene capacitor, Mouser 23PS212

C54, 47 nF, 25 VDC ceramic disk capacitor, Digikey P4307A-ND

Q8, J309 JFET, Newark

R15, 510 Ω resistor, Digikey 510QBK-ND

R17, 10 kΩ potentiometer, Mouser 314-1410-10K

R19, R21, 47 kΩ resistor, Digikey 47KQBK-ND

R20, 4.7 kΩ resistor, Digikey 4.7KQBK-ND

R23, 1.8 kΩ resistor, Digikey 1.8KQBK-ND

RFC2, 1 mH inductor, Mouser 43LS103

T68-7 powdered iron toroidal core (L9), Amidon

#28 AWG enamelled wire (L9)

nylon 0.5 inch, 6-32 round-head screw (L9)

nylon 6-32 nut (L9)

nylon #6 shoulder washer (L9)

BNC to mini-hook adapter (to counter)

30 inch BNC cable (to counter)

10:1 oscilloscope probe

banana to mini-hook test leads (2, to multimeter)

Problem 27 – Gain Limiting

R16, 1 kΩ potentiometer, Mouser 31CW301

U6, LM393N comparator, Digikey LM393N-ND

1.5 kΩ resistor, Digikey 1.5KQBK-ND

BNC to mini-hook adapter
30 inch BNC cable
10:1 oscilloscope probe
banana to mini-hook test leads (2, to multimeter)
shorting plug
thermometer
blow drier
plastic box

A hand-held blow (hair) drier is an effective heat source. For even heating and cooling of the VFO circuit, the board can be placed in a plastic box. The type used for food storage is suitable. Holes may drilled or cut in the box to allow hot air from the blow drier to flow in over one side of the board and exit over the other side. A hole should be drilled to allow proper thermometer placement.

Problem 28 – RF Mixer

C5, 10 nF, 25 V ceramic disk capacitor, Digikey P4300A-ND
C8, 47 nF, 25 VDC ceramic disk capacitor, Digikey P4307A-ND
C15, 2.2 μF, 25 VDC electrolytic capacitor, Mouser 140-XRL25V2.2
R2, 1 kΩ potentiometer, Mouser 31CW301
U1, U2, SA602AN mixer IC, Newark
30 inch BNC cable
10:1 oscilloscope probe

Problem 29 – Product Detector

C17, 8–50 pF, 250 VDC ceramic trimmer capacitor, Mouser 24AA024
C18, 270 pF, 50 VDC ceramic disk capacitor, Digikey P4029A-ND
X5, 4.91520 MHz crystal, 20 pF, HC-49 case, Digikey CTX050-ND
plastic crystal spacer, Bivar BI-CI-192-028-SR, Electronic Hardware Ltd.
3 kΩ resistor, Digikey 3KQBK-ND
BNC to mini-hook adapter
BNC barrel adapter
10:1 oscilloscope probe
banana to mini-hook test leads (2, to multimeter)
thermometer
blow drier
plastic box

Problem 30 – Transmit Mixer

C31, 5 pF, 50 VDC ceramic disk capacitor, Mouser 140-CD50S2-500J
C33, 47 nF, 25 VDC ceramic disk capacitor, Digikey P4307A-ND
C34, 8–50 pF, 250 VDC ceramic trimmer capacitor, Mouser 24AA024
C35, 270 pF, 50 VDC ceramic disk capacitor, Digikey P4029A-ND

L5, 18 μH inductor, Mouser 43LS185
U4, SA602AN mixer IC, Newark
X6, 4.91520 MHz crystal, 20 pF, HC-49 case, Digikey CTX050-ND
plastic crystal spacer, Bivar BI-CI-192-028-SR, Electronic Hardware Ltd.
BNC to mini-hook adapter
BNC 50 Ω termination
BNC tee
BNC barrel
30 inch BNC cable (2)
18 inch BNC cable (sync)
10:1 oscilloscope probe
keying relay cable
shorting plug

Problem 31 – Audio Amplifier

C20, C21, 100 nF, 100 VDC mylar capacitor, Mouser 140-PM2A104K
C22, C55, 10 nF, 25 V ceramic disk capacitor, Digikey P4300A-ND
C23, 2.2 μF, 25 VDC electrolytic capacitor, Mouser 140-XRL25V2.2
C27, C41, 100 μF, 25 VDC electrolytic capacitor
R7, 47 kΩ resistor, Digikey 47KQBK-ND
R22, 1.8 kΩ resistor, Digikey 1.8KQBK-ND
U3, LM386N-1 audio amplifier IC, Digikey LM386N-1-ND
5.6 Ω resistor, Digikey 5.6QBK-ND
8 Ω resistor, Digikey 8QBK-ND
1.5 kΩ resistor (2), Digikey 1.5KQBK-ND
BNC to mini-hook adapter (2)
30 inch BNC cable (2)
18 inch BNC cable (sync)
banana to mini-hook test leads (2, to multimeter)

Problem 32 – Automatic Gain Control

C29, 10 μF, 25 VDC electrolytic capacitor, Mouser 140-XRL25V10
C30, 2.2 μF, 25 VDC electrolytic capacitor, Mouser 140-XRL25V2.2
D5, D6, 1N5817 Schottky diode, Newark
Q2, Q3, J309 JFET, Newark
R5, 2.2 MΩ resistor network, Bourns 4608X-102-225, Newark
R6, 10 kΩ trimmer potentiometer, Mouser 36C14-ND
300 kΩ resistor, Digikey 300KQBK-ND
BNC to mini-hook adapter (2)
30 inch BNC cable (2)
18 inch BNC cable (sync)
banana to mini-hook test leads (2, to multimeter)

Problem 33 – Alignment

C19, 10 nF, 25 V ceramic disk capacitor, Digikey P4300A-ND
C26, 10 μF, 25 VDC electrolytic capacitor, Mouser 140-XRL25V10
C28, 100 nF, 100 VDC mylar capacitor, Mouser 140-PM2A104K
D1, D2, D3, D4, 1N4148/1N914A switching diode, Newark
J4, 3.5 mm stereo phone jack, Mouser 161-3500
R3, 150 kΩ resistor, Digikey 150KQBK-ND
R4, 8.2 MΩ resistor, Digikey 8.2MQBK-ND
R8, 500 Ω trimmer potentiometer, Digikey 36C53-ND
attenuator
BNC to mini-hook adapter (2)
BNC 50 Ω termination
BNC tee
30 inch BNC cable (3)
18 inch BNC cable (attenuator)
banana to mini-hook test leads (2, to multimeter)
keying switch

The keying switch can be any SPST switch wired to a 3.5 mm phone plug. A miniature toggle switch (Mouser 1055-TA1120 is suitable) is convenient and can be attached to a short section of wire with heat-shrink tubing.

Problem 34 – Receiver Response

attenuator
BNC to mini-hook adapter
BNC 50 Ω termination
BNC tee
BNC barrel
BNC to banana plug adapter
30 inch BNC cable (2)
18 inch BNC cable (2)
banana to mini-hook test leads (2, to multimeter)
keying switch
battery
antenna

For the Part A, the BNC to mini-hook adapter can be clipped to the speaker contacts, with a cable to a BNC tee at the multimeter (with BNC to banana plug adapter) and another cable running from the tee to the counter. Test leads can also be used for the multimeter. Substituting a BNC barrel for the short BNC cable between the source transceiver and attenuator may help reduce leakage. See Problem 10 for a description of the antenna.

Problem 35 – Intermodulation

attenuator
BNC 50 Ω termination
BNC tee
BNC barrel
30 inch BNC cable
18 inch BNC cable (2)
combiner
battery (2)

We use a ZFSC-2-4 2:1 power splitter/combiner for the intermodulation measurements. It is available from the Mini-Circuits Corporation, P.O. Box 350166, Brooklyn, New York 11235-0003, tel. 718.934.4500, http://www.minicircuits.com.

Problem 36 – Demonstration

Equipment for the demonstration requires some thought. Because of leakage around attenuators, it is difficult to provide a signal small enough for a reasonable MDS test, but large enough to measure on a counter. Using a completed transceiver as a source provides the proper range of frequencies. If the transmitter power is reduced until just measurable on a counter, a following attenuator should provide a reasonable test signal. To measure the frequency of the transceiver being demonstrated, the attenuation can be reduced until the counter reading is stable. An oscilloscope can be used for power measurement, or a power meter can be added at the output of the tested transceiver. The Diamond SX-200 power meter is quite suitable; it is available from Ham Radio Outlet, tel. 800.854.6046, http://www.hamradio.com.

Problem 39 – Listening

antenna
Microcraft Code Scanner for decoding Morse Code and radio teletype

It may be useful to provide cassette tape recorders to provide samples of Morse Code and to record received messages. The Code Scanner is designed to receive at 800 Hz and must be modified to operate at the frequency we use in our measurements, 620 Hz. The following resistor changes shift the frequency of the active filters.

Change R14 and R18 from 150 kΩ to 200 kΩ.
Change R15, R19, and R24 from 6.8 kΩ to 8.2 kΩ.
Change R16, R20, and R25 from 1.3 MΩ to 2 MΩ.
Change R17, R21, and R26 from 4.7 MΩ to 6.8 MΩ.
Change R23 from 100 kΩ to 150 kΩ.

Fourier Series

We can interpret the waveforms we see in terms of components at different frequencies. For example, we might consider a voltage $V(t)$ that is the sum of three cosine components:

$$V(t) = a_1 \cos(\omega_1 t) + a_2 \cos(\omega_2 t) + a_3 \cos(\omega_3 t). \tag{B.1}$$

We can include a DC term as a special case if we assume that one of the frequencies is zero. Representing a function in terms of its frequency components is helpful in understanding how the filter in a Class-C or Class-D amplifier operates. We also use the frequency components to define the relationship between DC and AC currents in oscillators and to predict mixer output frequencies.

B.1 FOURIER COEFFICIENTS

If a function is periodic, it has a special representation called a Fourier series, where each frequency component is a harmonic of the fundamental frequency. We will leave the discussion of why a function can be written in a Fourier series to a mathematics text, but we will see how to find the coefficients. We start by writing the function as an infinite sum of cosines and sines:

$$V(t) = a_0 + a_1 \cos(\omega t) + b_1 \sin(\omega t) + a_2 \cos(2\omega t) + b_2 \sin(2\omega t) + \cdots, \tag{B.2}$$

where a_0, a_1, b_1, a_2, and b_2 are the *Fourier coefficients*. For our functions we can simplify this sum. Notice that if we change the sign of t, the cosine terms do not change. We say that the cosine is an *even* function. In contrast, the sine is an *odd* function, and it changes sign. It turns out that the functions that we are interested in are even, and for this reason we do not need the sine terms. This means that we can write our series as

$$V(t) = \sum_{n=0}^{\infty} a_n \cos(n\omega t). \tag{B.3}$$

Now we can find a formula for the Fourier coefficients. To do this we need to use a remarkable property of the integrals of cosine products. Let us consider the integral I_{mn}, where

$$I_{mn} = \int_{-T/2}^{+T/2} \cos(m\omega t) \cos(n\omega t) \, dt, \tag{B.4}$$

where T is the period, given by

$$T = 2\pi/\omega. \tag{B.5}$$

There are several possibilities. If m and n are 0, the integrand is just 1, and the integral is T. If m and n are positive, and $m = n$, the integrand is $\cos^2(m\omega t)$, and the integral is $T/2$. If $m \neq n$, then we rewrite the product as a sum of cosines

$$I_{mn} = \frac{1}{2} \int_{-T/2}^{+T/2} (\cos[(m+n)\omega t] + \cos[(m-n)\omega t]) \, dt. \tag{B.6}$$

The integrals of cosines are sine functions. These are periodic, also, so that they have the same value at each limit, and the integrals will vanish. This means that if we integrate the product of two different harmonics over a period, the integral is zero. We say that different harmonics are *orthogonal*. We can summarize these results as follows:

$$I_{mn} = \begin{cases} T & \text{for } m = n = 0, \\ T/2 & \text{for } m = n > 0, \\ 0 & \text{for } m \neq n. \end{cases} \tag{B.7}$$

Now consider the integral V_n of the function $V(t)$, defined by

$$V_n = \int_{-T/2}^{+T/2} V(t) \cos(n\omega t) \, dt. \tag{B.8}$$

We substitute for $V(t)$ from Equation B.3 to obtain

$$V_n = \int_{-T/2}^{+T/2} \left(\sum_m a_m \cos(m\omega t) \right) \cos(n\omega t) \, dt. \tag{B.9}$$

I have used m as the index in the sum rather than n to keep them distinct. Let us bring the cosine factor $\cos(n\omega t)$ inside the sum. This gives us

$$V_n = \int_{-T/2}^{+T/2} \left(\sum_m a_m \cos(m\omega t) \cos(n\omega t) \right) dt. \tag{B.10}$$

We can do this integral by taking the integral of each of the terms in the sum separately and adding. We write

$$V_n = \sum_m a_m \left(\int_{-T/2}^{+T/2} \cos(m\omega t) \cos(n\omega t) \, dt \right) = \sum_m a_m I_{mn}. \tag{B.11}$$

The integrals are zero except where $m = n$. This means that we are only left with one term in the sum, given by

$$V_n = \begin{cases} Ta_0 & \text{if } n = 0, \\ Ta_n/2 & \text{if } n > 0. \end{cases} \tag{B.12}$$

I have replaced a_m with a_n, because $m = n$ for this term. We can invert this to find

$$a_0 = \frac{1}{T} \int_{-T/2}^{+T/2} V(t) \, dt. \tag{B.13}$$

This is the DC component, and we can think if it as just the average value of $V(t)$. For $n > 0$, we have

$$a_n = \frac{2}{T} \int_{-T/2}^{+T/2} V(t) \cos(n\omega t)\, dt. \tag{B.14}$$

Doing the integrals to find the Fourier coefficients requires practice. You should fill in the details of the following examples.

B.2 SQUARE WAVE

Now we find the Fourier coefficients for a square wave with voltages of $+1$ and -1 (Figure B.1a). The average value is zero, and so there is no DC component. The AC components are given by

$$a_n = \frac{2}{T} \int_{-T/2}^{+T/2} V(t) \cos(n\omega t)\, dt. \tag{B.15}$$

If n is even, the integral over the time the square wave is positive is zero, and so is the integral over the time that the square wave is negative. This means a_n is zero if n is even. If n is odd, the integrals over the positive part of the square wave are the same as the integrals over the positive part. You should sketch the cosines and the square wave to convince yourself of this. This means that we can write

$$a_n = \frac{4}{T} \left[\frac{\sin(n\omega t)}{n\omega} \right]_{-T/4}^{T/4} = \frac{2}{\pi} \left[\frac{\sin(n\omega t)}{n} \right]_{-T/4}^{T/4}. \tag{B.16}$$

The sines evaluate to $+1$ or -1, and we can write the first four coefficients as

$$a_1 = +\frac{4}{\pi}, \tag{B.17}$$

$$a_3 = -\frac{4}{3\pi}, \tag{B.18}$$

$$a_5 = +\frac{4}{5\pi}, \tag{B.19}$$

$$a_7 = -\frac{4}{7\pi}. \tag{B.20}$$

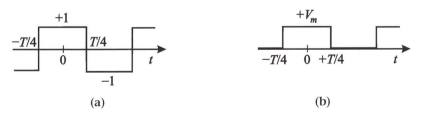

Figure B.1. Square waves for Fourier analysis (a), and pulse train with rectangular pulses with 50% duty cycle (b).

There are only odd harmonics, and the coefficients alternate in sign. The coefficients decrease as $1/n$. We can write the Fourier series for the square wave as

$$V(t) = \frac{4}{\pi}\left(\cos(\omega t) - \frac{\cos(3\omega t)}{3} + \frac{\cos(5\omega t)}{5} - \cdots\right). \tag{B.21}$$

We use these coefficients for studying mixers in Chapter 12.

We can also use this series to deduce the coefficients for rectangular pulses whose width is half the period (Figure B.1b). We say these pulses have a *duty cycle* of 50%. The DC component is half the pulse height V_m. The other components are the same as for the square wave, except that they need to be multiplied by $V_m/2$ to take the pulse height into account. This means that we can write the series for the rectangular pulses with a 50% duty cycle as

$$V(t) = \frac{V_m}{2} + \frac{2V_m}{\pi}\left(\cos(\omega t) - \frac{\cos(3\omega t)}{3} + \frac{\cos(5\omega t)}{5} - \cdots\right). \tag{B.22}$$

We use this series in analyzing a Class-D amplifier in Chapter 10 and in finding the channel spacing for pulsed transmissions in Chapter 12.

B.3 RECTIFIED COSINE

The voltage for the Class-C amplifier that we study in Chapter 10 looks like a rectified cosine (Figure B.2a). We use the first two terms of the Fourier series to find the relationship between the AC and DC components. We can write the DC term as

$$a_0 = \frac{1}{T}\int_{-T/4}^{T/4} V_m \cos(\omega t)\, dt = V_m/\pi \tag{B.23}$$

and the fundamental component as

$$a_1 = \frac{2}{T}\int_{-T/4}^{T/4} V_m \cos^2(\omega t)\, dt = V_m/2. \tag{B.24}$$

The fundamental frequency component is half of the original cosine. This makes sense, because the cosine is only on during half the cycle. The harmonics are

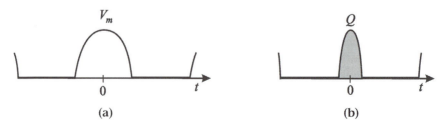

Figure B.2. Rectified cosine for Fourier analysis (a), and narrow pulses (b).

given by

$$a_n = \frac{2}{T} \int_{-T/4}^{T/4} V_m \cos(\omega t) \cos(n\omega t) \, dt$$

$$= \frac{V_m}{T} \int_{-T/4}^{T/4} (\cos[(n-1)\omega t] + \cos[(n+1)\omega t]) \, dt. \tag{B.25}$$

If n is odd and greater than 1, the integrals are zero. It may help to sketch the cosines to see this. If n is even, we can write

$$a_n = \frac{V_m}{2\pi} \left[\frac{\sin[(n-1)\omega t]}{n-1} + \frac{\sin[(n+1)\omega t]}{n+1} \right]_{-T/4}^{T/4}. \tag{B.26}$$

The sines evaluate to $+1$ or -1, and we can write the first four harmonics as

$$a_2 = +\frac{2V_m}{3\pi}, \tag{B.27}$$

$$a_4 = -\frac{2V_m}{15\pi}, \tag{B.28}$$

$$a_6 = +\frac{2V_m}{35\pi}, \tag{B.29}$$

$$a_8 = -\frac{2V_m}{63\pi}. \tag{B.30}$$

There are only even harmonics, and the coefficients alternate in sign. The coefficients decrease as $1/(n^2 - 1)$. We can write the series as

$$V(t) = \frac{V_m}{\pi} + \frac{V_m}{2} \cos(\omega t) + \frac{2V_m}{\pi} \left(\frac{\cos(2\omega t)}{3} - \frac{\cos(4\omega t)}{15} + \frac{\cos(6\omega t)}{35} - \cdots \right). \tag{B.31}$$

B.4 NARROW PULSES

As a final example, let us find the Fourier coefficients of narrow pulses of current (Figure B.2b). In Chapter 11, we use these coefficients to find the large-signal transconductance of a JFET and the output voltage in bipolar oscillators. We will let the total charge in each pulse be Q. We will assume that the pulses are narrow enough so that, for the harmonics we are interested in, the cosine will be equal to one over the entire pulse. We can write the DC term as

$$a_0 = \frac{1}{T} \int_{-T/2}^{+T/2} I(t) \, dt = Qf \tag{B.32}$$

and the AC terms as

$$a_n = \frac{2}{T} \int_{-T/2}^{+T/2} I(t) \cos(n\omega t) \, dt = 2Qf. \tag{B.33}$$

We write the series as

$$I(t) = Qf(1 + 2(\cos(\omega t) + \cos(2\omega t) + \cos(3\omega t) + \cdots)). \tag{B.34}$$

All harmonic components are present, and the coefficients are all the same. This is an idealization, and in practice the higher-order harmonics will begin to drop off when the pulse width is no longer narrower than the cosine. In the oscillators in Chapter 11, we compare the DC component a_0 and the fundamental AC component a_1. We can see that

$$a_1/a_0 = 2. \tag{B.35}$$

In words, the peak value of the fundamental component is twice the DC component. In measurements, we would likely use the peak-to-peak value of the fundamental, and this is *four* times the DC component.

FURTHER READING

The idea of representing functions in terms of frequency components is a central theme in electrical engineering, mathematics, and physics. These are called spectral representations, and they may be in the form of a series such as the Fourier series, or an integral such as the Laplace transform, or a combination of the two. *Signals and Systems,* by Oppenheim and Willsky, published by Prentice-Hall, gives a good overview of these series and transforms.

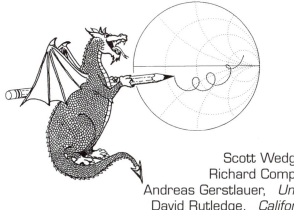

Puff 2.1

Scott Wedge, *Tanner Research, Inc.*
Richard Compton, *Lucent Technologies*
Andreas Gerstlauer, *University of California, Irvine*
David Rutledge, *California Institute of Technology*

Puff is a circuit simulator for linear circuits. It calculates scattering parameters and makes microstrip and stripline layouts. It also makes time-domain plots. The program is named after the magic dragon in the song by the popular American singing group Peter, Paul and Mary. *Puff* originated as a teaching tool for Caltech's microwave circuits course. It was created as an inexpensive and simple-to-use alternative to professional software whose high costs, copy protection schemes, and training requirements create difficulties in the academic environment. *Puff* uses a simple interactive schematic-capture type environment. After a circuit is laid out on the screen using cursor keys, a frequency or time domain analysis is available with a few keystrokes. This process is faster than using net lists, and errors are rare since the circuit is always visible on the screen. Intended for students and researchers, public distribution of the program began in 1987. *Puff* use, originally limited to Caltech, UCLA, and Cornell, has since spread to many other universities and colleges. The program has also become popular with working engineers, scientists, and amateur radio operators. Over 20,000 copies of versions 1.0, 1.5, 2.0, and 2.1 have been distributed worldwide, and translations have been made to Russian, Polish, and Japanese.

In this book, *Puff* is used for analyzing the filter and tuned transformer circuits in Chapters 5 and 6. Here we give an introduction to the use of the program, using the filters in Figures 5.4b and 5.8a as examples. Readers who are interested in more information about the program and its use in microwave circuits can get information about purchasing the complete manual at cost from our web site at http://www.its.caltech.edu/~mmic/puff.html. Lecture notes that give examples for using *Puff* in microwave circuits are also available at cost. You may write us at puff@caltech.edu or alternatively, write to Puff Distribution, MS 136-93, Caltech, Pasadena, CA, 91125.

C.1 WINDOWS

Puff is a DOS program, and for a computer running Windows95, *Puff* runs best in the Restart MS-DOS mode. Click on Restart in MS-DOS mode as you shut down windows. There is an alternate way to run *Puff* using your old version of DOS if you installed Windows95 as an upgrade: Boot your computer. When it is almost finished booting and just about ready to open the Windows95 page (timing is essential), push the *F8* button. This will give you a screen with choices. You should choose to run your previous version of DOS. Then follow the instructions in the next section.

Puff will run in the MS-DOS Prompt window within Windows95, but the program runs many times slower than under DOS. To preserve compatibility with MS-DOS programs, Windows95 allows the user to alter certain settings before launching the program. Older versions of Windows achieved this by the use of a PIF, or program information file. A PIF was required to run each DOS program. Windows95 condenses this feature into the shortcut links that are used to launch applications.

Two problems exist when using Puff with Windows95. The Print Screen button in Windows95 is configured to paste a copy of the screen to the clipboard instead of sending a screen dump to the printer. The shortcut link will have to altered to return the Print Screen key back to its original function. Also, the proper screen dump routine needs to be executed before *Puff* is started up. Windows95 allows the user to run a short program, called a batch file, before starting up the main application. The shortcut link should be configured to have the screen dump routine listed as its batch file.

To set up the shortcut link to *Puff*, follow these steps:

1. Using Windows Explorer, go to the directory that *Puff* is located in. Click on the `puff.exe` file using the right mouse button and select Create Shortcut. This will place a shortcut link to the executable in the *Puff* directory.
2. Now right click on this shortcut and select *Properties*. This will bring up the dialog window that allows you to alter the MS-DOS settings for the shortcut link.
3. Click on the *Program* tab near the top of the window. Go to the entry for *Batch file:* and type in the name for the required screen dump routine. Printers that are compatible with the HP Laserjets can use the routine named `vga2lasr.com`.
4. Make sure the *Close on exit* box is checked at the bottom.
5. Click on the *Screen* tab at the top. In the *Usage* section, select *Full-screen*. *Puff* doesn't run within a window.
6. Click on the *Misc* tab. In the *Windows shortcut keys* section, make sure that *PrtSc* should not be checked. When this box is checked, pressing the Print Screen button will dump an image of the screen to Window's clipboard instead of directing the output to the printer port.
7. Click on the *OK* box to save the settings and close the properties dialog window. *Puff* can now be run by simply clicking on this shortcut link. The screen dump

routine that was entered as the batch file will run before *Puff* is started, and the Print Screen key will allow you to send a screen dump to the LPT1 printer.

Some computers require that you follow this additional instruction: Click the right mouse button on the Puff.exe shortcut. Click on *Properties*; click on *Advanced*. This will open the Advanced properties window. Check the box *Prevent MS DOS programs from detecting Windows*. By default, the *Suggest MS DOS mode as necessary* box should also be checked.

To run *Puff* inside Windows NT, run the program in a full-screen DOS box. Some users also partition the hard drive and install DOS as the operating system on one of the partitions. When you boot your computer, the boot manager will ask you which operating system you wish to use. You must reboot when you switch between operating systems.

In Windows NT, the *PrtSc* key can be used to send an image to the clipboard when *Puff* is running in a window. *Alt Enter* will reduce the full screen to a window. You can paste the clipboard image into MSPaint or MS Photo Editor, and crop and invert the colors before printing.

C.2 GETTING STARTED

To run *Puff*, you need an IBM PC-compatible, running DOS 3.0 or later. In order to make a hard copy, you will need a screen-dump routine for your display and printer. Many printers are compatible with HP LaserJet printers. You should run the program `vgalasr.com` before you start *Puff*. A screen dump will then be possible using the *PrtSc* key. Alternatively, later DOS versions allow a graphics screen dump configured for your printer. For example, if your printer is compatible with an HP Deskjet, you can type `graphics DESKJET` to set up the screen dump.

To run the program, copy the files into a *Puff* directory on your hard disk. Put the copy in the logged disk drive, or with a hard disk, change to the *Puff* directory. Type `puff` and the return key. The computer will first display a system detect screen with a copyright notice and information concerning the hardware that *Puff* recognizes. After typing another key, the circuit file `setup.puf` will be loaded. *Puff* will automatically detect the type of display in use and set up a screen similar to Figure C.1. If something is wrong with the screen, edit the `setup.puf` file, listed in Table C.1, to try a different display type and override automatic selection. If it still does not work, check for the necessary graphics hardware. Once *Puff* has loaded, the word `setup` will appear on the screen to indicate that this file was used to start. You can specify a different starting file, say `yourfile.puf`, by typing `puff yourfile` when you start the program. *Puff* will add the extension `.puf` if none is given. If you specify a filename other than `setup` when starting, *Puff* will not show the system detect screen.

There are several files on the *Puff* diskette in addition to the screen dump programs. The program file `puff.exe` and overlay file `puff.ovr` must be present. In

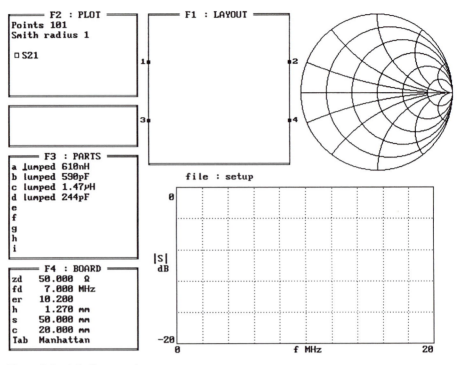

Figure C.1. A *Puff* screen dump.

addition, the setup.puf file should be available. Device files have the extension .dev, and there are several on your diskette, including a Fujitsu FHX04 HEMT transistor, fhx04.dev, a voltmeter, vmeter.dev, a voltage source, vsource.dev, a voltage-controlled current source, cs.dev, a unity gain differential amplifier, da.dev, and an operational amplifier with an open loop gain of 10,000, op.dev.

The *Puff* screen is subdivided into *windows*. Movement from window to window is accomplished by pressing function keys *F1* through *F4*. These keys are hot: They are always active. Other hot keys include *Esc*, which allows you to exit *Puff* and return to the operating system, *F10*, which toggles a small help window, and the cursor keys. At top center is the *Layout* window, where the circuit is constructed. This window is reached by typing the *F1* key. The numbers around the side of the circuit represent external connectors, or ports. There are two connectors on the left side, placed symmetrically above and below the center, and two on the right. The network appearing in the *Layout* window is analyzed from the *Plot* window. Located in the upper left corner of the screen, the *Plot* window is reached by typing *F2*. Here the scattering parameters are specified and their numerical values displayed. The rectangular and Smith chart plots are also controlled from the *Plot* window. The rectangular plot gives either a log-magnitude plot in the frequency domain or a linear impulse or step response in the time domain. The Smith chart, located at top right, gives a polar plot of the scattering coefficients. The circles inside the Smith chart are curves of constant resistance and reactance. Just below the *Plot* window, in the left center of the screen, is a three-line *Message*

Table C.1. Listing of `setup.puf`.

```
\b{oard} {setup.puf file for PUFF, version Electronics of Radio}
d   0    {display: 0 VGA or PUFF chooses, 1 EGA, 2 CGA, 3 Mono}
o   1    {artwork output: 0 dot-matrix, 1 LaserJet, 2 HPGL file}
t   2    {type: 0 microstrip, 1 stripline, 2 Manhattan}
zd    50.000 Ohms {normalizing impedance. zd>0}
fd     7.000 MHz  {design frequency. fd>0}
er    10.200      {dielectric constant. er>0}
h      1.270 mm   {dielectric thickness. h>0}
s     50.000 mm   {circuit-board side length. s>0}
c     20.000 mm   {connector separation. c>=0}
r      0.200 mm   {circuit resolution, r>0, Um=micrometers}
a      0.000 mm   {artwork width correction.}
mt     0.010 mm   {metal thickness, Um=micrometers.}
sr     0.000 Um   {metal surface roughness.}
lt     0.0E+00    {dielectric loss tangent.}
cd     5.8E+07    {conductivity of metal in mhos/meter.}
p      5.000      {photo reduction ratio. p<=203.2mm/s}
m      0.600      {mitering fraction. 0<=m<1}
\k{ey for plot window}
du    0   {upper dB-axis limit}
dl  -20   {lower dB-axis limit}
fl    0   {lower frequency limit. fl>=0}
fu   20   {upper frequency limit. fu>fl}
pts 101   {number of points, positive integer}
sr    1   {Smith-chart radius. sr>0}
S    21   {subscripts must be 1, 2, 3, or 4}
\p{arts window} {0=Ohms, D=degrees, U=micro, |=parallel}
lumped 610nH
lumped 590pF
lumped 1.47UH
lumped 244pF
```

Note: It is safest to edit a copy of this file, rather than the original. Be careful when you edit; if some of the board parameters are missing, the program will abort. Comments may be added at the end of the lines in braces, but they should not extend into the next line.

box, where error messages and requests for file names appear. Below the *Message* box is the *Parts* window, reached by typing *F3*. This window gives the current list of parts that may appear in the circuit. This list can only be edited from the *Parts* window. The equivalent physical dimensions of the *Layout* window are specified in the *Board* window, accessed by typing *F4*. The parameters given here also influence individual component dimensions. The two most important parameters are the normalizing impedance `zd`, and the design frequency `fd`. The normalizing impedance is used to calculate the scattering parameters. The design frequency is used to calculate the electrical length of transmission lines. Typing *F10* in any of the windows will bring up a small help screen. This screen will usually list the commands that are active for the current window. When in the *Board* window, however, *F10* will provide an explanation of the board parameters.

Maneuver through the *Puff* windows by pressing function keys *F1* through *F4*. In each of the windows, test the effects of typing and retyping the *F10* and *Tab* keys. When done, return *Puff* to its original state.

The spacing of the connectors in the *Layout* window is controlled by parameter c in the *Board* window. Hit *F4* to move to the *Board* window. Use the cursor keys and the number keys at the top of your keyboard to set c to zero. Type *F1* and see what happens to the connectors. When done, return c to its original value.

C.3 SCATTERING PARAMETERS

For computer simulations with *Puff*, we need to extend the idea of reflection and transmission coefficients to filters by defining scattering parameters. The voltage waves going into the filter are called *incident* waves, and we use a plus subscript for them. The voltage waves leaving the filter are called *scattered* waves, and we use a minus subscript for them. We call them incident and scattered waves instead of forward and reverse waves because it is not always clear which direction is forward. The input and output connections are called *ports*. *Puff* allows up to four ports in a simulation. This allows two filters to be run at the same time for comparison. The ports are indicated by the subscripts 1 to 4, and they are all given the same resistance. In *Puff*, the port resistance is equal to the parameter zd in the F4 *Board* window. We can use any of the ports as an input or output. For example, for a filter, we might use port 1 for the input and port 2 for the output. In Figure C.2, the input wave is V_{+1}, the reflected wave at the input is V_{-1}, and the output wave is V_{-2}. We can write the input and output powers in terms of these voltage waves. The input power P_+ is given by

$$P_+ = |V_{+1}|^2/(2R_s), \tag{C.1}$$

where R_s is the source resistance. This is the available power from the source. The reflected power P_- is given by

$$P_- = |V_{-1}|^2/(2R_s) \tag{C.2}$$

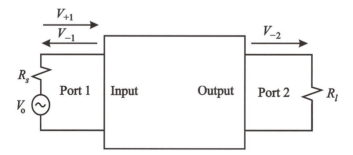

Figure C.2. A filter circuit with a source at the input (Port 1) and a load at the output (Port 2).

and the load power P is

$$P = |V_{-2}|^2/(2R_l), \tag{C.3}$$

where R_l is the load resistance.

The scattering parameters are complex quantities, and we have separate definitions for the magnitudes and phases. We write the magnitudes as

$$|s_{11}|^2 = P_-/P_+, \tag{C.4}$$
$$|s_{21}|^2 = P/P_+ \tag{C.5}$$

and the phases as

$$\angle s_{11} = \angle V_{-1} - \angle V_{+1}, \tag{C.6}$$
$$\angle s_{21} = \angle V_{-2} - \angle V_{+1}, \tag{C.7}$$

where s is a *scattering parameter,* and the subscripts are port numbers. Scattering parameters with equal indices like s_{ii} are reflection coefficients, and those like s_{ji}, where i and j are different port numbers, are transmission coefficients from port i to port j. $|s|^2$ gives the proportion of power transmitted or reflected, while $\angle s$ gives the phase difference between a scattered voltage wave and an incident voltage wave. *Puff* makes two plots of the scattering parameters. One is a Smith-chart plot in the complex plane. The other is a dB plot, given as

$$|s_{11}| = 10 \log(P_-/P_+)\,\text{dB}, \tag{C.8}$$
$$|s_{21}| = 10 \log(P/P_+)\,\text{dB}. \tag{C.9}$$

People use the same letter s whether it is expressed in dB or as a complex number. This makes it important to include the dB units, particularly since the complex scattering parameter has no units itself. Expressed in dB, $|s_{21}|$ is just the power gain, although we will say loss if it is negative.

C.4 EXAMPLES

As an example, we analyze the filter in Figure 5.4b. We begin with the screen in Figure C.1. Hit the *F3* key to make the *Parts* window active. This is indicated by the flashing cursor for part a, and the highlighted *F3* at the top of the window. Now hit *F1* to move to the *Layout* window. A large white × will appear in the center of the circuit board, and the first line in the *Parts* window will become highlighted. Part a is a lumped 610-nH inductor. Push ←, and *Puff* will draw a hollow blue rectangle, labeled a, to the left, and the × will move to the other end of the inductor. In the *Message* box, Δx –5.00mm will appear. This shows you the change in the x coordinate. When no length is given for a lumped part, *Puff* assumes it to be one tenth the size of the layout. Type 1 to connect to the first port. You should use the number key at the top of the keyboard rather than the numeric keypad. In fact, the *NumLock* key is disabled when *Puff* runs, to avoid mix ups between arrows and digits. *Puff* will make a gray outline path up and to the left to the first connector. Push →, and *Puff* will move back to the other end of the inductor. Type b to select the capacitor, and ↓ to draw it. Now hit the =

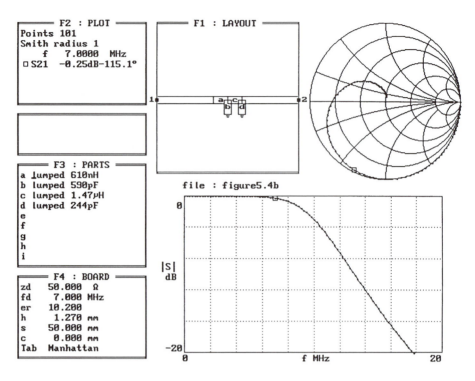

Figure C.3. Analyzing the filter in Figure 5.4b in *Puff*.

key to ground the end of the capacitor and type ↑ to move back to the other end. Without the ground, *Puff* thinks that the end of the capacitor is open-circuited. Now type c to select the second inductor, and → to draw it. Type 2 to connect to the second port. Type d to select the second capacitor, and ↓ to draw it. Hit the = key to ground the end of the capacitor. This completes the filter in Figure 5.4b.

When you type a key, *Puff* first checks for a valid keystroke. If it is not, *Puff* will beep and do nothing else. You can press z to hear the beep. Next, *Puff* checks to see if it can carry out the command. If it cannot, it will give an error message in the red *Message* box. For example, if you retype 1 in the present circuit, *Puff* will say, Port 1 is already joined, because multiple connections to the same port are not allowed.

This circuit is ready to be analyzed. Push *F2* to go to the *Plot* window, and type p to plot. The calculated s_{21} values will appear in complex form on a Smith chart as small dots joined by a cubic spline curve. A square marker indicates the transmission coefficient at the design frequency, and the numerical values are given in the *Plot* window. Any magnitude greater than 100 dB will be reported as ∞, and any magnitude as small as −100 dB will be listed as zero. When *Puff* finishes, the *Message* box indicates the time required for the calculation. The rectangular graph shows the loss of the filter. It is a low-pass filter with 3-dB frequency of 10 MHz. You can see this by pushing the *PageUp* key repeatedly to move the cursor to higher frequencies until the loss listed in the *Plot* window reaches 3 dB.

To plot the reflection coefficient, push *F2* to go to the *Plot* window. Use the ↑ and ↓ keys to move to the line below S21. For the input reflection coefficient, s_{11}, type 11. Now push p to plot the reflection and transmission coefficients. To find the input impedance, use the ↑ and ↓ keys to move to the S11 line, and type =. The real and imaginary parts of the input impedance, and the equivalent inductance or capacitance, will appear in the message box below the plot window.

Go into the *Layout* window and make sure that the × is at the right of part c. Hold down the shift key and type ←. Such a shift-cursor operation in the direction of a part will cause it to be erased. Hold down the shift key and type 1 to erase the path to the connector. What happens if the shift-cursor operation is not in the direction of a part?

Erase the entire layout by typing *Ctrl*-e from either the *Parts* or *Layout* window. Now we can analyze the band-pass filter in Figure 5.8a. Type *F3* to return to the *Parts* window. Use the *Delete* key and backspace keys to erase the part descriptions. Then type the following on line *a*:

    l 15μH + 34.5pF

The character μ is typed as *Alt*-m. The plus sign indicates a series connection. Any sequence of letters without numbers that starts with l and ends with a space indicates a lumped part. Now type the following on line *b*:

    l 86nH ∥ 6.0nF

The character ∥ is typed as *Alt*-p. It indicates a parallel connection. Return to the *Layout* window, and draw the circuit as it appears in Figure C.4. Now move to

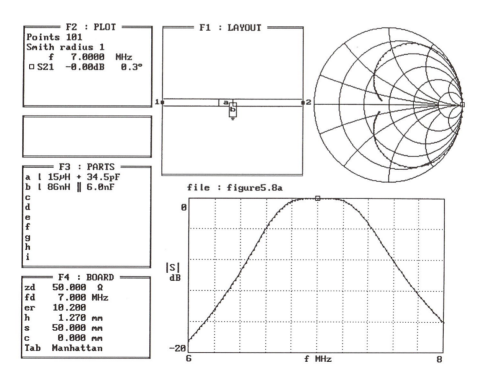

Figure C.4. Analyzing the filter in Figure 5.8a in *Puff*.

the *Plot* window. This is a band-pass filter, and we need to change the frequency scale to see its behavior clearly. To do this, use the up and down cursor keys to move the cursor to one of the frequencies at the bottom of the rectangular plot. Now type over the frequencies to narrow the plotting range from 6 MHz to 8 MHz. Push p again to plot. This should give the plot shown in Figure C.4. This shows the roll-off down to the 20-dB range. To see the behavior near the pass band, you can change the dB-scale to go from 0 to −3 in the same way that you changed the frequencies. Use the *PageUp* and *PageDown* keys to find the 3-dB bandwidth.

C.5 PARTS WINDOW

The program begins in the *Parts* window. The initial parts list is taken from the setup file. A previously saved circuit file can be read in by typing *Ctrl-r. Puff* will prompt for a file name and will add .puf if no extension is given. You can edit the parts list with the arrow keys, the backspace key, the carriage return (or enter) key, *Ins*, and *Del*. The first letter on each line, from a to i, identifies each part for use in the layout. By hitting the *Tab* key, the parts list may be doubled in size, allowing extra parts j through r to be defined. Hitting the *Tab* key a second time will shrink the parts list back to its normal size, unless an extra part was defined and used in the layout. The types of parts available in *Puff* are listed Table C.2, including examples of their use. Only the first letter of a part name is needed for it to be recognized: l is equivalent to lumped, t is equivalent to tline, q is equivalent to qline, etc. Free up space in the parts list by using single-character part descriptions. Commas and periods are both treated as decimal points. Spaces are ignored, except around device and indef file names. When you leave the *Parts* window, the list will be checked. There will be an error message if a part is longer than the board or shorter than the circuit resolution r given in the circuit file. *Puff* will redraw the circuit if it has changed and report an error if changes result in a part being drawn off the board.

The simplest part available in *Puff* is the attenuator, written as atten. It is drawn as an open blue square with red dots denoting its two ports. It is ideal; *Puff* will always consider it as matched to the normalizing impedance zd, and its attenuation will be frequency independent. The desired amount of attenuation is entered in dB. If no units are entered, *Puff* will assume dB. The atten part is reciprocal and symmetrical.

Plot the scattering parameters for an atten part that has a negative value of attenuation. Why is this different than that for an ideal amplifier?

The xformer part denotes a lossless and frequency-independent transformer. The only numerical parameter needed is the (unitless) turns ratio. *Puff* accepts a single real number for the turns ratio, not a ratio of two numbers. However, a colon may be used to write the number as a ratio to one. Therefore, 1.5:1 is a valid entry, whereas 3:2 is not. Because the transformer is antisymmetric, it is drawn in the layout as a trapezoid. Red dots are again used to denote the ports. For an $n : 1$ transformer, the wide end of the trapezoid represents the n side, while

Table C.2. Descriptions and Examples of the Parts Available in *Puff*.

Part	Description and Examples
atten	Ideal attenuator. Enter attenuation in dB: atten 3dB
xformer	Ideal transformer. Enter turns ratio: xformer 2:1
lumped	Resistor, capacitor, and/or inductor: lumped 50Ω‖1 nH‖1pF {parallel RLC} lumped 10+j10−j10Ω {series RLC resonant at fd}
tline	Ideal transmission line. Enter impedance and length: tline 50Ω 90° {λ/4 length at fd} tline1.0z 5mm +1.0h {artwork length correction added} tline! 50Ω 90° {analyze with advanced models}
qline	Finite Q tline: qline 50Ω 90° 75Qd {$Q = 75$ due to dielectric losses} qline 20mS 90° 50Qc {$Q = 50$ due to conductor losses}
clines	Ideal coupled transmission lines. Enter Z_e, Z_o, and length: clines 60Ω 90° {Z_e specified} clines 60Ω 40Ω 6mm {Z_e and Z_o specified} clines! 60Ω 90° {analyze with advanced models}
device	Read file with *s*-parameter data. Specify filename: device fhx04 {read filename fhx04.dev} device atf131.s2p {read file in EEsof format}
indef	Indefinite *s*-parameter generator. Converts *n* port *s*-parameter file to $n + 1$ port device. Specify filename as in device.

the narrow end represents the 1 side. If a negative turns ratio is entered, *Puff* will add a 180° phase shift to the transmission parameters.

Lay out and plot the scattering parameters for an ideal 180° phase shifter using an xformer with a -1:1 turns ratio. How would the schematic diagrams differ for a positive and negative turns ratio?

The lumped part is used to specify series and parallel combinations of resistance, capacitance, and/or inductance. It is drawn as an open blue rectangle. *Puff* understands four units for impedance and admittance: Ω (ohms, typed as *Alt*-o), S (siemens), z (normalized impedance), and y (normalized admittance). For example, a 100-Ω resistor may be specified as 100Ω, 0.01S, 2z, or 0.5y, assuming that zd is 50 Ω. Capacitance and inductance values are entered in units of farads (F) and henries (H), respectively. Note that these must be upper case. Resistance, capacitance, and inductance values may be positive, zero, or negative. Reactance values are specified by placing a j before or after a number, and using the impedance units Ω or z. For example, 25jΩ and 0.5jz specify a 25-Ω reactance at fd. *Puff* scales positive reactances proportionally and negative reactances inversely with frequency, interpreting them as inductive and capacitive reactances, respectively. This means that at 2fd, $j50\Omega$ becomes $j100\Omega$, and $-j50\Omega$ becomes $-j25\Omega$.

Specifications with the admittance units s and y are treated in a dual way: Positive susceptances scale proportionally with frequency, and the negative susceptances scale inversely.

Series circuits with two or three different lumped elements are specified by combining a real number, a positive imaginary number, and a negative imaginary number, all in the same lumped part. For example, 1+j10-j10z specifies a circuit that is resonant at the design frequency with a Q of 10. The resistance is equal to zd and the inductive and capacitive reactance are 10 zd at the design frequency. Note that the unit appears only once, after all the numbers. This type of lumped element entry is very convenient, for example, when taking values from filter tables. Parallel circuits are specified in a dual way using admittance units s and y. Series and parallel lumped elements that combine resistance and values of inductance and capacitance may also be formed. *Puff* has a special character for forcing a lumped part to be a parallel circuit when it does not contain admittance units. This character is ∥, typed as *Alt*-p, and it is referred to as the parallel sign. When used with a lumped part, the parallel sign forces the part's description to be interpreted as a parallel circuit, regardless of the unit. This allows part entries such as 50Ω∥1nH∥1pF for a parallel RLC circuit. Note that you can use symbols for the engineering prefixes *nano* and *pico*. These symbols may appear in front of units in the parts list, in .puf files, and in some *Board* window parameters. *Puff* recognizes the all the unit prefixes given in Table 1.1b. Note that each symbol is case sensitive (there is a big difference, for example, between 10mΩ and 10MΩ). The μ symbol is obtained by typing *Alt*-m.

Ground one terminal and compare the one-port scattering parameters for the following lumped parts:

```
lum 50Ω 1nH 1pF
lum 50Ω∥1nH∥1pF
lum 50Ω +1nH +1pF
lum 50Ω -1nH -1pF
lum 50Ω∥-1nH∥-1pF
```

Use *Alt*-p to obtain the ∥ sign and *Alt*-o to obtain the Ω sign. Use the *Tab* key, while in the *Plot* window, to align your plots with the Smith-chart circles.

The atten, xformer, and lumped parts do not have electrical length. When included in a circuit they are drawn in a default length, referred to here as the *Manhattan* length. An alternate length may be entered for the lumped part, although it is optional. This comes in handy when trying to align parts in the layout. The lumped length specification should come last, and the units must be in meters (m with an appropriate prefix). The *Manhattan* length is one tenth the size of the layout. All parts using *Manhattan* dimensions will be drawn with the same spacing between terminals.

The tline part is an ideal transmission line section. It is lossless and without dispersion. In the layout, it is drawn as a copper rectangle. The characteristic impedance or admittance of a tline may be specified in the same way as the resistance of a lumped part, except that it must be positive. Length units are required. Valid units include: meters (m with an appropriate prefix), h (the substrate thickness), and ° (degrees, typed as *Alt*-d). A transmission line specified with a 360° length will be one wavelength long at the design frequency fd. It is usually convenient to specify degrees other than, say millimeters (mm), but sometimes the physical units are useful for aligning a tline with other parts. You may also specify an artwork length correction to compensate for discontinuities. This is done by adding a plus or a minus sign and a second length. These corrections are made on the screen and in the artwork, but they do not affect the electrical length used in the analysis. For example, the length 90°-0.5h will be treated in the analysis as a quarter-wavelength section, but it will be drawn shorter on the screen and in the artwork. The h units are particularly useful here because open-circuit end corrections and phase shifts in tee junctions are proportional to the substrate thickness.

Often it happens that a desired electrical length results in a physical length larger than the size of the board or is perhaps negative for a de-imbedding problem. *Puff* provides for this eventuality by allowing you to force *Manhattan* dimensions upon the tline. This is done by placing an M (must be upper case) as the last character in the part's description. Then, regardless of the parameters specified, *Puff* will make the tline length one tenth the size of the layout. To find out the physical dimensions of a tline, place the cursor on the part, and hit the = key. *Puff* will calculate the dimensions for the part and list them in the *Message* box. The dimensions listed will include any artwork corrections or will be Manhattan dimensions if these were called for. Manhattan dimensions may be forced on all parts using the *Tab* key in the *Board* window.

The qline is similar to the tline but is lossy. In addition to impedance and length, you can add a value for the quality factor, or Q. The attenuation in the line is calculated by enforcing the given Q at the design frequency fd. Outside the design frequency, the attenuation is made to follow one of two models: Specify Qd or Q if you wish a dielectric loss model, and Qc if you prefer a conductor loss model. Manhattan dimensions are requested by placing an M (must be upper case) as the last character in the part's description. Artwork corrections are not allowed for the qline.

The clines part is a pair of coupled transmission lines. Its specification is similar to a tline, except that either one or two impedances or admittances may be given. If only one appears, then the specification looks the same as a tline. If this impedance is larger than zd, *Puff* interprets it as the even-mode impedance, and if smaller, as the odd-mode impedance. The other mode impedance is chosen to match the lines by forcing the product of even- and odd-mode impedances to be zd^2. If two impedances are given, the larger is the even-mode and the smaller the odd-mode impedance. As with the tline, use an upper case M at the end of a clines specification to force Manhattan dimensions.

The device part is used to read in a file containing multiport scattering parameters. Files may contain transistor data, measured data to be plotted, or parameters that define idealized parts, meters, or sources. A file name specifying the *s*-parameters must be given, preceded and followed by a space. *Puff* will assume a .dev extension if none is given. An optional length may be specified after the file name, as with the lumped part. The device is drawn in the *Layout* window as a

Table C.3. The File for the Fujitsu FHX04 HEMT transistor `fhx04.dev`. The frequencies are in GHz.

{FHX04 Fujitsu HEMT (89/90), f = 0 extrapolated; Vds = 2V, Ids = 10mA}							
f	s11		s21		s12		s22
0.0	1.000	0.0	4.375	180.0	0.000	0.0	0.625	0.0
1.0	0.982	−20.0	4.257	160.4	0.018	74.8	0.620	−15.2
2.0	0.952	−39.0	4.113	142.0	0.033	62.9	0.604	−28.9
3.0	0.910	−57.3	3.934	124.3	0.046	51.5	0.585	−42.4
4.0	0.863	−75.2	3.735	107.0	0.057	40.3	0.564	−55.8
5.0	0.809	−92.3	3.487	90.4	0.065	30.3	0.541	−69.2
6.0	0.760	−108.1	3.231	75.0	0.069	21.0	0.524	−82.0
7.0	0.727	−122.4	3.018	60.9	0.072	14.1	0.521	−93.6
8.0	0.701	−135.5	2.817	47.3	0.073	7.9	0.524	−104.7
9.0	0.678	−147.9	2.656	33.8	0.074	1.6	0.538	−115.4
10.0	0.653	−159.8	2.512	20.2	0.076	−4.0	0.552	−125.7
11.0	0.623	−171.1	2.367	7.1	0.076	−10.1	0.568	−136.4
12.0	0.601	178.5	2.245	−5.7	0.076	−15.9	0.587	−146.4
13.0	0.582	168.8	2.153	−18.4	0.076	−21.9	0.611	−156.2
14.0	0.564	160.2	2.065	−31.2	0.077	−28.6	0.644	−165.4
15.0	0.533	151.6	2.001	−44.5	0.079	−36.8	0.676	−174.8
16.0	0.500	142.8	1.938	−58.8	0.082	−48.5	0.707	174.2
17.0	0.461	134.3	1.884	−73.7	0.083	−61.7	0.733	163.6
18.0	0.424	126.6	1.817	−89.7	0.085	−77.9	0.758	150.9
19.0	0.385	121.7	1.708	−106.5	0.087	−97.2	0.783	139.1
20.0	0.347	119.9	1.613	−123.7	0.098	−119.9	0.793	126.6

blue arrowhead with red dots denoting the ports. The dot at the wide end of the arrowhead represents port 1 in the file, and the remaining ports follow at equal intervals along the arrow's axis.

If you wish to make your own `device` files, you should study the format of the file `fhx04.dev` given in Table C.3. There is an optional comment line in braces, followed by a template line. The f at the beginning of the template stands for frequency. If left out, *Puff* assumes that the scattering coefficients are independent of frequency. The scattering parameters that follow the frequency in the template may appear in any order, and *Puff* will assume that any parameters that are not given on the template line are zero. The program will figure out how many ports the device has from the highest port number that appears in the template line. *Puff* can handle up to four-port device files. The numbers that follow the template are separated by one or more spaces or carriage returns. The frequency is first (if it appears) followed by the magnitude (linear, not dB) and phase (in degrees) of each of the scattering parameters in turn. When *Puff* is calculating scattering parameters in the *Plot* window, it will interpolate linearly between points in the `device` file, as necessary. *Puff* will not extrapolate beyond a `device` file's frequency range and it will give an error when this is attempted. A previously saved `.puf` file (that includes saved scattering parameters) can also be recalled as a `device` file. Complex networks can then be formed by combining many smaller circuits.

The `indef` part is similar to the `device`, but it is used to generate *indefinite* scattering parameters from a file containing definite scattering parameter data. Indefinite parameters are those with an undefined ground terminal. If a one-port file is specified, `indef` will convert it to a two-port. If a two-port file is specified, `indef` will convert it to a three-port; and so on up to a four-port to five-port conversion. The `indef` part is most often used to model a transistor as a three-port. The *n* port to *n* + 1 port conversion is made possible by assuming that Kirchhoff's current law is valid, allowing what was the ground terminal to be converted to a port. The `indef` part is drawn on the screen in the same way as the `device` part, except that the port created from ground is drawn as a yellow dot. The extra port generally appears as the last port, except for the two-port to three-port `indef` where the extra port appears in the center of the part. Note that to turn an `indef` part into its `device` equivalent, the extra port should be shorted.

In addition to the `device` file format given in Table C.3, *Puff* can also read *s*-parameter data files in the EEsof format. This is done when the appropriate file extension is given for `device` or `indef` parts: A one-port file requires a `.S1P` extension, the two-port uses `.S2P`, and so on up to a four-port `.S4P`. *Puff*, however, cannot read every possible data format for these files. It is assumed that they possess the following format:

`# xHZ S MA R yy`

where *x* is the same engineering prefix used for `fd` in the *Board* window, and *yy* is the value for `zd`. *Puff* reads only scattering parameters given in the magnitude/angle format. These restrictions require some caution. If a device file contains frequencies given in GHz, *Puff* will give erroneous results if you try to plot data in MHz. The prefix for `fd` and the impedance of `zd` in the *Board* window *must* coincide with those used in all `device` and `indef` files. This goes for files in both EEsof and `.dev` formats. Noise parameters present in files are ignored.

Create a two-port circuit and plot all of its scattering parameters (s_{11}, s_{21}, s_{12}, and s_{22}). Use *Ctrl*-s from the *Plot* window to save it to a file. Erase the circuit, and then make a `device` part that will recall the scattering parameters from the previously saved file. Be sure to use the `.puf` extension. What happens if the number of points is different?

C.6 LAYOUT WINDOW

In the upper portion of the screen is the *Layout* window. The square represents the substrate, and the numbers on the sides represent connectors. Typing an arrow key will draw the selected part in the *Parts* window in the direction of the arrow. The *Message* box will show the change in the *x* and *y* coordinates. *Puff* starts out drawing part a, but you can select another part in the list by typing the letter for the part desired. The circuit can be grounded at any point by pushing the = key.

If there is already a part in the direction of the arrow key, *Puff* will move to the other end rather than draw over it. If the ends of two parts are closer together than the circuit resolution r, *Puff* will connect them together. *Puff* will stop you from drawing a part off the edge, but it will not stop you from crossing over a previous part. You can make a path to a connector by pushing one of the number keys 1, 2, 3, or 4 on the top row of the keyboard. Notice that *Puff* does this by first moving up or down and then right or left. The electrical length of a connector path is not taken into account in the analysis, and it is drawn in a different color to indicate this. You can erase the entire circuit and start over by pushing *Ctrl*-e. *Ctrl*-n moves the × to the nearest node. This can be useful if you are off the network and want to get back, or if you want to see if two nodes are connected.

The *Shift* key is used for erasing and moving around the layout. *Puff* will erase a part rather than move over it if you hold the *Shift* key down while pushing an arrow key. A ground can be removed with *Shift* =. The path to a connector is erased by holding the *Shift* key down while typing the connector number. If you are not at a connection to a port, this shift-number operation will move you to that port number without drawing the path. This is useful if you would rather start a circuit at a connector than in the center. The shift-arrow operation moves the × when there is no part to erase. The × moves half the length of the currently selected part. Half steps are used rather than full steps to allow centered and symmetrically laid out circuits. To move the full length of the part, step twice. It is important to note that this shift-arrow operation is actually drawing an invisible part. If you later change the size of an invisible part, it will have an impact on the layout, possibly resulting in a part being drawn off the board. If this happens, *Puff* will give you an error message that you may not believe. As a precaution, keep the number of invisible parts to a minimum.

There are special rules for drawing `clines`. Use the arrows to move along the lines and *Ctrl*-n to jump from one line to another. When connecting `clines` together, if you draw the second `clines` in the same direction as the first, the new lines will join the previous pair. This is the usual arrangement in a directional coupler. If you change directions, the `clines` will be staggered so that only one of the lines in each pair is connected. This is appropriate for a band-pass filter.

Typing *Ctrl*-a from the *Plot* window will activate *Puff* 's photographic artwork routines. The layout produced will be magnified by the photographic reduction ratio (p) in the circuit file. The artwork output parameter (o) in the circuit file allows you to specify dot-matrix or HP LaserJet printouts or the production of a Hewlett-Packard Graphics Language (HP-GL) file. *Puff* will prompt for titles to be placed atop the printout, or for an HP-GL file name. Only the `tlines`, `qlines`, and `clines` will appear in the artwork, and the corners will be mitered. You may adjust `tline` and `clines` lengths using discontinuity corrections in the parts list. Widths may be adjusted using the artwork width correction factor a in the circuit file.

You can inspect *Puff* 's layout calculations from the *Parts* window. Place the cursor in any `tline`, `qline`, or `clines` description and hit the = key. *Puff* will tell you the length and width for these parts, as well as the spacing for `clines`.

Lay out a simple circuit consisting of `tlines` and `clines`. Go to the *Board* window and use the *Tab* key to change the circuit type. Return to the *Layout* window and see how the circuit is affected. Repeat for microstrip, stripline, and Manhattan layouts. Can you make a microstrip circuit that is difficult to realize in stripline? What stripline circuits cannot be realized in microstrip?

Lay out a simple circuit. Go to the *Plot* window and save the circuit using *Ctrl*-s. Exit *Puff*; then use an ASCII editor to open the saved *Puff* file. Go to the section of the file that begins \c{ircuit}. *Puff* has saved your keystrokes in the *Layout* window in what we call a *keylist*. When reading a new file, *Puff* uses the keylist to redraw the circuit. What keystrokes are not saved? Is it better to erase all the parts using *Ctrl*-e or with repeated shift-arrow operations? Can you think of some advantages in keeping the keylist as short as possible?

The most common *Puff* layout errors involve invisible parts. Starting from a blank layout, select a `tline` part and make repeated shift-arrow operations to move the × cursor about the *Layout* window. Go to the *Parts* window, increase the length of the `tline` used, and then return to the *Layout* window. See how long you can make the `tline` before causing a layout error. What happens if you try to delete the `tline` from the parts list?

C.7 BOARD WINDOW

The relative dimensions of the circuit board in the *Layout* window are specified from the *Board* window. These dimensions set the scale used to draw the distributed components on the screen. Table C.4 gives a brief description of each of the parameters available. Access the *Board* window by pressing function key *F4*. Edit the parameters using the same keys as in the *Parts* window. *F10* brings up a help window explaining the board parameters.

The normalizing impedance `zd` is used to calculate the scattering parameters. It also defines the normalized impedance and admittance values (`z` and `y`) that may appear in the *Parts* window. Paths to connectors are drawn with transmission lines with impedance `zd`. The design frequency `fd` is used to determine the physical lengths of distributed components entered in degrees. The Hz units for `fd` may use any prefix that appears in the Chapter 2 table, although MHz and GHz are generally the most practical. The *Plot* window will inherit the frequency prefix. Change the prefix when you find yourself entering lots of zeros below the log-magnitude plot. Use caution. The prefixes are case sensitive. Also beware of frequency data in `device` and `indef` files that do not coincide with the `fd` prefix (e.g., *Puff* will give meaningless results if you use a device file with GHz frequency data when MHz is used for `fd`). Device scattering parameter data must also match the *Board* window's definition of `zd`.

The *Tab* key toggles between microstrip, Manhattan, and stripline layouts. Resort to the Manhattan mode when `tlines` or `clines` become too long or short. All distributed parts are then drawn with widths 1/20 the board size, and lengths 1/10 the board size. This mode also permits components with unrealizable values, such as `tlines` with negative electrical lengths. However, if *Puff* requires an

Table C.4. Description of Parameters that May Be Modified from the *Board* Window.

Parameter	Description
zd	Normalizing or characteristic impedance. Used in the calculation of scattering parameters. Units are Ωs with optional prefix.
fd	Design frequency. Used to compute electrical length of parts entered in degrees. Also the frequency used for the component sweep. Prefix given is carried over into the *Plot* window.
er	Relative dielectric constant of substrate. Used to calculate dimensions for microstrip and stripline components. Unitless.
h	Substrate thickness. One of three parameters that specify the equivalent dimensions for the *Layout* window. Significant in transmission line calculations.
s	Board size. Specifies the equivalent length of each side of the square circuit board that makes up the *Layout* window.
c	Connector separation. Sets the spacing between ports 1 and 3 and between ports 2 and 4 in the layout. Set to zero to create a centered two-port.
Tab	Circuit type. Use the *Tab* key to select a microstrip, stripline, or Manhattan layout.

electrical length calculation, stripline models will be used, and physically unrealizable parameters may not be allowed. If enabled, Manhattan dimensions will appear in the artwork, and artwork corrections will be ignored.

C.8 PLOT WINDOW

To reach the *Plot* window, push function key *F2*. The circuit in the *Layout* window is analyzed by typing p. If you type *Ctrl-p*, the previous plot will be drawn before the new plot, allowing a comparison of results. After an analysis has been completed, the *Plot* window lists the values of the scattering coefficients at design frequency fd. Use the *PgUp* and *PgDn* keys to move the markers and show the scattering coefficients at the other frequencies. The ↑ and ↓ keys can be used to move the cursor to various parameters. They cycle through a loop that includes the *Plot* window and the *x* and *y* axes on the rectangular plot. You can type over any parameter to change it. *Puff* can plot up to four different scattering parameters simultaneously. To select an *s*-parameter, move the cursor down toward the bottom of the *Plot* window, and a marker will appear, together with the letter s. Then type in the port numbers for the desired *s*-parameters. If you leave a line blank, it will be erased when you move the cursor. Those with EGA and VGA graphics can use the *Tab* key to change the Smith chart from an impedance chart to an admittance chart. Typing *Alt-s* with VGA graphics toggles an enlarged Smith chart.

The *Plot* window allows you to select the number of frequency `Points` to analyze. This must be a positive integer no greater than 500, assuming you have a full complement of memory. *Puff* interpolates between calculated points with a cubic spline. The interpolation is performed by splining the real and imaginary parts of the scattering coefficients separately. The independent variable for calculating the spline curve is the spacing on the Smith chart. This gives better results than using frequency as the splining parameter. If a curve kinks on the Smith chart, or gives erroneous ripples on the rectangular plot, it is an indication that the number of `Points` is too small.

You can plot an impulse response by typing `i`, or a step response by typing `s`. *Puff* will request a frequency interval specified by the ratio `fd/df`. It will then do a 256-point inverse fast Fourier transform of the scattering coefficients, and plot the results on a linear scale. The amplitude of the time-domain plot is the same as the Smith chart radius. The ratio `fd/df` will determine the time axis for the plot, which goes from $-1/8$`df` to $3/8$`df`. The upper and lower limits on the frequency axis of the magnitude plot are used for windowing the Fourier transform. The window is a raised cosine that goes to zero at the upper frequency limit. The scattering coefficients are set to zero outside the window.

A convenient way to see the impulse or step inputs is to draw an open-circuited connector path at an unused port. The reflection coefficient for this port is the input waveform. *Puff* normalizes the input waveforms so that the peak value is 1. The high-frequency limit controls the rise and fall times, and the low-frequency limit affects the ringing. Be aware that the time-domain waveforms are actually periodic, with period $1/$`df`, and aliasing from the previous pulse may affect the response. The step input is actually a square wave, and the response to the previous falling edge will affect the rising step that follows.

To save a network in a circuit file, type *Ctrl-s* and give a file name. The *Parts* window and the data in the *Plot* window will be saved along with the circuit. The `.puf` extension will be added to the file name if one is not specified. Typing *Ctrl-a* from the *Plot* window will activate *Puff*'s photographic artwork routines. The layout produced will be magnified by the photographic reduction ratio (p) in the circuit file. Only the `tlines`, `qlines`, and `clines` will appear in the artwork, and the corners will be mitered. The artwork output parameter (o) in the circuit file allows you to specify dot-matrix or HP LaserJet printouts or the production of a Hewlett-Packard Graphics Language (HP-GL) file. *Puff* will prompt for titles to be placed atop the printout, or for an HP-GL file name. The HP-GL file will be created with the `.hpg` extension. The *Puff* diskette includes the program `hpg2com`. Use this to dump `.hpg` files to a serial plotter connected at port COM1.

C.9 COMPONENT SWEEP

The simple optimizer included in *Puff* is called the *component sweep*. Instead of sweeping with respect to frequency, a circuit's scattering coefficients may be swept with respect to a changing component parameter. This feature is invoked by placing a question mark (?) in front of the parameter to be swept in the appropriate position of a part's description. For example, to find the optimum value for a tuning capacitor, one could specify a part as `lumped ?5pF`. When plotting in the *Plot* window, the frequency will then be held constant at the design frequency `fd`, and the values specified in the x axis of the rectangular plot will be substituted for (in the above example) capacitance values in picofarads. In this manner, any parameter used in the parts list may be designated as a sweep parameter, but only one parameter may be swept at a time. Swept lumped elements are restricted to single resistors, capacitors, or inductors. A description such as `lumped ?1+5j−5jΩ` is not allowed since it is a series RLC circuit. In addition, the parallel sign ‖ may not be present in the lumped specification. The unit and prefix given in the part's description (following the `?`) is inherited by the component sweep.

Component Data

This appendix gives data for most of the components that are used in the NorCal 40A. We appreciate the manufacturers giving us permission to copy data sheets. Often data sheets have a great deal of information, but it is important to make your own measurements to check them. Manufacturers vary greatly in how conservatively they rate their devices. In addition, they may be testing the devices under conditions that are different from yours. Often several manufacturers sell a device with same part number, and the performance between different manufacturers varies. These data sheets are no substitute for the data books and Web sites that give the complete list of products that a manufacturer sells. Many of these devices are part of a wide line of products that will cover a range of frequencies, functions, and power levels. For much more component information, see *Data Book for Homebrewers and QRPers,* by Paul Harden, published by Quicksilver Printing. The book is available from Five Watt Press, 740 Galena Street, Aurora, CO 80010-3922, email: qrpbook@aol.com.

Table D.1. Resistors, inductors and capacitors
Table D.2. Iron and ferrite cores

Motorola – http://mot2.indirect.com

1N5817 Schottky barrier rectifier
MVAM108 silicon tuning diode
P2N2222A general purpose npn silicon transistor
2N3553 2.5-W high frequency npn silicon transistor
2N3906 general purpose pnp silicon transistor
2N4124 general purpose npn silicon transistor
J309 n-channel VHF/UHF JFET
MC78L08AC three-terminal low-current, positive-voltage regulator

National Semiconductor – http://www.national.com

LM386N-1 National Semiconductor low-voltage audio power amplifier

Philips – http://www.semiconductors.philips.com

1N4148 high-speed diode
SA602AN double-balanced mixer and oscillator

Table D.1. Characteristics of the Resistors, Inductors and Capacitors in the NorCal 40A.

Type	±%	Range	ppm/°C	Q	Use
carbon film resistor	5	1 Ω–1 MΩ	−240		dividers, damping
inductors	5	220 nH–1 mH	+1,200	50 at 7 MHz	chokes, filters
small ceramic	5	10–1,000 pF	−800	600 at 7 MHz	RF filters
NP0 ceramic	5	18–220 pF	−30	800 at 2 MHz	VFO resonator
large ceramic	20	1–47 nF	−30,000	20 at 7 MHz	RF bypass
polystyrene	5	100 pF–10 nF	−150	250 at 2 MHz	VFO resonator
polyester film	5	1–470 nF	+900	240 at 1 kHz	audio coupling
aluminum electrolytic	20	220 nF–10 mF	+2,000	6 at 1 kHz	audio filtering, supply filtering
air variable		2–24 pF	+50	1,000 at 2 MHz	VFO adjustment
ceramic trimmer		7–70 pF	−1,600	200 at 7 MHz	filter adjustment, crystal oscillator
varactor MVAM108		30–500 pF	+500	150 at 2 MHz (3 V bias)	VFO tuning

Note: ±% gives the tolerance. Parts with a tolerance of 5% are made with standard values, where the first two digits come from this list: 10, 11, 12, 13, 15, 16, 18, 20, 22, 24, 27, 30, 33, 36, 39, 43, 47, 51, 56, 62, 68, 75, 82, 91. For 20% parts, the standard values are typically 10, 22, 33, 47, and 68. The range of values are those listed in a catalog for the particular line of parts that we use. For variable components, "range" gives the extreme adjustable values that we measured. The column "ppm/°C" gives our measured values of the temperature coefficient for a representative part; "ppm" is parts per million. "NP0" designates a part with a low-temperature coefficient. The Qs are also our measured values. They vary with frequency. Perhaps it is surprising that even the air variable capacitor has a temperature coefficient, but you should remember that metals expand as the temperature goes up.

Table D.2. The Cores in the NorCal 40A Transceiver.

Core	A_l, nH/turn²	Q	ppm/°C	Material	Paint	Use
T37-2	4.0 (28 turns)	170	+100	iron powder	red	filter
T68-7	5.0 (60 turns)	200	+50	iron powder	white	oscillator
FT37-43	160 (14 turns)	1	−30	nickel–zinc ferrite	orange spot	transformer
FT37-61	66 (1 turns)	50	+500	nickel–zinc ferrite	none	tuned transformer

Note: There is a lot of information in the core number itself. For example, in FT37-43, "F" indicates a ferrite, "T" a toroidal core, "37" the outside diameter in hundredths of an inch, and "61" the manufacturing recipe. These are our measured values for A_l, Q, and the temperature coefficient. The values of A_l vary ±10% from lot to lot. All measurements are at 7 MHz, except for the T68-7 core, which is at 2 MHz. The values vary with frequency, and if the specific value of the inductance is critical, you should measure the inductance constant at the frequency you are interested in. Temperature coefficients are given for ferrites for completeness, but it is a poor idea to use ferrites in an application where temperature stability is important, because characteristics differ greatly over even a modest range of temperature.

MOTOROLA
SEMICONDUCTOR TECHNICAL DATA

Order this document
by 1N5817/D

Axial Lead Rectifiers

. . . employing the Schottky Barrier principle in a large area metal–to–silicon power diode. State–of–the–art geometry features chrome barrier metal, epitaxial construction with oxide passivation and metal overlap contact. Ideally suited for use as rectifiers in low–voltage, high–frequency inverters, free wheeling diodes, and polarity protection diodes.

- Extremely Low v_F
- Low Stored Charge, Majority Carrier Conduction
- Low Power Loss/High Efficiency

Mechanical Characteristics

- Case: Epoxy, Molded
- Weight: 0.4 gram (approximately)
- Finish: All External Surfaces Corrosion Resistant and Terminal Leads are Readily Solderable
- Lead and Mounting Surface Temperature for Soldering Purposes: 220°C Max. for 10 Seconds, 1/16″ from case
- Shipped in plastic bags, 1000 per bag.
- Available Tape and Reeled, 5000 per reel, by adding a "RL" suffix to the part number
- Polarity: Cathode Indicated by Polarity Band
- Marking: 1N5817, 1N5818, 1N5819

1N5817
1N5818
1N5819

1N5817 and 1N5819 are
Motorola Preferred Devices

**SCHOTTKY BARRIER
RECTIFIERS
1 AMPERE
20, 30 and 40 VOLTS**

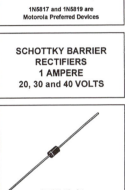

CASE 59–04

MAXIMUM RATINGS

Rating	Symbol	1N5817	1N5818	1N5819	Unit
Peak Repetitive Reverse Voltage Working Peak Reverse Voltage DC Blocking Voltage	V_{RRM} V_{RWM} V_R	20	30	40	V
Non–Repetitive Peak Reverse Voltage	V_{RSM}	24	36	48	V
RMS Reverse Voltage	$V_{R(RMS)}$	14	21	28	V
Average Rectified Forward Current (2) ($V_{R(equiv)} \leq 0.2 V_R(dc)$, $T_L = 90°C$, $R_{\theta JA} = 80°C/W$, P.C. Board Mounting, see Note 2, $T_A = 55°C$)	I_O		1.0		A
Ambient Temperature (Rated $V_R(dc)$, $P_{F(AV)} = 0$, $R_{\theta JA} = 80°C/W$)	T_A	85	80	75	°C
Non–Repetitive Peak Surge Current (Surge applied at rated load conditions, half–wave, single phase 60 Hz, $T_L = 70°C$)	I_{FSM}		25 (for one cycle)		A
Operating and Storage Junction Temperature Range (Reverse Voltage applied)	T_J, T_{stg}		–65 to +125		°C
Peak Operating Junction Temperature (Forward Current applied)	$T_{J(pk)}$		150		°C

THERMAL CHARACTERISTICS (2)

Characteristic	Symbol	Max	Unit
Thermal Resistance, Junction to Ambient	$R_{\theta JA}$	80	°C/W

ELECTRICAL CHARACTERISTICS ($T_L = 25°C$ unless otherwise noted) (2)

Characteristic		Symbol	1N5817	1N5818	1N5819	Unit
Maximum Instantaneous Forward Voltage (1)	(i_F = 0.1 A) (i_F = 1.0 A) (i_F = 3.0 A)	v_F	0.32 0.45 0.75	0.33 0.55 0.875	0.34 0.6 0.9	V
Maximum Instantaneous Reverse Current @ Rated dc Voltage (1)	(T_L = 25°C) (T_L = 100°C)	I_R	1.0 10	1.0 10	1.0 10	mA

(1) Pulse Test: Pulse Width = 300 μs, Duty Cycle = 2.0%.
(2) Lead Temperature reference is cathode lead 1/32″ from case.

Preferred devices are Motorola recommended choices for future use and best overall value.

Rev 3

 **MOTOROLA**

© Motorola, Inc. 1996

1N5817 1N5818 1N5819

NOTE 1 — DETERMINING MAXIMUM RATINGS

Reverse power dissipation and the possibility of thermal runaway must be considered when operating this rectifier at reverse voltages above 0.1 V_{RWM}. Proper derating may be accomplished by use of equation (1).

$$T_{A(max)} = T_{J(max)} - R_{\theta JA}P_{F(AV)} - R_{\theta JA}P_{R(AV)} \quad (1)$$

where $T_{A(max)}$ = Maximum allowable ambient temperature
$T_{J(max)}$ = Maximum allowable junction temperature (125°C or the temperature at which thermal runaway occurs, whichever is lowest)
$P_{F(AV)}$ = Average forward power dissipation
$P_{R(AV)}$ = Average reverse power dissipation
$R_{\theta JA}$ = Junction–to–ambient thermal resistance

Figures 1, 2, and 3 permit easier use of equation (1) by taking reverse power dissipation and thermal runaway into consideration. The figures solve for a reference temperature as determined by equation (2).

$$T_R = T_{J(max)} - R_{\theta JA}P_{R(AV)} \quad (2)$$

Substituting equation (2) into equation (1) yields:

$$T_{A(max)} = T_R - R_{\theta JA}P_{F(AV)} \quad (3)$$

Inspection of equations (2) and (3) reveals that T_R is the ambient temperature at which thermal runaway occurs or where T_J = 125°C, when forward power is zero. The transition from one boundary condition to the other is evident on the curves of Figures 1, 2, and 3 as a difference in the rate of change of the slope in the vicinity of 115°C. The data of Figures 1, 2, and 3 are based upon dc conditions. For use in common rectifier circuits, Table 1 indicates suggested factors for an equivalent dc voltage to use for conservative design, that is:

$$V_{R(equiv)} = V_{in(PK)} \times F \quad (4)$$

The factor F is derived by considering the properties of the various rectifier circuits and the reverse characteristics of Schottky diodes.

EXAMPLE: Find $T_{A(max)}$ for 1N5818 operated in a 12–volt dc supply using a bridge circuit with capacitive filter such that I_{DC} = 0.4 A ($I_{F(AV)}$ = 0.5 A), $I_{(FM)}/I_{(AV)}$ = 10, Input Voltage = 10 V$_{(rms)}$, $R_{\theta JA}$ = 80°C/W.

Step 1. Find $V_{R(equiv)}$. Read F = 0.65 from Table 1,
  ∴ $V_{R(equiv)}$ = (1.41)(10)(0.65) = 9.2 V.
Step 2. Find T_R from Figure 2. Read T_R = 109°C
  @ V_R = 9.2 V and $R_{\theta JA}$ = 80°C/W.
Step 3. Find $P_{F(AV)}$ from Figure 4. **Read $P_{F(AV)}$ = 0.5 W

$$@ \frac{I_{(FM)}}{I_{(AV)}} = 10 \text{ and } I_{F(AV)} = 0.5 \text{ A.}$$

Step 4. Find $T_{A(max)}$ from equation (3).
  $T_{A(max)}$ = 109 − (80) (0.5) = 69°C.

**Values given are for the 1N5818. Power is slightly lower for the 1N5817 because of its lower forward voltage, and higher for the 1N5819.

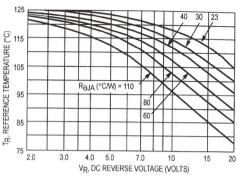

Figure 1. Maximum Reference Temperature 1N5817

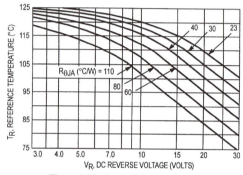

Figure 2. Maximum Reference Temperature 1N5818

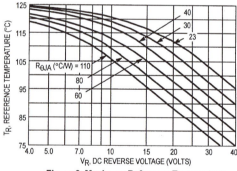

Figure 3. Maximum Reference Temperature 1N5819

Table 1. Values for Factor F

Circuit	Half Wave		Full Wave, Bridge		Full Wave, Center Tapped*†	
Load	Resistive	Capacitive*	Resistive	Capacitive	Resistive	Capacitive
Sine Wave	0.5	1.3	0.5	0.65	1.0	1.3
Square Wave	0.75	1.5	0.75	0.75	1.5	1.5

*Note that $V_{R(PK)} \approx 2.0\ V_{in(PK)}$. †Use line to center tap voltage for V_{in}.

1N5817 1N5818 1N5819

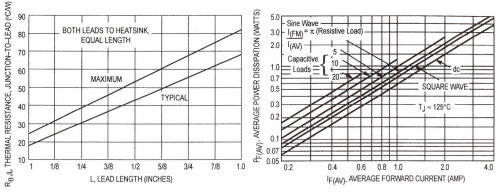

Figure 4. Steady–State Thermal Resistance

**Figure 5. Forward Power Dissipation
1N5817–19**

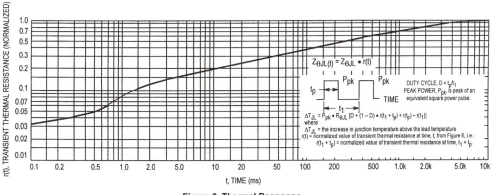

Figure 6. Thermal Response

NOTE 2 — MOUNTING DATA

Data shown for thermal resistance junction–to–ambient ($R_{\theta JA}$) for the mountings shown are to be used as typical guideline values for preliminary engineering, or in case the tie point temperature cannot be measured.

TYPICAL VALUES FOR $R_{\theta JA}$ IN STILL AIR

Mounting Method	Lead Length, L (in)				$R_{\theta JA}$
	1/8	1/4	1/2	3/4	
1	52	65	72	85	°C/W
2	67	80	87	100	°C/W
3	50				°C/W

Mounting Method 1
P.C. Board with
1–1/2″ x 1–1/2″
copper surface.

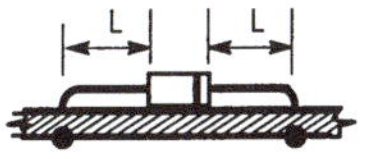

Mounting Method 2

VECTOR PIN MOUNTING

Mounting Method 3
P.C. Board with
1–1/2″ x 1–1/2″
copper surface.

L = 3/8″

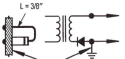

BOARD GROUND
PLANE

1N5817 1N5818 1N5819

NOTE 3 — THERMAL CIRCUIT MODEL
(For heat conduction through the leads)

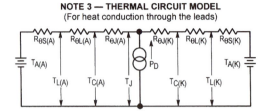

Use of the above model permits junction to lead thermal resistance for any mounting configuration to be found. For a given total lead length, lowest values occur when one side of the rectifier is brought as close as possible to the heatsink. Terms in the model signify:

T_A = Ambient Temperature T_C = Case Temperature
T_L = Lead Temperature T_J = Junction Temperature
$R_{\theta S}$ = Thermal Resistance, Heatsink to Ambient
$R_{\theta L}$ = Thermal Resistance, Lead to Heatsink
$R_{\theta J}$ = Thermal Resistance, Junction to Case
P_D = Power Dissipation

(Subscripts A and K refer to anode and cathode sides, respectively.)
Values for thermal resistance components are:
$R_{\theta L}$ = 100°C/W/in typically and 120°C/W/in maximum
$R_{\theta J}$ = 36°C/W typically and 46°C/W maximum.

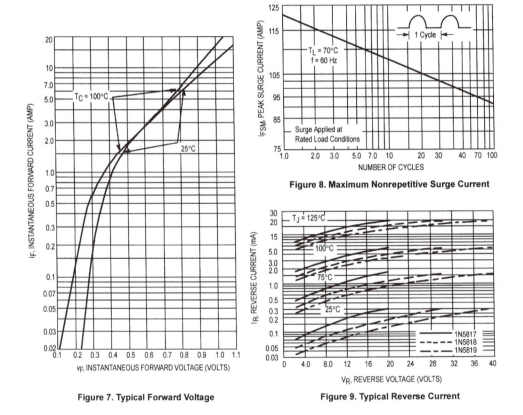

Figure 7. Typical Forward Voltage

Figure 8. Maximum Nonrepetitive Surge Current

Figure 9. Typical Reverse Current

1N5817 1N5818 1N5819

NOTE 4 — HIGH FREQUENCY OPERATION

Since current flow in a Schottky rectifier is the result of majority carrier conduction, it is not subject to junction diode forward and reverse recovery transients due to minority carrier injection and stored charge. Satisfactory circuit analysis work may be performed by using a model consisting of an ideal diode in parallel with a variable capacitance. (See Figure 10.)

Rectification efficiency measurements show that operation will be satisfactory up to several megahertz. For example, relative waveform rectification efficiency is approximately 70 percent at 2.0 MHz, e.g., the ratio of dc power to RMS power in the load is 0.28 at this frequency, whereas perfect rectification would yield 0.406 for sine wave inputs. However, in contrast to ordinary junction diodes, the loss in waveform efficiency is not indicative of power loss: it is simply a result of reverse current flow through the diode capacitance, which lowers the dc output voltage.

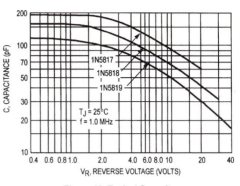

Figure 10. Typical Capacitance

1N5817 1N5818 1N5819

PACKAGE DIMENSIONS

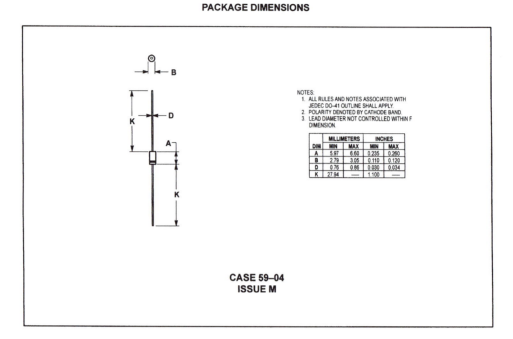

NOTES:
1. ALL RULES AND NOTES ASSOCIATED WITH JEDEC DO–41 OUTLINE SHALL APPLY.
2. POLARITY DENOTED BY CATHODE BAND.
3. LEAD DIAMETER NOT CONTROLLED WITHIN F DIMENSION.

DIM	MILLIMETERS		INCHES	
	MIN	MAX	MIN	MAX
A	5.97	6.60	0.235	0.260
B	2.79	3.05	0.110	0.120
D	0.76	0.86	0.030	0.034
K	27.94	—	1.100	—

CASE 59–04
ISSUE M

Mfax is a trademark of Motorola, Inc.

How to reach us:
USA / EUROPE / Locations Not Listed: Motorola Literature Distribution;
P.O. Box 5405, Denver, Colorado 80217. 303–675–2140 or 1–800–441–2447

Mfax™: RMFAX0@email.sps.mot.com – TOUCHTONE 602–244–6609
INTERNET: http://Design–NET.com

JAPAN: Nippon Motorola Ltd.; Tatsumi–SPD–JLDC, 6F Seibu–Butsuryu–Center, 3–14–2 Tatsumi Koto–Ku, Tokyo 135, Japan. 81–3–3521–8315

ASIA/PACIFIC: Motorola Semiconductors H.K. Ltd.; 8B Tai Ping Industrial Park, 51 Ting Kok Road, Tai Po, N.T., Hong Kong. 852–26629298

 **MOTOROLA**

◊

1N5817/D

SILICON TUNING DIODES

. . . designed for electronic tuning of AM receivers and high capacitance, high tuning ratio applications.

- High Capacitance Ratio — C_R = 15 (Min), MVAM108, 115, 125
- Guaranteed Diode Capacitance — C_t = 440 pF (Min) — 560 pF (Max) @ V_R = 1.0 Vdc, f = 1.0 MHz, MVAM108, MVAM115, MVAM125
- Guaranteed Figure of Merit — Q = 150 (Min) @ V_R = 1.0 Vdc, f = 1.0 MHz

MVAM108★
MVAM109★
MVAM115★
MVAM125★

CASE 182-02, STYLE 1
(TO-226AC)

2 O————|◁—————O 1
Cathode   Anode

**TUNING DIODES
WITH VERY HIGH
CAPACITANCE RATIO**

★These are Motorola
designated preferred devices.

MAXIMUM RATINGS

Rating		Symbol	Value	Unit
Reverse Voltage	MVAM108	V_R	12	Volts
	MVAM109		15	
	MVAM115		18	
	MVAM125		28	
Forward Current		I_F	50	mA
Power Dissipation @ T_A = 25°C		P_D	280	mW
Derate above 25°C			2.8	mW/°C
Operating and Storage Junction Temperature Range		T_J, T_{stg}	−55 to +125	°C

ELECTRICAL CHARACTERISTICS (T_A = 25°C unless otherwise noted, Each Device)

Characteristic		Symbol	Min	Typ	Max	Unit
Breakdown Voltage		$V_{(BR)R}$				Vdc
(I_R = 10 μAdc)	MVAM108		12	—	—	
	MVAM109		15	—	—	
	MVAM115		18	—	—	
	MVAM125		28	—	—	
Reverse Current		I_R				nAdc
(V_R = 8.0 V)	MVAM108		—	—	100	
(V_R = 9.0 V)	MVAM109		—	—	100	
(V_R = 15 V)	MVAM115		—	—	100	
(V_R = 25 V)	MVAM125		—	—	100	
Diode Capacitance Temperature Coefficient (1) (V_R = 1.0 Vdc, f = 1.0 MHz, T_A = −40°C to +85°C)		TC_C	—	435	—	ppm/°C
Case Capacitance (f = 1.0 MHz, Lead Length 1/16")		C_C	—	0.18	—	pF
Diode Capacitance (2)		C_t				pF
(V_R = 1.0 Vdc, f = 1.0 MHz)	MVAM108, 115, 125		440	500	560	
	MVAM109		400	460	520	
Figure of Merit (f = 1.0 MHz, Lead Length 1/16", V_R = 1.0 Vdc)		Q	150	—	—	—
Capacitance Ratio						—
(f = 1.0 MHz)	MVAM108	C1/C8	15	—	—	
	MVAM109	C1/C9	12	—	—	
	MVAM115	C1/C15	15	—	—	
	MVAM125	C1/C25	15	—	—	

NOTES:
1. The effect of increasing temperature 1.0°C, at any operating point, is equivalent to lowering the effective tuning voltage 1.25 mV. The percent change of capacitance per °C is nearly constant from −40°C to +100°C.
2. Upon request, diodes are available in matched sets. All diodes in a set can be matched for capacitance to 3% or 2.0 pF (whichever is greater) at all points along the specified tuning range.

MVAM108, MVAM109, MVAM115, MVAM125

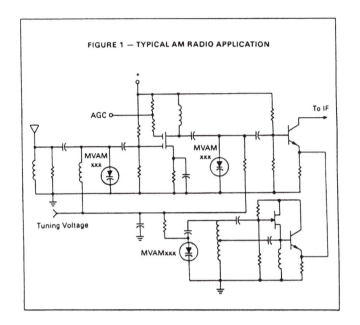

FIGURE 1 — TYPICAL AM RADIO APPLICATION

FIGURE 2 — CAPACITANCE versus REVERSE VOLTAGE

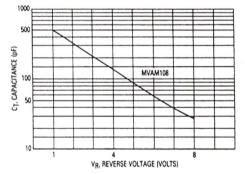

FIGURE 3 — CAPACITANCE versus REVERSE VOLTAGE

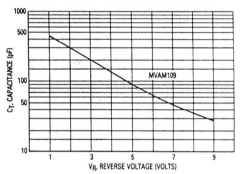

FIGURE 4 — CAPACITANCE versus REVERSE VOLTAGE

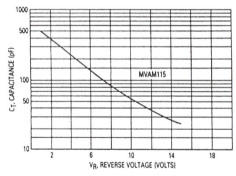

FIGURE 5 — CAPACITANCE versus REVERSE VOLTAGE

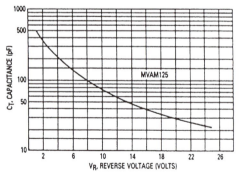

MOTOROLA SMALL-SIGNAL TRANSISTORS, FETs AND DIODES

MOTOROLA
SEMICONDUCTOR TECHNICAL DATA

Order this document
by P2N2222A/D

Amplifier Transistors

NPN Silicon

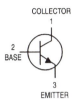

COLLECTOR
1

2
BASE

3
EMITTER

P2N2222A

CASE 29–04, STYLE 17
TO–92 (TO–226AA)

MAXIMUM RATINGS

Rating	Symbol	Value	Unit
Collector–Emitter Voltage	V_{CEO}	40	Vdc
Collector–Base Voltage	V_{CBO}	75	Vdc
Emitter–Base Voltage	V_{EBO}	6.0	Vdc
Collector Current — Continuous	I_C	600	mAdc
Total Device Dissipation @ $T_A = 25°C$ Derate above 25°C	P_D	625 5.0	mW mW/°C
Total Device Dissipation @ $T_C = 25°C$ Derate above 25°C	P_D	1.5 12	Watts mW/°C
Operating and Storage Junction Temperature Range	T_J, T_{stg}	−55 to +150	°C

THERMAL CHARACTERISTICS

Characteristic	Symbol	Max	Unit
Thermal Resistance, Junction to Ambient	$R_{\theta JA}$	200	°C/W
Thermal Resistance, Junction to Case	$R_{\theta JC}$	83.3	°C/W

ELECTRICAL CHARACTERISTICS ($T_A = 25°C$ unless otherwise noted)

Characteristic	Symbol	Min	Max	Unit
OFF CHARACTERISTICS				
Collector–Emitter Breakdown Voltage ($I_C = 10$ mAdc, $I_B = 0$)	$V_{(BR)CEO}$	40	—	Vdc
Collector–Base Breakdown Voltage ($I_C = 10$ μAdc, $I_E = 0$)	$V_{(BR)CBO}$	75	—	Vdc
Emitter–Base Breakdown Voltage ($I_E = 10$ μAdc, $I_C = 0$)	$V_{(BR)EBO}$	6.0	—	Vdc
Collector Cutoff Current ($V_{CE} = 60$ Vdc, $V_{EB(off)} = 3.0$ Vdc)	I_{CEX}	—	10	nAdc
Collector Cutoff Current ($V_{CB} = 60$ Vdc, $I_E = 0$) ($V_{CB} = 60$ Vdc, $I_E = 0$, $T_A = 150°C$)	I_{CBO}	— —	0.01 10	μAdc
Emitter Cutoff Current ($V_{EB} = 3.0$ Vdc, $I_C = 0$)	I_{EBO}	—	10	nAdc
Collector Cutoff Current ($V_{CE} = 10$ V)	I_{CEO}	—	10	nAdc
Base Cutoff Current ($V_{CE} = 60$ Vdc, $V_{EB(off)} = 3.0$ Vdc)	I_{BEX}	—	20	nAdc

 **MOTOROLA**

P2N2222A

ELECTRICAL CHARACTERISTICS (T_A = 25°C unless otherwise noted) (Continued)

Characteristic	Symbol	Min	Max	Unit
ON CHARACTERISTICS				
DC Current Gain	h_{FE}			—
(I_C = 0.1 mAdc, V_{CE} = 10 Vdc)		35	—	
(I_C = 1.0 mAdc, V_{CE} = 10 Vdc)		50	—	
(I_C = 10 mAdc, V_{CE} = 10 Vdc)		75	—	
(I_C = 10 mAdc, V_{CE} = 10 Vdc, T_A = −55°C)		35	—	
(I_C = 150 mAdc, V_{CE} = 10 Vdc)[1]		100	300	
(I_C = 150 mAdc, V_{CE} = 1.0 Vdc)[1]		50	—	
(I_C = 500 mAdc, V_{CE} = 10 Vdc)[1]		40	—	
Collector−Emitter Saturation Voltage[1]	$V_{CE(sat)}$			Vdc
(I_C = 150 mAdc, I_B = 15 mAdc)		—	0.3	
(I_C = 500 mAdc, I_B = 50 mAdc)		—	1.0	
Base−Emitter Saturation Voltage[1]	$V_{BE(sat)}$			Vdc
(I_C = 150 mAdc, I_B = 15 mAdc)		0.6	1.2	
(I_C = 500 mAdc, I_B = 50 mAdc)		—	2.0	
SMALL−SIGNAL CHARACTERISTICS				
Current−Gain — Bandwidth Product[2]	f_T	300	—	MHz
(I_C = 20 mAdc, V_{CE} = 20 Vdc, f = 100 MHz)				
Output Capacitance	C_{obo}	—	8.0	pF
(V_{CB} = 10 Vdc, I_E = 0, f = 1.0 MHz)				
Input Capacitance	C_{ibo}	—	25	pF
(V_{EB} = 0.5 Vdc, I_C = 0, f = 1.0 MHz)				
Input Impedance	h_{ie}			$k\Omega$
(I_C = 1.0 mAdc, V_{CE} = 10 Vdc, f = 1.0 kHz)		2.0	8.0	
(I_C = 10 mAdc, V_{CE} = 10 Vdc, f = 1.0 kHz)		0.25	1.25	
Voltage Feedback Ratio	h_{re}			$\times 10^{-4}$
(I_C = 1.0 mAdc, V_{CE} = 10 Vdc, f = 1.0 kHz)		—	8.0	
(I_C = 10 mAdc, V_{CE} = 10 Vdc, f = 1.0 kHz)		—	4.0	
Small−Signal Current Gain	h_{fe}			—
(I_C = 1.0 mAdc, V_{CE} = 10 Vdc, f = 1.0 kHz)		50	300	
(I_C = 10 mAdc, V_{CE} = 10 Vdc, f = 1.0 kHz)		75	375	
Output Admittance	h_{oe}			$\mu mhos$
(I_C = 1.0 mAdc, V_{CE} = 10 Vdc, f = 1.0 kHz)		5.0	35	
(I_C = 10 mAdc, V_{CE} = 10 Vdc, f = 1.0 kHz)		25	200	
Collector Base Time Constant	$rb'C_c$	—	150	ps
(I_E = 20 mAdc, V_{CB} = 20 Vdc, f = 31.8 MHz)				
Noise Figure	N_F	—	4.0	dB
(I_C = 100 μAdc, V_{CE} = 10 Vdc, R_S = 1.0 kΩ, f = 1.0 kHz)				

SWITCHING CHARACTERISTICS

Delay Time	(V_{CC} = 30 Vdc, $V_{BE(off)}$ = −2.0 Vdc,	t_d	—	10	ns
Rise Time	I_C = 150 mAdc, I_{B1} = 15 mAdc) (Figure 1)	t_r	—	25	ns
Storage Time	(V_{CC} = 30 Vdc, I_C = 150 mAdc,	t_s	—	225	ns
Fall Time	I_{B1} = I_{B2} = 15 mAdc) (Figure 2)	t_f	—	60	ns

1. Pulse Test: Pulse Width ≤ 300 μs, Duty Cycle ≤ 2.0%.
2. f_T is defined as the frequency at which $|h_{fe}|$ extrapolates to unity.

P2N2222A

SWITCHING TIME EQUIVALENT TEST CIRCUITS

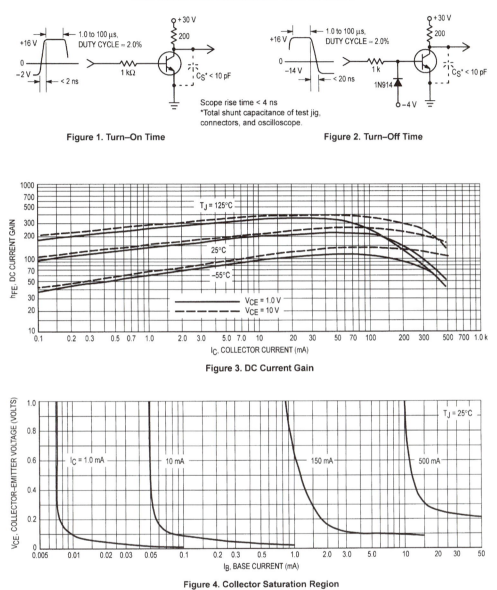

Figure 1. Turn–On Time

Scope rise time < 4 ns
*Total shunt capacitance of test jig, connectors, and oscilloscope.

Figure 2. Turn–Off Time

Figure 3. DC Current Gain

Figure 4. Collector Saturation Region

P2N2222A

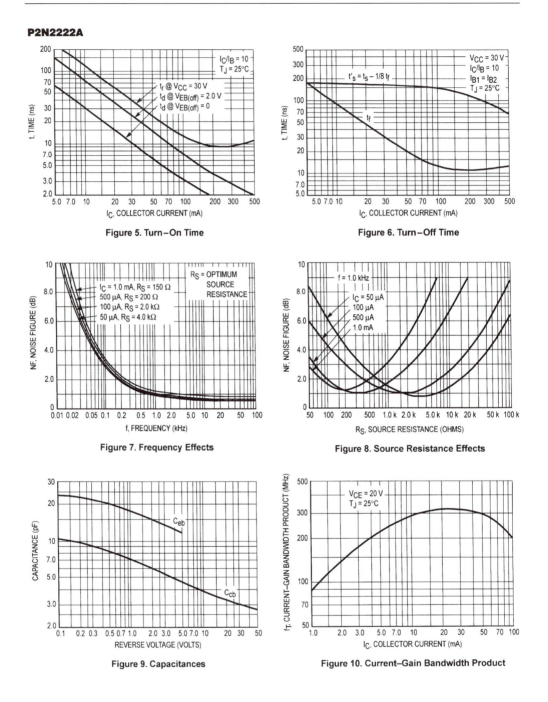

Figure 5. Turn–On Time

Figure 6. Turn–Off Time

Figure 7. Frequency Effects

Figure 8. Source Resistance Effects

Figure 9. Capacitances

Figure 10. Current–Gain Bandwidth Product

Motorola Small–Signal Transistors, FETs and Diodes Device Data

P2N2222A

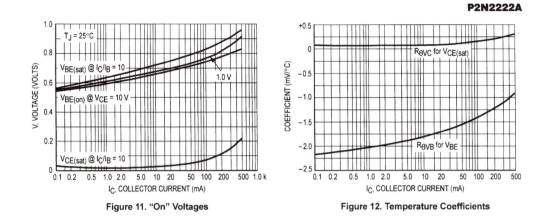

Figure 11. "On" Voltages   Figure 12. Temperature Coefficients

P2N2222A

PACKAGE DIMENSIONS

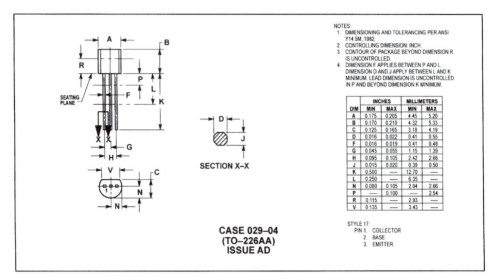

NOTES:
1. DIMENSIONING AND TOLERANCING PER ANSI Y14.5M, 1982.
2. CONTROLLING DIMENSION: INCH.
3. CONTOUR OF PACKAGE BEYOND DIMENSION R IS UNCONTROLLED.
4. DIMENSION F APPLIES BETWEEN P AND L. DIMENSION D AND J APPLY BETWEEN L AND K MINIMUM. LEAD DIMENSION IS UNCONTROLLED IN P AND BEYOND DIMENSION K MINIMUM.

DIM	INCHES		MILLIMETERS	
	MIN	MAX	MIN	MAX
A	0.175	0.205	4.45	5.20
B	0.170	0.210	4.32	5.33
C	0.125	0.165	3.18	4.19
D	0.016	0.022	0.41	0.55
F	0.016	0.019	0.41	0.48
G	0.045	0.055	1.15	1.39
H	0.095	0.105	2.42	2.66
J	0.015	0.020	0.39	0.50
K	0.500	—	12.70	—
L	0.250	—	6.35	—
N	0.080	0.105	2.04	2.66
P	—	0.100	—	2.54
R	0.115	—	2.93	—
V	0.135	—	3.43	—

SECTION X–X

CASE 029–04
(TO–226AA)
ISSUE AD

STYLE 17:
PIN 1. COLLECTOR
2. BASE
3. EMITTER

How to reach us:
USA/EUROPE/Locations Not Listed: Motorola Literature Distribution; P.O. Box 20912; Phoenix, Arizona 85036. 1–800–441–2447 or 602–303–5454

MFAX: RMFAX0@email.sps.mot.com – TOUCHTONE 602–244–6609
INTERNET: http://Design–NET.com

JAPAN: Nippon Motorola Ltd.; Tatsumi–SPD–JLDC, 6F Seibu–Butsuryu–Center, 3–14–2 Tatsumi Koto–Ku, Tokyo 135, Japan. 03–81–3521–8315

ASIA/PACIFIC: Motorola Semiconductors H.K. Ltd.; 8B Tai Ping Industrial Park, 51 Ting Kok Road, Tai Po, N.T., Hong Kong. 852–26629298

 **MOTOROLA**

◊

P2N2222A/D

MOTOROLA
■ **SEMICONDUCTOR** ■
TECHNICAL DATA

2N3553

The RF Line

2.5 W — 175 MHz

HIGH FREQUENCY
TRANSISTOR

NPN SILICON

NPN SILICON HIGH-FREQUENCY TRANSISTOR

. . . designed for amplifier and oscillator applications in military and industrial equipment. Suitable for use as output, driver or pre-driver stages in VHF equipment.

* Specified 175 MHz, 28 Vdc Characteristics —
  Output Power = 2.5 Watts
  Minimum Gain = 10 dB
  Efficiency = 50%

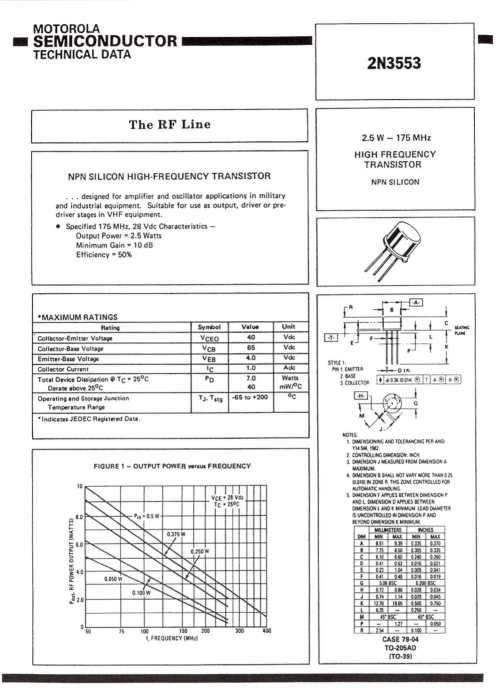

*MAXIMUM RATINGS

Rating	Symbol	Value	Unit
Collector-Emitter Voltage	V_{CEO}	40	Vdc
Collector-Base Voltage	V_{CB}	65	Vdc
Emitter-Base Voltage	V_{EB}	4.0	Vdc
Collector Current	I_C	1.0	Adc
Total Device Dissipation @ T_C = 25°C	P_D	7.0	Watts
Derate above 25°C		40	mW/°C
Operating and Storage Junction Temperature Range	T_J, T_{stg}	-65 to +200	°C

*Indicates JEDEC Registered Data.

STYLE 1:
PIN 1. EMITTER
2. BASE
3. COLLECTOR

NOTES:
1. DIMENSIONING AND TOLERANCING PER ANSI Y14.5M, 1982.
2. CONTROLLING DIMENSION: INCH.
3. DIMENSION J MEASURED FROM DIMENSION A MAXIMUM.
4. DIMENSION B SHALL NOT VARY MORE THAN 0.25 (0.010) IN ZONE R. THIS ZONE CONTROLLED FOR AUTOMATIC HANDLING.
5. DIMENSION F APPLIES BETWEEN DIMENSION P AND L. DIMENSION D APPLIES BETWEEN DIMENSION L AND K MINIMUM. LEAD DIAMETER IS UNCONTROLLED IN DIMENSION P AND BEYOND DIMENSION K MINIMUM.

DIM	MILLIMETERS MIN	MILLIMETERS MAX	INCHES MIN	INCHES MAX
A	8.51	9.39	0.335	0.370
B	7.75	8.50	0.305	0.335
C	6.10	6.60	0.240	0.260
D	0.41	0.53	0.016	0.021
E	0.23	1.04	0.009	0.041
F	0.41	0.48	0.016	0.019
G	5.08 BSC		0.200 BSC	
H	0.72	0.86	0.028	0.034
J	0.74	1.14	0.029	0.045
K	12.70	19.05	0.500	0.750
L	6.35	—	0.250	—
M	45° BSC		45° BSC	
P	—	1.27	—	0.050
R	2.54	—	0.100	—

CASE 79-04
TO-205AD
(TO-39)

FIGURE 1 — OUTPUT POWER versus FREQUENCY

V_{CE} = 28 Vdc
T_C = 25°C

P_{in} = 0.5 W
0.375 W
0.250 W
0.100 W
0.050 W

P_{out}, RF POWER OUTPUT (WATTS)

f, FREQUENCY (MHz)

2N3553

*ELECTRICAL CHARACTERISTICS (T_A = 25°C unless otherwise noted)

Characteristic	Symbol	Min	Typ	Max	Unit
OFF CHARACTERISTICS					
Collector-Emitter Sustaining Voltage (1) (I_C = 200 mAdc, I_B = 0)	$V_{CEO(sus)}$	40	–	–	Vdc
Emitter-Base Breakdown Voltage (I_E = 0.1 mAdc, I_C = 0)	$V_{(BR)EBO}$	4.0	–	–	Vdc
Collector Cutoff Current (V_{CE} = 30 Vdc, I_B = 0)	I_{CEO}	–	–	0.1	mAdc
Collector Cutoff Current (V_{CE} = 30 Vdc, $V_{BE(off)}$ = 1.5 Vdc, T_C = 200°C) (V_{CE} = 65 Vdc, $V_{BE(off)}$ = 1.5 Vdc)	I_{CEX}	– –	– –	5.0 1.0	mAdc
Emitter Cutoff Current (V_{BE} = 4.0 Vdc, I_C = 0)	I_{EBO}	–	–	0.1	mAdc
ON CHARACTERISTICS					
DC Current Gain (I_C = 250 mAdc, V_{CE} = 5.0 Vdc)	h_{FE}	10	–	–	–
Collector-Emitter Saturation Voltage (I_C = 250 mAdc, I_B = 50 mAdc)	$V_{CE(sat)}$	–	–	1.0	Vdc
DYNAMIC CHARACTERISTICS					
Current-Gain–Bandwidth Product (I_C = 100 mAdc, V_{CE} = 28 Vdc, f = 100 MHz)	f_T	–	500	–	MHz
Output Capacitance (V_{CB} = 30 Vdc, I_E = 0, f = 100 kHz)	C_{ob}	–	8.0	10	pF
FUNCTIONAL TESTS					
Power Input (V_{CE} = 28 Vdc, P_{out} = 2.5 Watts, f = 175 MHz)	P_{in}	–	–	0.25	Watt
Common-Emitter Amplifier Power Gain (V_{CE} = 28 Vdc, P_{out} = 2.5 Watts, f = 175 MHz)	G_{pe}	10	–	–	dB
Collector Efficiency (V_{CE} = 28 Vdc, P_{out} = 2.5 Watts, f = 175 MHz)	η	50	–	–	%

*Indicates JEDEC Registered Data
(1) Pulsed thru a 25 mH inductor.

FIGURE 2 – 175 MHz TEST CIRCUIT SCHEMATIC

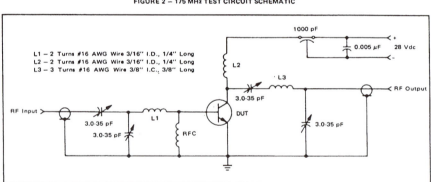

L1 – 2 Turns #16 AWG Wire 3/16" I.D., 1/4" Long
L2 – 2 Turns #16 AWG Wire 3/16" I.D., 1/4" Long
L3 – 3 Turns #16 AWG Wire 3/8" I.C., 3/8" Long

MOTOROLA
SEMICONDUCTOR TECHNICAL DATA

Order this document
by 2N3905/D

General Purpose Transistors
PNP Silicon

**2N3905
2N3906***

*Motorola Preferred Device

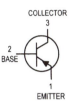

COLLECTOR
3

2
BASE

1
EMITTER

**CASE 29–04, STYLE 1
TO–92 (TO–226AA)**

MAXIMUM RATINGS

Rating	Symbol	Value	Unit
Collector–Emitter Voltage	V_{CEO}	40	Vdc
Collector–Base Voltage	V_{CBO}	40	Vdc
Emitter–Base Voltage	V_{EBO}	5.0	Vdc
Collector Current — Continuous	I_C	200	mAdc
Total Device Dissipation @ T_A = 25°C Derate above 25°C	P_D	625 5.0	mW mW/°C
Total Power Dissipation @ T_A = 60°C	P_D	250	mW
Total Device Dissipation @ T_C = 25°C Derate above 25°C	P_D	1.5 12	Watts mW/°C
Operating and Storage Junction Temperature Range	T_J, T_{stg}	–55 to +150	°C

THERMAL CHARACTERISTICS[1]

Characteristic	Symbol	Max	Unit
Thermal Resistance, Junction to Ambient	$R_{\theta JA}$	200	°C/W
Thermal Resistance, Junction to Case	$R_{\theta JC}$	83.3	°C/W

ELECTRICAL CHARACTERISTICS (T_A = 25°C unless otherwise noted)

Characteristic	Symbol	Min	Max	Unit
OFF CHARACTERISTICS				
Collector–Emitter Breakdown Voltage [2] (I_C = 1.0 mAdc, I_B = 0)	$V_{(BR)CEO}$	40	—	Vdc
Collector–Base Breakdown Voltage (I_C = 10 µAdc, I_E = 0)	$V_{(BR)CBO}$	40	—	Vdc
Emitter–Base Breakdown Voltage (I_E = 10 µAdc, I_C = 0)	$V_{(BR)EBO}$	5.0	—	Vdc
Base Cutoff Current (V_{CE} = 30 Vdc, V_{EB} = 3.0 Vdc)	I_{BL}	—	50	nAdc
Collector Cutoff Current (V_{CE} = 30 Vdc, V_{EB} = 3.0 Vdc)	I_{CEX}	—	50	nAdc

1. Indicates Data in addition to JEDEC Requirements.
2. Pulse Test: Pulse Width ≤ 300 µs; Duty Cycle ≤ 2.0%.

Preferred devices are Motorola recommended choices for future use and best overall value.

REV 2

 **MOTOROLA**

2N3905 2N3906

ELECTRICAL CHARACTERISTICS (T_A = 25°C unless otherwise noted) (Continued)

Characteristic		Symbol	Min	Max	Unit
ON CHARACTERISTICS(1)					
DC Current Gain		h_{FE}			—
(I_C = 0.1 mAdc, V_{CE} = 1.0 Vdc)	2N3905		30	—	
	2N3906		60	—	
(I_C = 1.0 mAdc, V_{CE} = 1.0 Vdc)	2N3905		40	—	
	2N3906		80	—	
(I_C = 10 mAdc, V_{CE} = 1.0 Vdc)	2N3905		50	150	
	2N3906		100	300	
(I_C = 50 mAdc, V_{CE} = 1.0 Vdc)	2N3905		30	—	
	2N3906		60	—	
(I_C = 100 mAdc, V_{CE} = 1.0 Vdc)	2N3905		15	—	
	2N3906		30	—	
Collector–Emitter Saturation Voltage		$V_{CE(sat)}$			Vdc
(I_C = 10 mAdc, I_B = 1.0 mAdc)			—	0.25	
(I_C = 50 mAdc, I_B = 5.0 mAdc			—	0.4	
Base–Emitter Saturation Voltage		$V_{BE(sat)}$			Vdc
(I_C = 10 mAdc, I_B = 1.0 mAdc)			0.65	0.85	
(I_C = 50 mAdc, I_B = 5.0 mAdc)			—	0.95	
SMALL–SIGNAL CHARACTERISTICS					
Current–Gain — Bandwidth Product		f_T			MHz
(I_C = 10 mAdc, V_{CE} = 20 Vdc, f = 100 MHz)	2N3905		200	—	
	2N3906		250	—	
Output Capacitance		C_{obo}	—	4.5	pF
(V_{CB} = 5.0 Vdc, I_E = 0, f = 1.0 MHz)					
Input Capacitance		C_{ibo}	—	10.0	pF
(V_{EB} = 0.5 Vdc, I_C = 0, f = 1.0 MHz)					
Input Impedance		h_{ie}			k Ω
(I_C = 1.0 mAdc, V_{CE} = 10 Vdc, f = 1.0 kHz)	2N3905		0.5	8.0	
	2N3906		2.0	12	
Voltage Feedback Ratio		h_{re}			X 10^{-4}
(I_C = 1.0 mAdc, V_{CE} = 10 Vdc, f = 1.0 kHz)	2N3905		0.1	5.0	
	2N3906		0.1	10	
Small–Signal Current Gain		h_{fe}			—
(I_C = 1.0 mAdc, V_{CE} = 10 Vdc, f = 1.0 kHz)	2N3905		50	200	
	2N3906		100	400	
Output Admittance		h_{oe}			µmhos
(I_C = 1.0 mAdc, V_{CE} = 10 Vdc, f = 1.0 kHz)	2N3905		1.0	40	
	2N3906		3.0	60	
Noise Figure		NF			dB
(I_C = 100 µAdc, V_{CE} = 5.0 Vdc, R_S = 1.0 k Ω, f = 1.0 kHz)	2N3905		—	5.0	
	2N3906		—	4.0	

SWITCHING CHARACTERISTICS

			Symbol	Min	Max	Unit
Delay Time	(V_{CC} = 3.0 Vdc, V_{BE} = 0.5 Vdc,		t_d	—	35	ns
Rise Time	I_C = 10 mAdc, I_{B1} = 1.0 mAdc)		t_r	—	35	ns
Storage Time		2N3905	t_s	—	200	ns
		2N3906		—	225	
Fall Time	(V_{CC} = 3.0 Vdc, I_C = 10 mAdc, I_{B1} = I_{B2} = 1.0 mAd	2N3905	t_f	—	60	ns
		2N3906		—	75	

1. Pulse Test: Pulse Width ≤ 300 µs; Duty Cycle ≤ 2.0%.

2N3905 2N3906

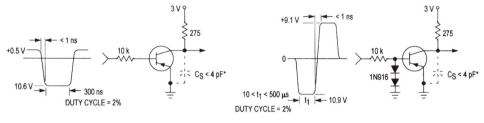

* Total shunt capacitance of test jig and connectors

**Figure 1. Delay and Rise Time
Equivalent Test Circuit**

**Figure 2. Storage and Fall Time
Equivalent Test Circuit**

TYPICAL TRANSIENT CHARACTERISTICS

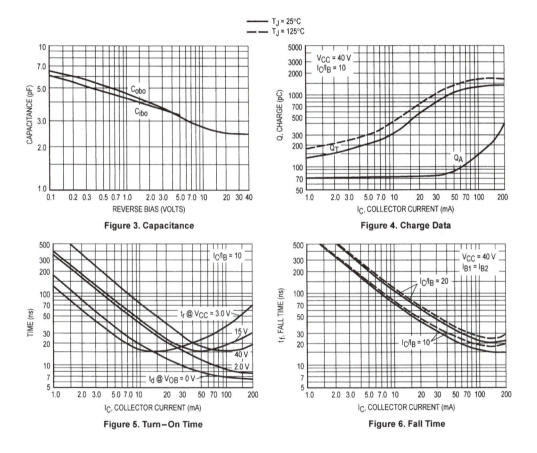

Figure 3. Capacitance

Figure 4. Charge Data

Figure 5. Turn–On Time

Figure 6. Fall Time

2N3905 2N3906

TYPICAL AUDIO SMALL–SIGNAL CHARACTERISTICS
NOISE FIGURE VARIATIONS
($V_{CE} = -5.0$ Vdc, $T_A = 25°C$, Bandwidth = 1.0 Hz)

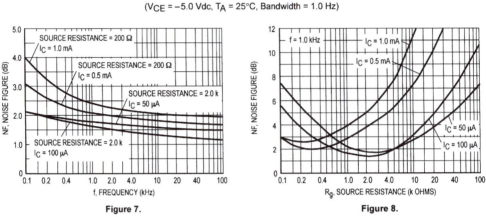

Figure 7.

Figure 8.

h PARAMETERS
($V_{CE} = -10$ Vdc, $f = 1.0$ kHz, $T_A = 25°C$)

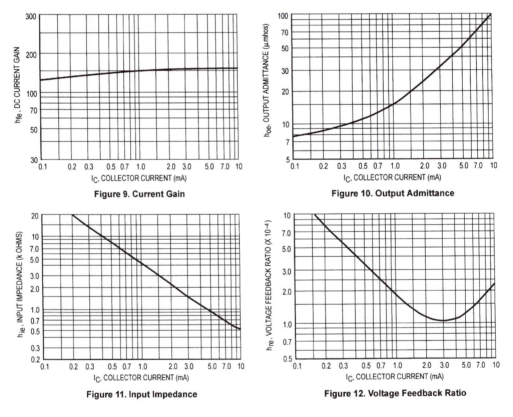

Figure 9. Current Gain

Figure 10. Output Admittance

Figure 11. Input Impedance

Figure 12. Voltage Feedback Ratio

Motorola Small–Signal Transistors, FETs and Diodes Device Data

2N3905 2N3906

TYPICAL STATIC CHARACTERISTICS

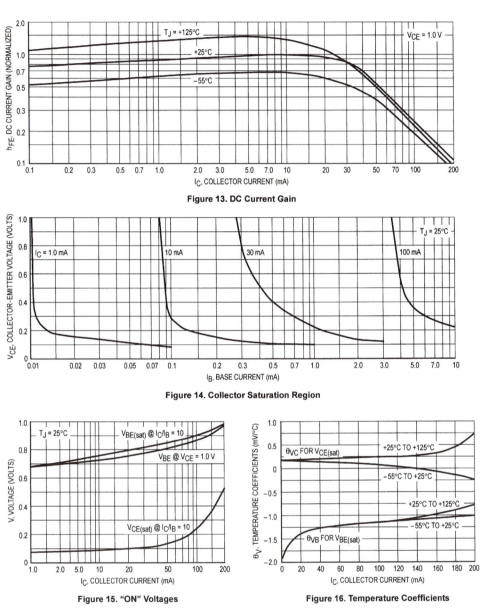

Figure 13. DC Current Gain

Figure 14. Collector Saturation Region

Figure 15. "ON" Voltages

Figure 16. Temperature Coefficients

2N3905 2N3906

PACKAGE DIMENSIONS

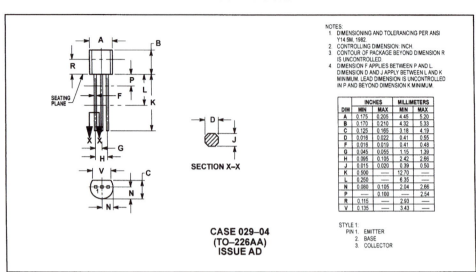

SECTION X-X

CASE 029-04
(TO-226AA)
ISSUE AD

NOTES:
1. DIMENSIONING AND TOLERANCING PER ANSI Y14.5M, 1982.
2. CONTROLLING DIMENSION: INCH.
3. CONTOUR OF PACKAGE BEYOND DIMENSION R IS UNCONTROLLED.
4. DIMENSION F APPLIES BETWEEN P AND L. DIMENSION D AND J APPLY BETWEEN L AND K MINIMUM. LEAD DIMENSION IS UNCONTROLLED IN P AND BEYOND DIMENSION K MINIMUM.

DIM	INCHES MIN	INCHES MAX	MILLIMETERS MIN	MILLIMETERS MAX
A	0.175	0.205	4.45	5.20
B	0.170	0.210	4.32	5.33
C	0.125	0.165	3.18	4.19
D	0.016	0.022	0.41	0.55
F	0.016	0.019	0.41	0.48
G	0.045	0.055	1.15	1.39
H	0.095	0.105	2.42	2.66
J	0.015	0.020	0.39	0.50
K	0.500	—	12.70	—
L	0.250	—	6.35	—
N	0.080	0.105	2.04	2.66
P	—	0.100	—	2.54
R	0.115	—	2.93	—
V	0.135	—	3.43	—

STYLE 1:
PIN 1. EMITTER
2. BASE
3. COLLECTOR

How to reach us:
USA/EUROPE/Locations Not Listed: Motorola Literature Distribution; P.O. Box 20912; Phoenix, Arizona 85036. 1–800–441–2447 or 602–303–5454

MFAX: RMFAX0@email.sps.mot.com – TOUCHTONE 602–244–6609
INTERNET: http://Design–NET.com

JAPAN: Nippon Motorola Ltd.; Tatsumi–SPD–JLDC, 6F Seibu–Butsuryu–Center, 3–14–2 Tatsumi Koto–Ku, Tokyo 135, Japan. 03–81–3521–8315

ASIA/PACIFIC: Motorola Semiconductors H.K. Ltd.; 8B Tai Ping Industrial Park, 51 Ting Kok Road, Tai Po, N.T., Hong Kong. 852–26629298

 **MOTOROLA**

2N3905/D

MOTOROLA
SEMICONDUCTOR TECHNICAL DATA

Order this document
by 2N4123/D

General Purpose Transistors

NPN Silicon

2N4123
2N4124

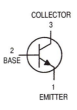

COLLECTOR
3

2
BASE

1
EMITTER

CASE 29–04, STYLE 1
TO–92 (TO–226AA)

MAXIMUM RATINGS

Rating	Symbol	2N4123	2N4124	Unit
Collector–Emitter Voltage	V_{CEO}	30	25	Vdc
Collector–Base Voltage	V_{CBO}	40	30	Vdc
Emitter–Base Voltage	V_{EBO}	5.0		Vdc
Collector Current — Continuous	I_C	200		mAdc
Total Device Dissipation @ T_A = 25°C Derate above 25°C	P_D	625 5.0		mW mW/°C
Total Device Dissipation @ T_C = 25°C Derate above 25°C	P_D	1.5 12		Watts mW/°C
Operating and Storage Junction Temperature Range	T_J, T_{stg}	−55 to +150		°C

THERMAL CHARACTERISTICS

Characteristic	Symbol	Max	Unit
Thermal Resistance, Junction to Ambient	$R_{\theta JA}$	200	°C/W
Thermal Resistance, Junction to Case	$R_{\theta JC}$	83.3	°C/W

ELECTRICAL CHARACTERISTICS (T_A = 25°C unless otherwise noted)

Characteristic	Symbol	Min	Max	Unit
OFF CHARACTERISTICS				
Collector–Emitter Breakdown Voltage[1] (I_C = 1.0 mAdc, I_E = 0) 2N4123 2N4124	$V_{(BR)CEO}$	 30 25	 — —	Vdc
Collector–Base Breakdown Voltage (I_C = 10 µAdc, I_E = 0) 2N4123 2N4124	$V_{(BR)CBO}$	 40 30	 — —	Vdc
Emitter–Base Breakdown Voltage (I_E = 10 µAdc, I_C = 0)	$V_{(BR)EBO}$	5.0	—	Vdc
Collector Cutoff Current (V_{CB} = 20 Vdc, I_E = 0)	I_{CBO}	—	50	nAdc
Emitter Cutoff Current (V_{EB} = 3.0 Vdc, I_C = 0)	I_{EBO}	—	50	nAdc

1. Pulse Test: Pulse Width = 300 µs, Duty Cycle = 2.0%.

 **MOTOROLA**

2N4123 2N4124

ELECTRICAL CHARACTERISTICS (T_A = 25°C unless otherwise noted) (Continued)

Characteristic		Symbol	Min	Max	Unit
ON CHARACTERISTICS					
DC Current Gain[1]		h_{FE}			—
(I_C = 2.0 mAdc, V_{CE} = 1.0 Vdc) 2N4123			50	150	
2N4124			120	360	
(I_C = 50 mAdc, V_{CE} = 1.0 Vdc) 2N4123			25	—	
2N4124			60	—	
Collector–Emitter Saturation Voltage[1]		$V_{CE(sat)}$	—	0.3	Vdc
(I_C = 50 mAdc, I_B = 5.0 mAdc)					
Base–Emitter Saturation Voltage[1]		$V_{BE(sat)}$	—	0.95	Vdc
(I_C = 50 mAdc, I_B = 5.0 mAdc)					
SMALL–SIGNAL CHARACTERISTICS					
Current–Gain — Bandwidth Product		f_T			MHz
(I_C = 10 mAdc, V_{CE} = 20 Vdc, f = 100 MHz) 2N4123			250	—	
2N4124			300	—	
Input Capacitance		C_{ibo}	—	8.0	pF
(V_{EB} = 0.5 Vdc, I_C = 0, f = 1.0 MHz)					
Collector–Base Capacitance		C_{cb}	—	4.0	pF
(I_E = 0, V_{CB} = 5.0 V, f = 1.0 MHz)					
Small–Signal Current Gain		h_{fe}			—
(I_C = 2.0 mAdc, V_{CE} = 10 Vdc, R_S = 10 k ohm, f = 1.0 kHz) 2N4123			50	200	
2N4124			120	480	
Current Gain — High Frequency		$|h_{fe}|$			—
(I_C = 10 mAdc, V_{CE} = 20 Vdc, f = 100 MHz) 2N4123			2.5	—	
2N4124			3.0	—	
(I_C = 2.0 mAdc, V_{CE} = 10 V, f = 1.0 kHz) 2N4123			50	200	
(I_C = 2.0 mAdc, V_{CE} = 10 V, f = 1.0 kHz) 2N4124			120	480	
Noise Figure		NF			dB
(I_C = 100 μAdc, V_{CE} = 5.0 Vdc, R_S = 1.0 k ohm, f = 1.0 kHz) 2N4123			—	6.0	
2N4124			—	5.0	

1. Pulse Test: Pulse Width = 300 μs, Duty Cycle = 2.0%.

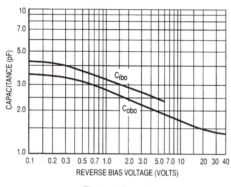

Figure 1. Capacitance

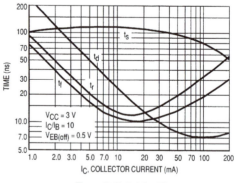

Figure 2. Switching Times

AUDIO SMALL–SIGNAL CHARACTERISTICS
NOISE FIGURE
(V_{CE} = 5 Vdc, T_A = 25°C)
Bandwidth = 1.0 Hz

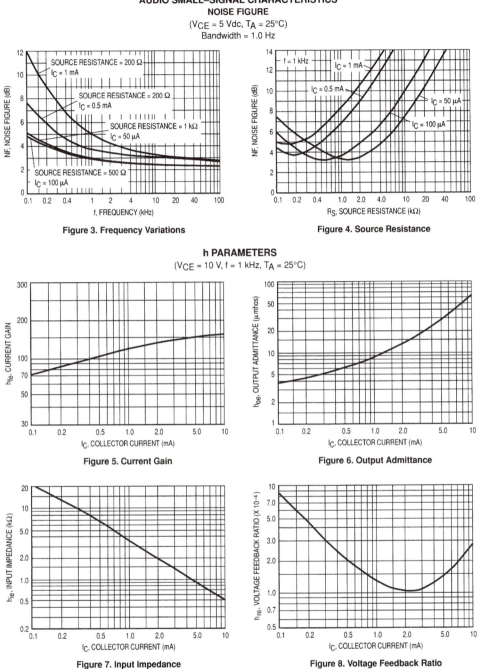

Figure 3. Frequency Variations

Figure 4. Source Resistance

h PARAMETERS
(V_{CE} = 10 V, f = 1 kHz, T_A = 25°C)

Figure 5. Current Gain

Figure 6. Output Admittance

Figure 7. Input Impedance

Figure 8. Voltage Feedback Ratio

2N4123 2N4124

STATIC CHARACTERISTICS

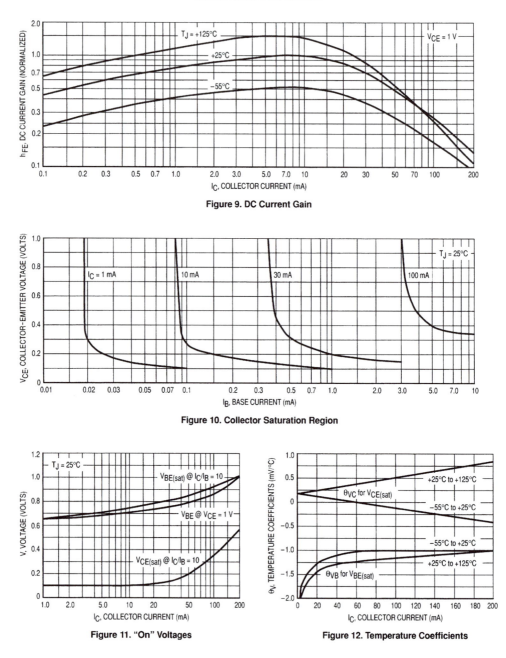

Figure 9. DC Current Gain

Figure 10. Collector Saturation Region

Figure 11. "On" Voltages

Figure 12. Temperature Coefficients

2N4123 2N4124

PACKAGE DIMENSIONS

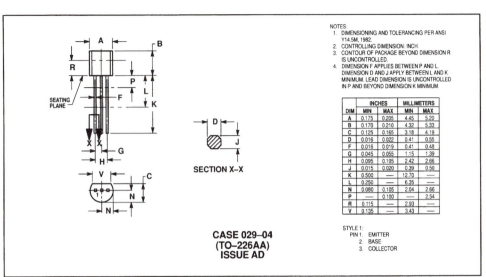

NOTES:
1. DIMENSIONING AND TOLERANCING PER ANSI Y14.5M, 1982.
2. CONTROLLING DIMENSION: INCH.
3. CONTOUR OF PACKAGE BEYOND DIMENSION R IS UNCONTROLLED.
4. DIMENSION F APPLIES BETWEEN P AND L. DIMENSION D AND J APPLY BETWEEN L AND K MINIMUM. LEAD DIMENSION IS UNCONTROLLED IN P AND BEYOND DIMENSION K MINIMUM.

DIM	INCHES		MILLIMETERS	
	MIN	MAX	MIN	MAX
A	0.175	0.205	4.45	5.20
B	0.170	0.210	4.32	5.33
C	0.125	0.165	3.18	4.19
D	0.016	0.022	0.41	0.55
F	0.016	0.019	0.41	0.48
G	0.045	0.055	1.15	1.39
H	0.095	0.105	2.42	2.66
J	0.015	0.020	0.39	0.50
K	0.500	—	12.70	—
L	0.250	—	6.35	—
N	0.080	0.105	2.04	2.66
P	—	0.100	—	2.54
R	0.115	—	2.93	—
V	0.135	—	3.43	—

SECTION X–X

CASE 029–04
(TO–226AA)
ISSUE AD

STYLE 1:
PIN 1. EMITTER
2. BASE
3. COLLECTOR

2N4123 2N4124

How to reach us:
USA/EUROPE: Motorola Literature Distribution;
P.O. Box 20912; Phoenix, Arizona 85036. 1–800–441–2447

MFAX: RMFAX0@email.sps.mot.com – TOUCHTONE (602) 244–6609
INTERNET: http://Design–NET.com

JAPAN: Nippon Motorola Ltd.; Tatsumi–SPD–JLDC, Toshikatsu Otsuki,
6F Seibu–Butsuryu–Center, 3–14–2 Tatsumi Koto–Ku, Tokyo 135, Japan. 03–3521–8315

HONG KONG: Motorola Semiconductors H.K. Ltd.; 8B Tai Ping Industrial Park,
51 Ting Kok Road, Tai Po, N.T., Hong Kong. 852–26629298

◊

2N4123/D

MOTOROLA
SEMICONDUCTOR TECHNICAL DATA

Order this document
by J308/D

JFET VHF/UHF Amplifiers
N–Channel — Depletion

1 DRAIN

3
GATE

2 SOURCE

J308
J309
J310

Motorola Preferred Devices

1
2
3

CASE 29–04, STYLE 5
TO–92 (TO–226AA)

MAXIMUM RATINGS

Rating	Symbol	Value	Unit
Drain–Source Voltage	V_{DS}	25	Vdc
Gate–Source Voltage	V_{GS}	25	Vdc
Forward Gate Current	I_{GF}	10	mAdc
Total Device Dissipation @ T_A = 25°C Derate above 25°C	P_D	350 2.8	mW mW/°C
Junction Temperature Range	T_J	−65 to +125	°C
Storage Temperature Range	T_{stg}	−65 to +150	°C

ELECTRICAL CHARACTERISTICS (T_A = 25°C unless otherwise noted)

Characteristic	Symbol	Min	Typ	Max	Unit
OFF CHARACTERISTICS					
Gate–Source Breakdown Voltage (I_G = −1.0 µAdc, V_{DS} = 0)	$V_{(BR)GSS}$	−25	—	—	Vdc
Gate Reverse Current (V_{GS} = −15 Vdc, V_{DS} = 0, T_A = 25°C) (V_{GS} = −15 Vdc, V_{DS} = 0, T_A = +125°C)	I_{GSS}	 — —	 — —	 −1.0 −1.0	 nAdc µAdc
Gate Source Cutoff Voltage (V_{DS} = 10 Vdc, I_D = 1.0 nAdc) J308 J309 J310	$V_{GS(off)}$	 −1.0 −1.0 −2.0	 — — —	 −6.5 −4.0 −6.5	Vdc
ON CHARACTERISTICS					
Zero–Gate–Voltage Drain Current[1] (V_{DS} = 10 Vdc, V_{GS} = 0) J308 J309 J310	I_{DSS}	 12 12 24	 — — —	 60 30 60	mAdc
Gate–Source Forward Voltage (V_{DS} = 0, I_G = 1.0 mAdc)	$V_{GS(f)}$	—	—	1.0	Vdc
SMALL–SIGNAL CHARACTERISTICS					
Common–Source Input Conductance (V_{DS} = 10 Vdc, I_D = 10 mAdc, f = 100 MHz) J308 J309 J310	$Re(y_{is})$	 — — —	 0.7 0.7 0.5	 — — —	mmhos
Common–Source Output Conductance (V_{DS} = 10 Vdc, I_D = 10 mAdc, f = 100 MHz)	$Re(y_{os})$	—	0.25	—	mmhos
Common–Gate Power Gain (V_{DS} = 10 Vdc, I_D = 10 mAdc, f = 100 MHz)	G_{pg}	—	16	—	dB

1. Pulse Test: Pulse Width ≤ 300 µs, Duty Cycle ≤ 3.0%.

 **MOTOROLA**

J308 J309 J310

ELECTRICAL CHARACTERISTICS (T_A = 25°C unless otherwise noted) (Continued)

Characteristic		Symbol	Min	Typ	Max	Unit
SMALL–SIGNAL CHARACTERISTICS (continued)						
Common–Source Forward Transconductance (V_{DS} = 10 Vdc, I_D = 10 mAdc, f = 100 MHz)		$Re(y_{fs})$	—	12	—	mmhos
Common–Gate Input Conductance (V_{DS} = 10 Vdc, I_D = 10 mAdc, f = 100 MHz)		$Re(y_{ig})$	—	12	—	mmhos
Common–Source Forward Transconductance (V_{DS} = 10 Vdc, I_D = 10 mAdc, f = 1.0 kHz)	J308 J309 J310	g_{fs}	8000 10000 8000	— — —	20000 20000 18000	µmhos
Common–Source Output Conductance (V_{DS} = 10 Vdc, I_D = 10 mAdc, f = 1.0 kHz)		g_{os}	—	—	250	µmhos
Common–Gate Forward Transconductance (V_{DS} = 10 Vdc, I_D = 10 mAdc, f = 1.0 kHz)	J308 J309 J310	g_{fg}	— — —	13000 13000 12000	— — —	µmhos
Common–Gate Output Conductance (V_{DS} = 10 Vdc, I_D = 10 mAdc, f = 1.0 kHz)	J308 J309 J310	g_{og}	— — —	150 100 150	— — —	µmhos
Gate–Drain Capacitance (V_{DS} = 0, V_{GS} = –10 Vdc, f = 1.0 MHz)		C_{gd}	—	1.8	2.5	pF
Gate–Source Capacitance (V_{DS} = 0, V_{GS} = –10 Vdc, f = 1.0 MHz)		C_{gs}	—	4.3	5.0	pF
FUNCTIONAL CHARACTERISTICS						
Noise Figure (V_{DS} = 10 Vdc, I_D = 10 mAdc, f = 450 MHz)		NF	—	1.5	—	dB
Equivalent Short–Circuit Input Noise Voltage (V_{DS} = 10 Vdc, I_D = 10 mAdc, f = 100 Hz)		$\bar{e}_n$	—	10	—	$nV/\sqrt{Hz}$

J308 J309 J310

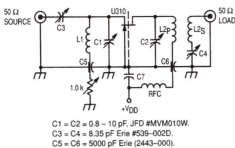

C1 = C2 = 0.8 – 10 pF, JFD #MVM010W.
C3 = C4 = 8.35 pF Erie #539–002D.
C5 = C6 = 5000 pF Erie (2443–000).
C7 = 1000 pF, Allen Bradley #FA5C.
RFC = 0.33 μH Miller #9230–30.
L1 = One Turn #16 Cu, 1/4″ I.D. (Air Core).
L2p = One Turn #16 Cu, 1/4″ I.D. (Air Core).
L2S = One Turn #16 Cu, 1/4″ I.D. (Air Core).

Figure 1. 450 MHz Common–Gate Amplifier Test Circuit

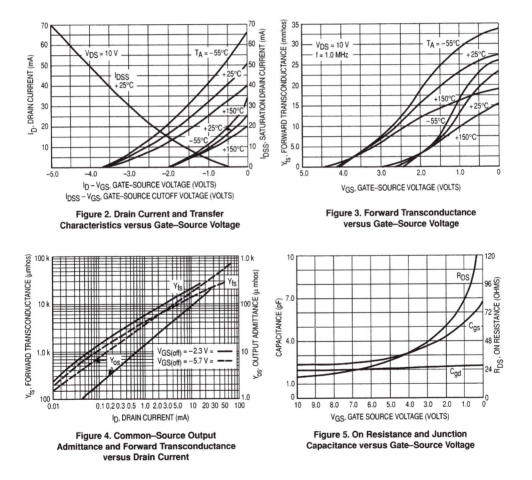

Figure 2. Drain Current and Transfer Characteristics versus Gate–Source Voltage

Figure 3. Forward Transconductance versus Gate–Source Voltage

Figure 4. Common–Source Output Admittance and Forward Transconductance versus Drain Current

Figure 5. On Resistance and Junction Capacitance versus Gate–Source Voltage

J308 J309 J310

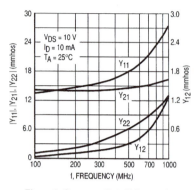

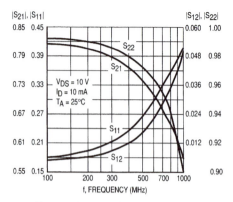

Figure 6. Common–Gate Y Parameter Magnitude versus Frequency

Figure 7. Common–Gate S Parameter Magnitude versus Frequency

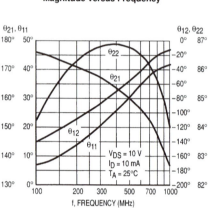

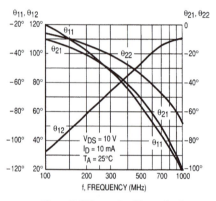

Figure 8. Common–Gate Y Parameter Phase–Angle versus Frequency

Figure 9. S Parameter Phase–Angle versus Frequency

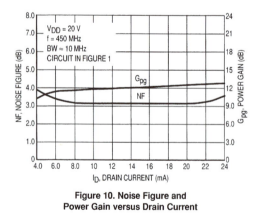

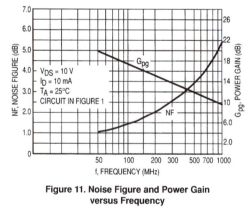

Figure 10. Noise Figure and Power Gain versus Drain Current

Figure 11. Noise Figure and Power Gain versus Frequency

J308 J309 J310

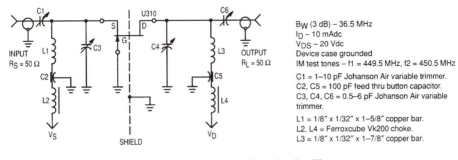

B_W (3 dB) – 36.5 MHz
I_D – 10 mAdc
V_{DS} – 20 Vdc
Device case grounded
IM test tones – f1 = 449.5 MHz, f2 = 450.5 MHz

C1 = 1–10 pF Johanson Air variable trimmer.
C2, C5 = 100 pF feed thru button capacitor.
C3, C4, C6 = 0.5–6 pF Johanson Air variable trimmer.

L1 = 1/8″ x 1/32″ x 1–5/8″ copper bar.
L2, L4 = Ferroxcube Vk200 choke.
L3 = 1/8″ x 1/32″ x 1–7/8″ copper bar.

Figure 12. 450 MHz IMD Evaluation Amplifier

Amplifier power gain and IMD products are a function of the load impedance. For the amplifier design shown above with C4 and C6 adjusted to reflect a load to the drain resulting in a nominal power gain of 9 dB, the 3rd order intercept point (IP) value is 29 dBm. Adjusting C4, C6 to provide larger load values will result in higher gain, smaller bandwidth and lower IP values. For example, a nominal gain of 13 dB can be achieved with an intercept point of 19 dBm.

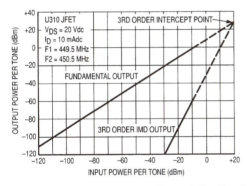

Example of intercept point plot use:
Assume two in–band signals of –20 dBm at the amplifier input. They will result in a 3rd-order IMD signal at the output of –90 dBm. Also, each signal level at the output will be –11 dBm, showing an amplifier gain of 9.0 dB and an intermodulation ratio (IMR) capability of 79 dB. The gain and IMR values apply only for signal levels below comparison.

Figure 13. Two-Tone 3rd-Order Intercept Point

J308 J309 J310

PACKAGE DIMENSIONS

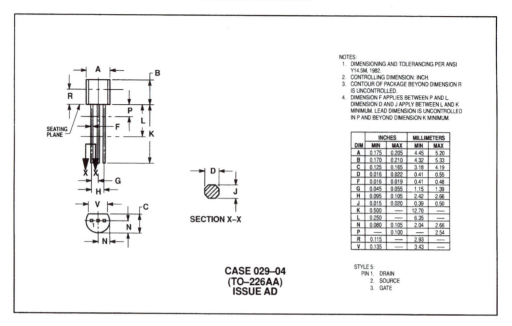

NOTES:
1. DIMENSIONING AND TOLERANCING PER ANSI Y14.5M, 1982.
2. CONTROLLING DIMENSION: INCH.
3. CONTOUR OF PACKAGE BEYOND DIMENSION R IS UNCONTROLLED.
4. DIMENSION F APPLIES BETWEEN P AND L. DIMENSION D AND J APPLY BETWEEN L AND K MINIMUM. LEAD DIMENSION IS UNCONTROLLED IN P AND BEYOND DIMENSION K MINIMUM.

DIM	INCHES MIN	INCHES MAX	MILLIMETERS MIN	MILLIMETERS MAX
A	0.175	0.205	4.45	5.20
B	0.170	0.210	4.32	5.33
C	0.125	0.165	3.18	4.19
D	0.016	0.022	0.41	0.55
F	0.016	0.019	0.41	0.48
G	0.045	0.055	1.15	1.39
H	0.095	0.105	2.42	2.66
J	0.015	0.020	0.39	0.50
K	0.500	—	12.70	—
L	0.250	—	6.35	—
N	0.080	0.105	2.04	2.66
P	—	0.100	—	2.54
R	0.115	—	2.93	—
V	0.135	—	3.43	—

SECTION X–X

CASE 029–04
(TO–226AA)
ISSUE AD

STYLE 5:
PIN 1. DRAIN
     2. SOURCE
     3. GATE

Order this document by MC78L00/D

MOTOROLA

Three-Terminal Low Current Positive Voltage Regulators

MC78L00, A Series

The MC78L00, A Series of positive voltage regulators are inexpensive, easy–to–use devices suitable for a multitude of applications that require a regulated supply of up to 100 mA. Like their higher powered MC7800 and MC78M00 Series cousins, these regulators feature internal current limiting and thermal shutdown making them remarkably rugged. No external components are required with the MC78L00 devices in many applications.

These devices offer a substantial performance advantage over the traditional zener diode–resistor combination, as output impedance and quiescent current are substantially reduced.

- Wide Range of Available, Fixed Output Voltages
- Low Cost
- Internal Short Circuit Current Limiting
- Internal Thermal Overload Protection
- No External Components Required
- Complementary Negative Regulators Offered (MC79L00 Series)
- Available in either ±5% (AC) or ±10% (C) Selections

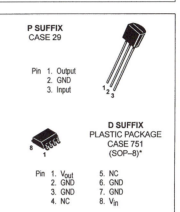

P SUFFIX
CASE 29

Pin 1. Output
    2. GND
    3. Input

D SUFFIX
PLASTIC PACKAGE
CASE 751
(SOP–8)*

Pin 1. V_{out} 5. NC
    2. GND   6. GND
    3. GND   7. GND
    4. NC   8. V_{in}

* SOP–8 is an internally modified SO–8 package. Pins 2, 3, 6, and 7 are electrically common to the die attach flag. This internal lead frame modification decreases package thermal resistance and increases power dissipation capability when appropriately mounted on a printed circuit board. SOP–8 conforms to all external dimensions of the standard SO–8 package.

Representative Schematic Diagram

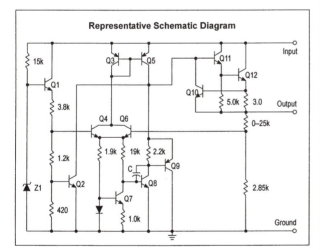

Standard Application

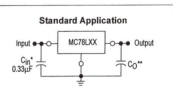

A common ground is required between the input and the output voltages. The input voltage must remain typically 2.0 V above the output voltage even during the low point on the input ripple voltage.

*C_{in} is required if regulator is located an appreciable distance from power supply filter.
**C_O is not needed for stability; however, it does improve transient response.

ORDERING INFORMATION

Device	Operating Temperature Range	Package
MC78LXXACD*	T_J = 0° to +125°C	SOP–8
MC78LXXACP		Plastic Power
MC78LXXCP		Plastic Power
MC78LXXABD*	T_J = –40° to +125°C	SOP–8
MC78LXXABP*		Plastic Power

XX indicates nominal voltage
*Available in 5, 8, 9, 12 and 15 V devices.

DEVICE TYPE/NOMINAL VOLTAGE

10%	5%	Voltage
MC78L05C	MC78L05AC	5.0
MC78L08C	MC78L08AC	8.0
MC78L09C	MC78L09AC	9.0
MC78L12C	MC78L12AC	12
MC78L15C	MC78L15AC	15
MC78L18C	MC78L18AC	18
MC78L24C	MC78L24AC	24

Rev 3

MC78L00, A Series

MAXIMUM RATINGS (T_A = +125°C, unless otherwise noted.)

Rating	Symbol	Value	Unit
Input Voltage (2.6 V–8.0 V) (12 V–18 V) (24 V)	V_I	30 35 40	Vdc
Storage Temperature Range	T_{stg}	–65 to +150	°C
Operating Junction Temperature Range	T_J	0 to +150	°C

ELECTRICAL CHARACTERISTICS (V_I = 10 V, I_O = 40 mA, C_I = 0.33 μF, C_O = 0.1 μF, –40°C < T_J < +125°C (for MC78LXXAB), 0°C < T_J < +125°C (for MC78LXXAC), unless otherwise noted.)

Characteristics	Symbol	MC78L05AC, AB			MC78L05C			Unit
		Min	Typ	Max	Min	Typ	Max	
Output Voltage (T_J = +25°C)	V_O	4.8	5.0	5.2	4.6	5.0	5.4	Vdc
Line Regulation (T_J = +25°C, I_O = 40 mA) 7.0 Vdc ≤ V_I ≤ 20 Vdc 8.0 Vdc ≤ V_I ≤ 20 Vdc	Reg$_{line}$	 – –	 55 45	 150 100	 – –	 55 45	 200 150	mV
Load Regulation (T_J = +25°C, 1.0 mA ≤ I_O ≤ 100 mA) (T_J = +25°C, 1.0 mA ≤ I_O ≤ 40 mA)	Reg$_{load}$	 – –	 11 5.0	 60 30	 – –	 11 5.0	 60 30	mV
Output Voltage (7.0 Vdc ≤ V_I ≤ 20 Vdc, 1.0 mA ≤ I_O ≤ 40 mA) (V_I = 10 V, 1.0 mA ≤ I_O ≤ 70 mA)	V_O	 4.75 4.75	 – –	 5.25 5.25	 4.5 4.5	 – –	 5.5 5.5	Vdc
Input Bias Current (T_J = +25°C) (T_J = +125°C)	I_{IB}	 – –	 3.8 –	 6.0 5.5	 – –	 3.8 –	 6.0 5.5	mA
Input Bias Current Change (8.0 Vdc ≤ V_I ≤ 20 Vdc) (1.0 mA ≤ I_O ≤ 40 mA)	ΔI_{IB}	 – –	 – –	 1.5 0.1	 – –	 – –	 1.5 0.2	mA
Output Noise Voltage (T_A = +25°C, 10 Hz ≤ f ≤ 100 kHz)	V_n	–	40	–	–	40	–	μV
Ripple Rejection (I_O = 40 mA, f = 120 Hz, 8.0 Vdc ≤ V_I ≤ 18 V, T_J = +25°C)	RR	41	49	–	40	49	–	dB
Dropout Voltage (T_J = +25°C)	$V_I - V_O$	–	1.7	–	–	1.7	–	Vdc

ELECTRICAL CHARACTERISTICS (V_I = 14 V, I_O = 40 mA, C_I = 0.33 μF, C_O = 0.1 μF, –40°C < T_J < +125°C (for MC78LXXAB), 0°C < T_J < +125°C (for MC78LXXAC), unless otherwise noted.)

Characteristics	Symbol	MC78L08AC, AB			MC78L08C			Unit
		Min	Typ	Max	Min	Typ	Max	
Output Voltage (T_J = +25°C)	V_O	7.7	8.0	8.3	7.36	8.0	8.64	Vdc
Line Regulation (T_J = +25°C, I_O = 40 mA) 10.5 Vdc ≤ V_I ≤ 23 Vdc 11 Vdc ≤ V_I ≤ 23 Vdc	Reg$_{line}$	 – –	 20 12	 175 125	 – –	 20 12	 200 150	mV
Load Regulation (T_J = +25°C, 1.0 mA ≤ I_O ≤ 100 mA) (T_J = +25°C, 1.0 mA ≤ I_O ≤ 40 mA)	Reg$_{load}$	 – –	 15 8.0	 80 40	 – –	 15 6.0	 80 40	mV
Output Voltage (10.5 Vdc ≤ V_I ≤ 23 Vdc, 1.0 mA ≤ I_O ≤ 40 mA) (V_I = 14 V, 1.0 mA ≤ I_O ≤ 70 mA)	V_O	 7.6 7.6	 – –	 8.4 8.4	 7.2 7.2	 – –	 8.8 8.8	Vdc
Input Bias Current (T_J = +25°C) (T_J = +125°C)	I_{IB}	 – –	 3.0 –	 6.0 5.5	 – –	 3.0 –	 6.0 5.5	mA
Input Bias Current Change (11 Vdc ≤ V_I ≤ 23 Vdc) (1.0 mA ≤ I_O ≤ 40 mA)	ΔI_{IB}	 – –	 – –	 1.5 0.1	 – –	 – –	 1.5 0.2	mA
Output Noise Voltage (T_A = +25°C, 10 Hz ≤ f ≤ 100 kHz)	V_n	–	60	–	–	52	–	μV
Ripple Rejection (I_O = 40 mA, f = 120 Hz, 12 V ≤ V_I ≤ 23 V, T_J = +25°C)	RR	37	57	–	36	55	–	dB
Dropout Voltage (T_J = +25°C)	$V_I - V_O$	–	1.7	–	–	1.7	–	Vdc

MC78L00, A Series

ELECTRICAL CHARACTERISTICS (V_I = 15 V, I_O = 40 mA, C_I = 0.33 μF, C_O = 0.1 μF, −40°C < T_J < +125°C (for MC78LXXAB), 0°C < T_J < +125°C (for MC78LXXAC), unless otherwise noted.)

Characteristics	Symbol	MC78L09AC, AB			MC78L09C			Unit
		Min	Typ	Max	Min	Typ	Max	
Output Voltage (T_J = +25°C)	V_O	8.6	9.0	9.4	8.3	9.0	9.7	Vdc
Line Regulation	Reg$_{line}$							mV
(T_J = +25°C, I_O = 40 mA)								
11.5 Vdc ≤ V_I ≤ 24 Vdc		−	20	175	−	20	200	
12 Vdc ≤ V_I ≤ 24 Vdc		−	12	125	−	12	150	
Load Regulation	Reg$_{load}$							mV
(T_J = +25°C, 1.0 mA ≤ I_O ≤ 100 mA)		−	15	90	−	15	90	
(T_J = +25°C, 1.0 mA ≤ I_O ≤ 40 mA)		−	8.0	40	−	6.0	40	
Output Voltage	V_O							Vdc
(11.5 Vdc ≤ V_I ≤ 24 Vdc, 1.0 mA ≤ I_O ≤ 40 mA)		8.5	−	9.5	8.1	−	9.9	
(V_I = 15 V, 1.0 mA ≤ I_O ≤ 70 mA)		8.5	−	9.5	8.1	−	9.9	
Input Bias Current	I_{IB}							mA
(T_J = +25°C)		−	3.0	6.0	−	3.0	6.0	
(T_J = +125°C)		−	−	5.5	−	−	5.5	
Input Bias Current Change	ΔI_{IB}							mA
(11 Vdc ≤ V_I ≤ 23 Vdc)		−	−	1.5	−	−	1.5	
(1.0 mA ≤ I_O ≤ 40 mA)		−	−	0.1	−	−	0.2	
Output Noise Voltage	V_n	−	60	−	−	52	−	μV
(T_A = +25°C, 10 Hz ≤ f ≤ 100 kHz)								
Ripple Rejection (I_O = 40 mA,	RR	37	57	−	36	55	−	dB
f = 120 Hz, 13 V ≤ V_I ≤ 24 V, T_J = +25°C)								
Dropout Voltage	$V_I − V_O$	−	1.7	−	−	1.7	−	Vdc
(T_J = +25°C)								

ELECTRICAL CHARACTERISTICS (V_I = 19 V, I_O = 40 mA, C_I = 0.33 μF, C_O = 0.1 μF, −40°C < T_J < +125°C (for MC78LXXAB), 0°C < T_J < +125°C (for MC78LXXAC), unless otherwise noted.)

Characteristics	Symbol	MC78L12AC, AB			MC78L12C			Unit
		Min	Typ	Max	Min	Typ	Max	
Output Voltage (T_J = +25°C)	V_O	11.5	12	12.5	11.1	12	12.9	Vdc
Line Regulation	Reg$_{line}$							mV
(T_J = +25°C, I_O = 40 mA)								
14.5 Vdc ≤ V_I ≤ 27 Vdc		−	120	250	−	120	250	
16 Vdc ≤ V_I ≤ 27 Vdc		−	100	200	−	100	200	
Load Regulation	Reg$_{load}$							mV
(T_J = +25°C, 1.0 mA ≤ I_O ≤ 100 mA)		−	20	100	−	20	100	
(T_J = +25°C, 1.0 mA ≤ I_O ≤ 40 mA)		−	10	50	−	10	50	
Output Voltage	V_O							Vdc
(14.5 Vdc ≤ V_I ≤ 27 Vdc, 1.0 mA ≤ I_O ≤ 40 mA)		11.4	−	12.6	10.8	−	13.2	
(V_I = 19 V, 1.0 mA ≤ I_O ≤ 70 mA)		11.4	−	12.6	10.8	−	13.2	
Input Bias Current	I_{IB}							mA
(T_J = +25°C)		−	4.2	6.5	−	4.2	6.5	
(T_J = +125°C)		−	−	6.0	−	−	6.0	
Input Bias Current Change	ΔI_{IB}							mA
(16 Vdc ≤ V_I ≤ 27 Vdc)		−	−	1.5	−	−	1.5	
(1.0 mA ≤ I_O ≤ 40 mA)		−	−	0.1	−	−	0.2	
Output Noise Voltage	V_n	−	80	−	−	80	−	μV
(T_A = +25°C, 10 Hz ≤ f ≤ 100 kHz)								
Ripple Rejection (I_O = 40 mA,	RR	37	42	−	36	42	−	dB
f = 120 Hz, 15 V ≤ V_I ≤ 25 V, T_J = +25°C)								
Dropout Voltage	$V_I − V_O$	−	1.7	−	−	1.7	−	Vdc
(T_J = +25°C)								

MC78L00, A Series

ELECTRICAL CHARACTERISTICS (V_I = 23 V, I_O = 40 mA, C_I = 0.33 µF, C_O = 0.1 µF, −40°C < T_J < +125°C (for MC78LXXAB), 0°C < T_J < +125°C (for MC78LXXAC), unless otherwise noted.)

Characteristics	Symbol	MC78L15AC, AB			MC78L15C			Unit
		Min	Typ	Max	Min	Typ	Max	
Output Voltage (T_J = +25°C)	V_O	14.4	15	15.6	13.8	15	16.2	Vdc
Line Regulation (T_J = +25°C, I_O = 40 mA)	Reg_{line}							mV
17.5 Vdc ≤ V_I ≤ 30 Vdc		–	130	300	–	130	300	
20 Vdc ≤ V_I ≤ 30 Vdc		–	110	250	–	110	250	
Load Regulation	Reg_{load}							mV
(T_J = +25°C, 1.0 mA ≤ I_O ≤ 100 mA)		–	25	150	–	25	150	
(T_J = +25°C, 1.0 mA ≤ I_O ≤ 40 mA)		–	12	75	–	12	75	
Output Voltage	V_O							Vdc
(17.5 Vdc ≤ V_I ≤ 30 Vdc, 1.0 mA ≤ I_O ≤ 40 mA)		14.25	–	15.75	13.5	–	16.5	
(V_I = 23 V, 1.0 mA ≤ I_O ≤ 70 mA)		14.25	–	15.75	13.5	–	16.5	
Input Bias Current	I_{IB}							mA
(T_J = +25°C)		–	4.4	6.5	–	4.4	6.5	
(T_J = +125°C)		–	–	6.0	–	–	6.0	
Input Bias Current Change	ΔI_{IB}							mA
(20 Vdc ≤ V_I ≤ 30 Vdc)		–	–	1.5	–	–	1.5	
(1.0 mA ≤ I_O ≤ 40 mA)		–	–	0.1	–	–	0.2	
Output Noise Voltage (T_A = +25°C, 10 Hz ≤ f ≤ 100 kHz)	V_n	–	90	–	–	90	–	µV
Ripple Rejection (I_O = 40 mA, f = 120 Hz, 18.5 V ≤ V_I ≤ 28.5 V, T_J = +25°C)	RR	34	39	–	33	39	–	dB
Dropout Voltage (T_J = +25°C)	$V_I - V_O$	–	1.7	–	–	1.7	–	Vdc

ELECTRICAL CHARACTERISTICS (V_I = 27 V, I_O = 40 mA, C_I = 0.33 µF, C_O = 0.1 µF, 0°C < T_J < +125°C, unless otherwise noted.)

Characteristics	Symbol	MC78L18AC			MC78L18C			Unit
		Min	Typ	Max	Min	Typ	Max	
Output Voltage (T_J = +25°C)	V_O	17.3	18	18.7	16.6	18	19.4	Vdc
Line Regulation (T_J = +25°C, I_O = 40 mA)	Reg_{line}							mV
21.4 Vdc ≤ V_I ≤ 33 Vdc		–	45	325	–	32	325	
20.7 Vdc ≤ V_I ≤ 33 Vdc								
22 Vdc ≤ V_I ≤ 33 Vdc		–	35	275	–	27	275	
21 Vdc ≤ V_I ≤ 33 Vdc								
Load Regulation	Reg_{load}							mV
(T_J = +25°C, 1.0 mA ≤ I_O ≤ 100 mA)		–	30	170	–	30	170	
(T_J = +25°C, 1.0 mA ≤ I_O ≤ 40 mA)		–	15	85	–	15	85	
Output Voltage	V_O							Vdc
(21.4 Vdc ≤ V_I ≤ 33 Vdc, 1.0 mA ≤ I_O ≤ 40 mA)					16.2	–	19.8	
(20.7 Vdc ≤ V_I ≤ 33 Vdc, 1.0 mA ≤ I_O ≤ 40 mA)		17.1	–	18.9				
(V_I = 27 V, 1.0 mA ≤ I_O ≤ 70 mA)					16.2	–	19.8	
(V_I = 27 V, 1.0 mA ≤ I_O ≤ 70 mA)		17.1	–	18.9				
Input Bias Current	I_{IB}							mA
(T_J = +25°C)		–	3.1	6.5	–	3.1	6.5	
(T_J = +125°C)		–	–	6.0	–	–	6.0	
Input Bias Current Change	ΔI_{IB}							mA
(22 Vdc ≤ V_I ≤ 33 Vdc)					–	–	1.5	
(21 Vdc ≤ V_I ≤ 33 Vdc)		–	–	1.5				
(1.0 mA ≤ I_O ≤ 40 mA)		–	–	0.1	–	–	0.2	
Output Noise Voltage (T_A = +25°C, 10 Hz ≤ f ≤ 100 kHz)	V_n	–	150	–	–	150	–	µV
Ripple Rejection (I_O = 40 mA, f = 120 Hz, 23 V ≤ V_I ≤ 33 V, T_J = +25°C)	RR	33	48	–	32	46	–	dB
Dropout Voltage (T_J = +25°C)	$V_I - V_O$	–	1.7	–	–	1.7	–	Vdc

MOTOROLA ANALOG IC DEVICE DATA

MC78L00, A Series

ELECTRICAL CHARACTERISTICS (V_I = 33 V, I_O = 40 mA, C_I = 0.33 µF, C_O = 0.1 µF, 0°C < T_J < +125°C, unless otherwise noted.)

Characteristics	Symbol	MC78L24AC			MC78L24C			Unit
		Min	Typ	Max	Min	Typ	Max	
Output Voltage (T_J = +25°C)	V_O	23	24	25	22.1	24	25.9	Vdc
Line Regulation (T_J = +25°C, I_O = 40 mA) 27.5 Vdc ≤ V_I ≤ 38 Vdc 28 Vdc ≤ V_I ≤ 80 Vdc 27 Vdc ≤ V_I ≤ 38 Vdc	Reg$_{line}$	 – – –	 – 50 60	 – 300 350	 – – –	 35 30 –	 350 300 –	mV
Load Regulation (T_J = +25°C, 1.0 mA ≤ I_O ≤ 100 mA) (T_J = +25°C, 1.0 mA ≤ I_O ≤ 40 mA)	Reg$_{load}$	 – –	 40 20	 200 100	 – –	 40 20	 200 100	mV
Output Voltage (28 Vdc ≤ V_I ≤ 38 Vdc, 1.0 mA ≤ I_O ≤ 40 mA) (27 Vdc ≤ V_I ≤ 38 Vdc, 1.0 mA ≤ I_O ≤ 40 mA) (28 Vdc ≤ V_I = 33 Vdc, 1.0 mA ≤ I_O ≤ 70 mA) (27 Vdc ≤ V_I ≤ 33 Vdc, 1.0 mA ≤ I_O ≤ 70 mA)	V_O	 22.8 22.8	 – –	 25.2 25.2	 21.6 21.6 	 – – 	 26.4 26.4 	Vdc
Input Bias Current (T_J = +25°C) (T_J = +125°C)	I_{IB}	 – –	 3.1 –	 6.5 6.0	 – –	 3.1 –	 6.5 6.0	mA
Input Bias Current Change (28 Vdc ≤ V_I ≤ 38 Vdc) (1.0 mA ≤ I_O ≤ 40 mA)	ΔI_{IB}	 – –	 – –	 1.5 0.1	 – –	 – –	 1.5 0.2	mA
Output Noise Voltage (T_A = +25°C, 10 Hz ≤ f ≤ 100 kHz)	V_n	–	200	–	–	200	–	µV
Ripple Rejection (I_O = 40 mA, f = 120 Hz, 29 V ≤ V_I ≤ 35 V, T_J = +25°C)	RR	31	45	–	30	43	–	dB
Dropout Voltage (T_J = +25°C)	$V_I - V_O$	–	1.7	–	–	1.7	–	Vdc

MC78L00, A Series

Figure 1. Dropout Characteristics

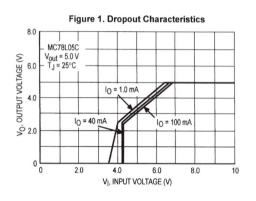

Figure 2. Dropout Voltage versus Junction Temperature

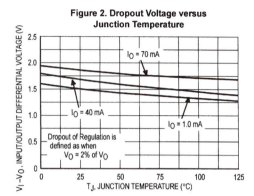

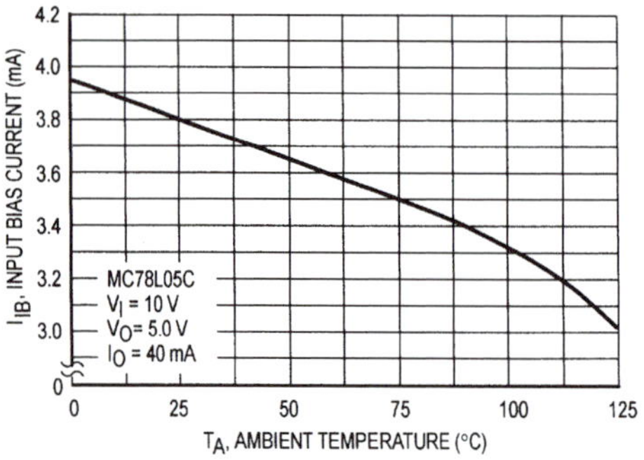

Figure 3. Input Bias Current versus Ambient Temperature

Figure 4. Input Bias Current versus Input Voltage

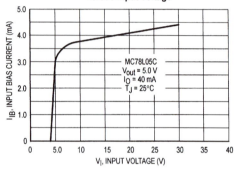

Figure 5. Maximum Average Power Dissipation versus Ambient Temperature – TO–92 Type Package

Figure 6. SOP–8 Thermal Resistance and Maximum Power Dissipation versus P.C.B. Copper Length

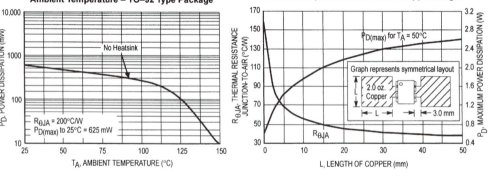

MC78L00, A Series

APPLICATIONS INFORMATION

Design Considerations

The MC78L00 Series of fixed voltage regulators are designed with Thermal Overload Protection that shuts down the circuit when subjected to an excessive power overload condition. Internal Short Circuit Protection limits the maximum current the circuit will pass.

In many low-current applications, compensation capacitors are not required. However, it is recommended that the regulator input be bypassed with a capacitor if the regulator is connected to the power supply filter with long wire lengths, or if the output load capacitance is large. The input bypass capacitor should be selected to provide good high–frequency characteristics to insure stable operation under all load conditions. A 0.33 μF or larger tantalum, mylar, or other capacitor having low internal impedance at high frequencies should be chosen. The bypass capacitor should be mounted with the shortest possible leads directly across the regulators input terminals. Good construction techniques should be used to minimize ground loops and lead resistance drops since the regulator has no external sense lead. Bypassing the output is also recommended.

Figure 7. Current Regulator

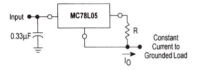

The MC78L00 regulators can also be used as a current source when connected as above. In order to minimize dissipation the MC78L05C is chosen in this application. Resistor R determines the current as follows:

$$I_O = \frac{5.0\,V}{R} + I_B$$

$I_B = 3.8$ mA over line and load changes

For example, a 100 mA current source would require R to be a 50 Ω, 1/2 W resistor and the output voltage compliance would be the input voltage less 7 V.

Figure 8. ± 15 V Tracking Voltage Regulator

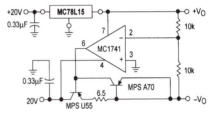

Figure 9. Positive and Negative Regulator

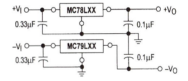

MC78L00, A Series

OUTLINE DIMENSIONS

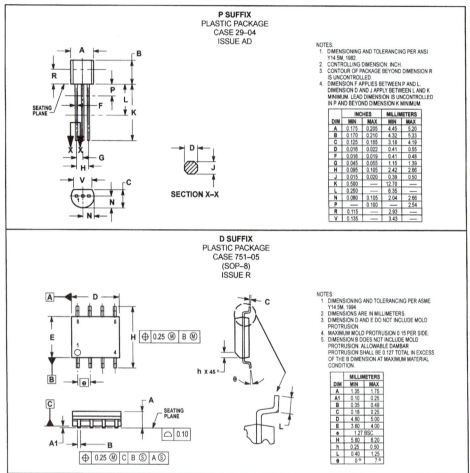

P SUFFIX
PLASTIC PACKAGE
CASE 29–04
ISSUE AD

NOTES:
1. DIMENSIONING AND TOLERANCING PER ANSI Y14.5M, 1982.
2. CONTROLLING DIMENSION: INCH.
3. CONTOUR OF PACKAGE BEYOND DIMENSION R IS UNCONTROLLED.
4. DIMENSION F APPLIES BETWEEN P AND L. DIMENSION D AND J APPLY BETWEEN L AND K MINIMUM. LEAD DIMENSION IS UNCONTROLLED IN P AND BEYOND DIMENSION K MINIMUM.

SECTION X–X

DIM	INCHES MIN	INCHES MAX	MILLIMETERS MIN	MILLIMETERS MAX
A	0.175	0.205	4.45	5.20
B	0.170	0.210	4.32	5.33
C	0.125	0.165	3.18	4.19
D	0.016	0.022	0.41	0.55
F	0.016	0.019	0.41	0.48
G	0.045	0.055	1.15	1.39
H	0.095	0.105	2.42	2.66
J	0.015	0.020	0.39	0.50
K	0.500	—	12.70	—
L	0.250	—	6.35	—
N	0.080	0.105	2.04	2.66
P	—	0.100	—	2.54
R	0.115	—	2.93	—
V	0.135	—	3.43	—

D SUFFIX
PLASTIC PACKAGE
CASE 751–05
(SOP–8)
ISSUE R

NOTES:
1. DIMENSIONING AND TOLERANCING PER ASME Y14.5M, 1994.
2. DIMENSIONS ARE IN MILLIMETERS.
3. DIMENSION D AND E DO NOT INCLUDE MOLD PROTRUSION.
4. MAXIMUM MOLD PROTRUSION 0.15 PER SIDE.
5. DIMENSION B DOES NOT INCLUDE MOLD PROTRUSION. ALLOWABLE DAMBAR PROTRUSION SHALL BE 0.127 TOTAL IN EXCESS OF THE B DIMENSION AT MAXIMUM MATERIAL CONDITION.

DIM	MILLIMETERS MIN	MILLIMETERS MAX
A	1.35	1.75
A1	0.10	0.25
B	0.35	0.49
C	0.18	0.25
D	4.80	5.00
E	3.80	4.00
e	1.27 BSC	
H	5.80	6.20
h	0.25	0.50
L	0.40	1.25
θ	0°	7°

SEATING PLANE

How to reach us:
USA / EUROPE / Locations Not Listed: Motorola Literature Distribution; P.O. Box 20912; Phoenix, Arizona 85036. 1–800–441–2447 or 602–303–5454

MFAX: RMFAX0@email.sps.mot.com – TOUCHTONE 602–244–6609
INTERNET: http://Design–NET.com

JAPAN: Nippon Motorola Ltd.; Tatsumi–SPD–JLDC, 6F Seibu–Butsuryu–Center, 3–14–2 Tatsumi Koto–Ku, Tokyo 135, Japan. 03–81–3521–8315

ASIA/PACIFIC: Motorola Semiconductors H.K. Ltd.; 8B Tai Ping Industrial Park, 51 Ting Kok Road, Tai Po, N.T., Hong Kong. 852–26629298

MOTOROLA

◇

MC78L00/D

September 1997

N *National Semiconductor*

LM386
Low Voltage Audio Power Amplifier

General Description

The LM386 is a power amplifier designed for use in low voltage consumer applications. The gain is internally set to 20 to keep external part count low, but the addition of an external resistor and capacitor between pins 1 and 8 will increase the gain to any value up to 200.

The inputs are ground referenced while the output is automatically biased to one half the supply voltage. The quiescent power drain is only 24 milliwatts when operating from a 6 volt supply, making the LM386 ideal for battery operation.

Features

■ Battery operation
■ Minimum external parts
■ Wide supply voltage range: 4V–12V or 5V–18V
■ Low quiescent current drain: 4 mA
■ Voltage gains from 20 to 200
■ Ground referenced input
■ Self-centering output quiescent voltage
■ Low distortion
■ Available in 8 pin MSOP package

Applications

■ AM-FM radio amplifiers
■ Portable tape player amplifiers
■ Intercoms
■ TV sound systems
■ Line drivers
■ Ultrasonic drivers
■ Small servo drivers
■ Power converters

Equivalent Schematic and Connection Diagrams

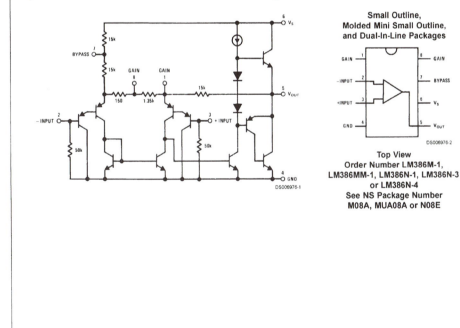

Small Outline,
Molded Mini Small Outline,
and Dual-In-Line Packages

DS006976-2

Top View
Order Number LM386M-1,
LM386MM-1, LM386N-1, LM386N-3
or LM386N-4
See NS Package Number
M08A, MUA08A or N08E

Absolute Maximum Ratings (Note 2)

If Military/Aerospace specified devices are required, please contact the National Semiconductor Sales Office/ Distributors for availability and specifications.

Supply Voltage	
(LM386N-1, -3, LM386M-1)	15V
Supply Voltage (LM386N-4)	22V
Package Dissipation (Note 3)	
(LM386N)	1.25W
(LM386M)	0.73W
(LM386MM-1)	0.595W
Input Voltage	±0.4V
Storage Temperature	65 C to +150 C
Operating Temperature	0 C to +70 C
Junction Temperature	+150 C
Soldering Information	

Dual-In-Line Package	
Soldering (10 sec)	+260 C
Small Outline Package	
(SOIC and MSOP)	
Vapor Phase (60 sec)	+215 C
Infrared (15 sec)	+220 C

See AN-450 "Surface Mounting Methods and Their Effect on Product Reliability" for other methods of soldering surface mount devices.

Thermal Resistance	
θ_{JC} (DIP)	37 C/W
θ_{JA} (DIP)	107 C/W
θ_{JC} (SO Package)	35 C/W
θ_{JA} (SO Package)	172 C/W
θ_{JA} (MSOP)	210 C/W
θ_{JC} (MSOP)	56 C/W

Electrical Characteristics(Notes 1, 2)

T_A = 25 C

Parameter	Conditions	Min	Typ	Max	Units
Operating Supply Voltage (V_S)					
LM386N-1, -3, LM386M-1, LM386MM-1		4		12	V
LM386N-4		5		18	V
Quiescent Current (I_Q)	V_S = 6V, V_{IN} = 0		4	8	mA
Output Power (P_{OUT})					
LM386N-1, LM386M-1, LM386MM-1	V_S = 6V, R_L = 8Ω, THD = 10%	250	325		mW
LM386N-3	V_S = 9V, R_L = 8Ω, THD = 10%	500	700		mW
LM386N-4	V_S = 16V, R_L = 32Ω, THD = 10%	700	1000		mW
Voltage Gain (A_V)	V_S = 6V, f = 1 kHz		26		dB
	10 μF from Pin 1 to 8		46		dB
Bandwidth (BW)	V_S = 6V, Pins 1 and 8 Open		300		kHz
Total Harmonic Distortion (THD)	V_S = 6V, R_L = 8Ω, P_{OUT} = 125 mW		0.2		%
	f = 1 kHz, Pins 1 and 8 Open				
Power Supply Rejection Ratio (PSRR)	V_S = 6V, f = 1 kHz, C_{BYPASS} = 10 μF		50		dB
	Pins 1 and 8 Open, Referred to Output				
Input Resistance (R_{IN})			50		kΩ
Input Bias Current (I_{BIAS})	V_S = 6V, Pins 2 and 3 Open		250		nA

Note 1: All voltages are measured with respect to the ground pin, unless otherwise specified.

Note 2: Absolute Maximum Ratings indicate limits beyond which damage to the device may occur. Operating Ratings indicate conditions for which the device is functional, but do not guarantee specific performance limits. Electrical Characteristics state DC and AC electrical specifications under particular test conditions which guarantee specific performance limits. This assumes that the device is within the Operating Ratings. Specifications are not guaranteed for parameters where no limit is given, however, the typical value is a good indication of device performance.

Note 3: For operation in ambient temperatures above 25 C, the device must be derated based on a 150 C maximum junction temperature and 1) a thermal resistance of 80 C/W junction to ambient for the dual-in-line package and 2) a thermal resistance of 170 C/W for the small outline package.

Application Hints

GAIN CONTROL

To make the LM386 a more versatile amplifier, two pins (1 and 8) are provided for gain control. With pins 1 and 8 open the 1.35 kΩ resistor sets the gain at 20 (26 dB). If a capacitor is put from pin 1 to 8, bypassing the 1.35 kΩ resistor, the gain will go up to 200 (46 dB). If a resistor is placed in series with the capacitor, the gain can be set to any value from 20 to 200. Gain control can also be done by capacitively coupling a resistor (or FET) from pin 1 to ground.

Additional external components can be placed in parallel with the internal feedback resistors to tailor the gain and frequency response for individual applications. For example, we can compensate poor speaker bass response by frequency shaping the feedback path. This is done with a series RC from pin 1 to 5 (paralleling the internal 15 kΩ resistor). For 6 dB effective bass boost: R ≅ 15 kΩ, the lowest value for good stable operation is R = 10 kΩ if pin 8 is open. If pins 1 and 8 are bypassed then R as low as 2 kΩ can be used. This restriction is because the amplifier is only compensated for closed-loop gains greater than 9.

INPUT BIASING

The schematic shows that both inputs are biased to ground with a 50 kΩ resistor. The base current of the input transistors is about 250 nA, so the inputs are at about 12.5 mV when left open. If the dc source resistance driving the LM386 is higher than 250 kΩ it will contribute very little additional offset (about 2.5 mV at the input, 50 mV at the output). If the dc source resistance is less than 10 kΩ, then shorting the unused input to ground will keep the offset low (about 2.5 mV at the input, 50 mV at the output). For dc source resistances between these values we can eliminate excess offset by putting a resistor from the unused input to ground, equal in value to the dc source resistance. Of course all offset problems are eliminated if the input is capacitively coupled.

When using the LM386 with higher gains (bypassing the 1.35 kΩ resistor between pins 1 and 8) it is necessary to bypass the unused input, preventing degradation of gain and possible instabilities. This is done with a 0.1 μF capacitor or a short to ground depending on the dc source resistance on the driven input.

Typical Performance Characteristics

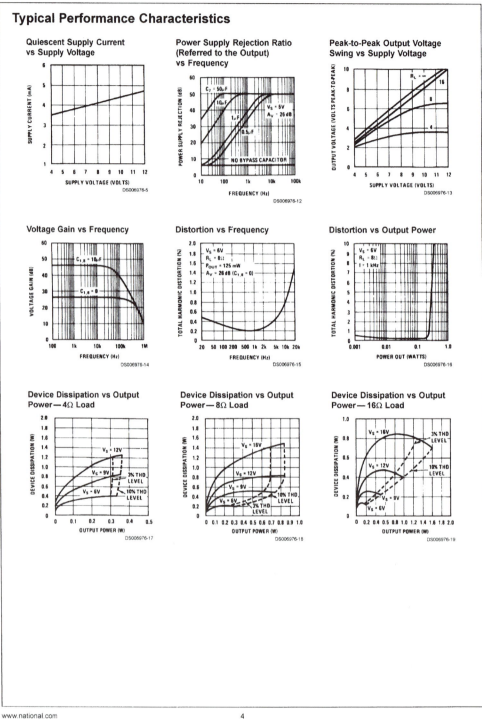

Quiescent Supply Current vs Supply Voltage

Power Supply Rejection Ratio (Referred to the Output) vs Frequency

Peak-to-Peak Output Voltage Swing vs Supply Voltage

Voltage Gain vs Frequency

Distortion vs Frequency

Distortion vs Output Power

Device Dissipation vs Output Power — 4Ω Load

Device Dissipation vs Output Power — 8Ω Load

Device Dissipation vs Output Power — 16Ω Load

Typical Applications

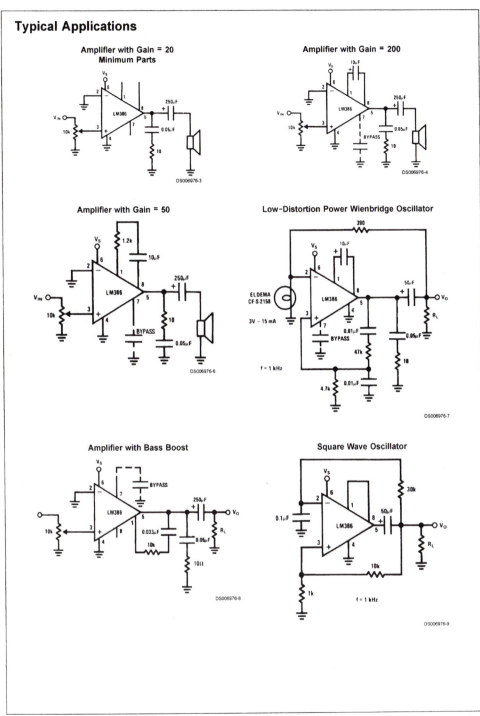

Amplifier with Gain = 20
Minimum Parts

DS006976-3

Amplifier with Gain = 200

DS006976-4

Amplifier with Gain = 50

DS006976-6

Low-Distortion Power Wienbridge Oscillator

f = 1 kHz

DS006976-7

Amplifier with Bass Boost

DS006976-8

Square Wave Oscillator

f = 1 kHz

DS006976-9

Typical Applications (Continued)

Frequency Response with Bass Boost

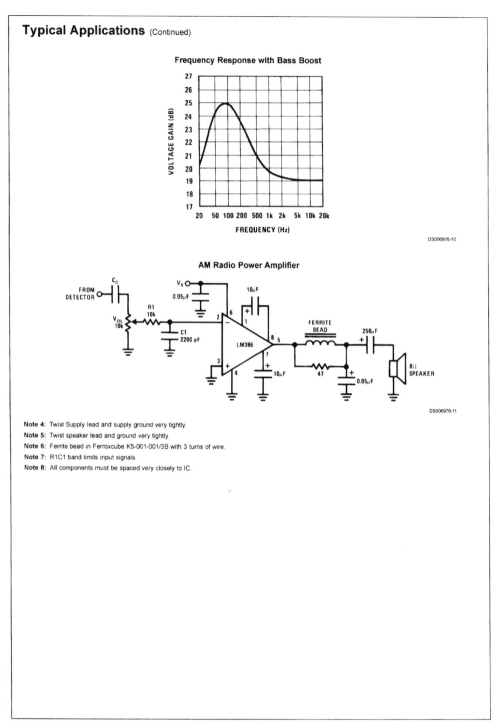

DS006976-10

AM Radio Power Amplifier

DS006976-11

Note 4: Twist Supply lead and supply ground very tightly.

Note 5: Twist speaker lead and ground very tightly.

Note 6: Ferrite bead in Ferroxcube K5-001-001/3B with 3 turns of wire.

Note 7: R1C1 band limits input signals.

Note 8: All components must be spaced very closely to IC.

Physical Dimensions inches (millimeters) unless otherwise noted

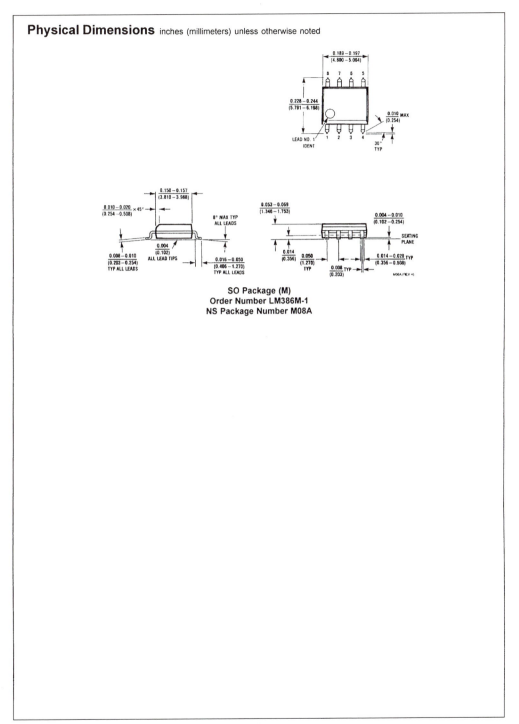

SO Package (M)
Order Number LM386M-1
NS Package Number M08A

Physical Dimensions inches (millimeters) unless otherwise noted (Continued)

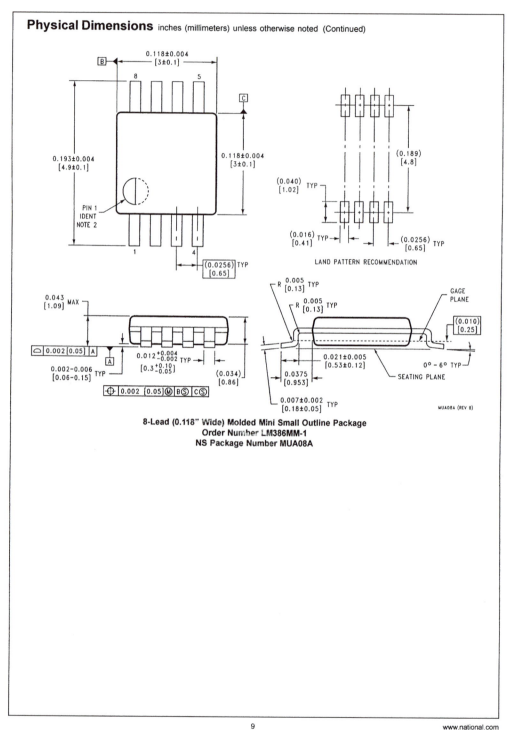

LAND PATTERN RECOMMENDATION

8-Lead (0.118" Wide) Molded Mini Small Outline Package
Order Number LM386MM-1
NS Package Number MUA08A

MUA08A (REV B)

LM386 Low Voltage Audio Power Amplifier

Physical Dimensions inches (millimeters) unless otherwise noted (Continued)

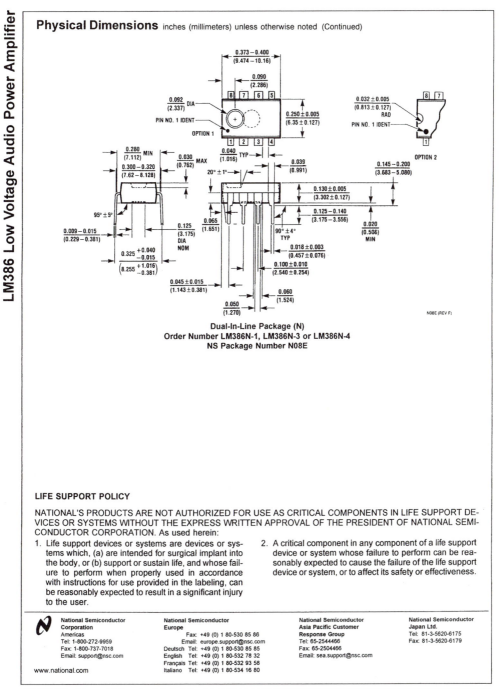

Dual-In-Line Package (N)
Order Number LM386N-1, LM386N-3 or LM386N-4
NS Package Number N08E

N08E (REV F)

LIFE SUPPORT POLICY

NATIONAL'S PRODUCTS ARE NOT AUTHORIZED FOR USE AS CRITICAL COMPONENTS IN LIFE SUPPORT DE-
VICES OR SYSTEMS WITHOUT THE EXPRESS WRITTEN APPROVAL OF THE PRESIDENT OF NATIONAL SEMI-
CONDUCTOR CORPORATION. As used herein:

1. Life support devices or systems are devices or systems which, (a) are intended for surgical implant into
the body, or (b) support or sustain life, and whose failure to perform when properly used in accordance
with instructions for use provided in the labeling, can
be reasonably expected to result in a significant injury
to the user.

2. A critical component in any component of a life support
device or system whose failure to perform can be reasonably expected to cause the failure of the life support
device or system, or to affect its safety or effectiveness.

**National Semiconductor
Corporation
Americas**
Tel: 1-800-272-9959
Fax: 1-800-737-7018
Email: support@nsc.com
www.national.com

**National Semiconductor
Europe**
Fax: +49 (0) 1 80-530 85 86
Email: europe.support@nsc.com
Deutsch Tel: +49 (0) 1 80-530 85 85
English   Tel: +49 (0) 1 80-532 78 32
Français Tel: +49 (0) 1 80-532 93 58
Italiano Tel: +49 (0) 1 80-534 16 80

**National Semiconductor
Asia Pacific Customer
Response Group**
Tel: 65-2544466
Fax: 65-2504466
Email: sea.support@nsc.com

**National Semiconductor
Japan Ltd.**
Tel: 81-3-5620-6175
Fax: 81-3-5620-6179

DISCRETE SEMICONDUCTORS

DATA SHEET

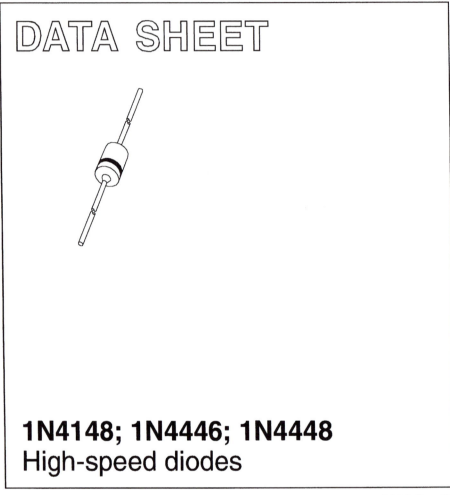

1N4148; 1N4446; 1N4448
High-speed diodes

Product specification   1996 Sep 03
Supersedes data of April 1996
File under Discrete Semiconductors, SC01

Philips
Semiconductors

PHILIPS

Philips Semiconductors

Product specification

High-speed diodes

1N4148; 1N4446; 1N4448

FEATURES

- Hermetically sealed leaded glass SOD27 (DO-35) package
- High switching speed: max. 4 ns
- General application
- Continuous reverse voltage: max. 75 V
- Repetitive peak reverse voltage: max. 75 V
- Repetitive peak forward current: max. 450 mA.

APPLICATIONS

- High-speed switching.

DESCRIPTION

The 1N4148, 1N4446, 1N4448 are high-speed switching diodes fabricated in planar technology, and encapsulated in hermetically sealed leaded glass SOD27 (DO-35) packages.

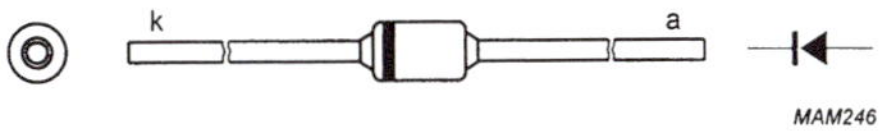

The diodes are type branded.

Fig.1 Simplified outline (SOD27; DO-35) and symbol.

LIMITING VALUES

In accordance with the Absolute Maximum Rating System (IEC 134).

SYMBOL	PARAMETER	CONDITIONS	MIN.	MAX.	UNIT
V_{RRM}	repetitive peak reverse voltage		–	75	V
V_R	continuous reverse voltage		–	75	V
I_F	continuous forward current	see Fig.2; note 1	–	200	mA
I_{FRM}	repetitive peak forward current		–	450	mA
I_{FSM}	nonrepetitive peak forward current	square wave; T_j = 25 °C prior to surge; see Fig.4			
		t = 1 μs	–	4	A
		t = 1 ms	–	1	A
		t = 1 s	–	0.5	A
P_{tot}	total power dissipation	T_{amb} = 25 °C; note 1	–	500	mW
T_{stg}	storage temperature		–65	+200	°C
T_j	junction temperature		–	200	°C

Note

1. Device mounted on an FR4 printed circuit-board; lead length 10 mm.

2

Philips Semiconductors

Product specification

High-speed diodes

1N4148; 1N4446; 1N4448

ELECTRICAL CHARACTERISTICS

$T_j = 25\ °C$; unless otherwise specified.

SYMBOL	PARAMETER	CONDITIONS	MIN.	MAX.	UNIT
V_F	forward voltage	see Fig.3			
	1N4148	$I_F = 10$ mA	–	1.0	V
	1N4446	$I_F = 20$ mA	–	1.0	V
	1N4448	$I_F = 5$ mA	0.62	0.72	V
		$I_F = 100$ mA	–	1.0	V
I_R	reverse current	$V_R = 20$ V; see Fig.5		25	nA
		$V_R = 20$ V; $T_j = 150\ °C$; see Fig.5	–	50	µA
I_R	reverse current; 1N4448	$V_R = 20$ V; $T_j = 100\ °C$; see Fig.5	–	3	µA
C_d	diode capacitance	$f = 1$ MHz; $V_R = 0$; see Fig.6		4	pF
t_{rr}	reverse recovery time	when switched from $I_F = 10$ mA to $I_R = 60$ mA; $R_L = 100\ \Omega$; measured at $I_R = 1$ mA; see Fig.7		4	ns
V_{fr}	forward recovery voltage	when switched from $I_F = 50$ mA; $t_r = 20$ ns; see Fig.8	–	2.5	V

THERMAL CHARACTERISTICS

SYMBOL	PARAMETER	CONDITIONS	VALUE	UNIT
$R_{th\ j\text{-}tp}$	thermal resistance from junction to tie-point	lead length 10 mm	240	K/W
$R_{th\ j\text{-}a}$	thermal resistance from junction to ambient	lead length 10 mm; note 1	350	K/W

Note

1. Device mounted on a printed circuit-board without metallization pad.

Philips Semiconductors

Product specification

High-speed diodes

1N4148; 1N4446; 1N4448

GRAPHICAL DATA

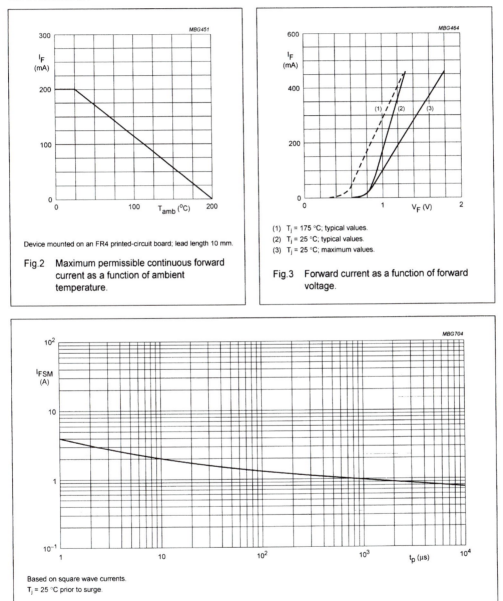

Device mounted on an FR4 printed-circuit board; lead length 10 mm.

Fig.2 Maximum permissible continuous forward current as a function of ambient temperature.

(1) T_j = 175 °C; typical values.
(2) T_j = 25 °C; typical values.
(3) T_j = 25 °C; maximum values.

Fig.3 Forward current as a function of forward voltage.

Based on square wave currents.
T_j = 25 °C prior to surge.

Fig.4 Maximum permissible nonrepetitive peak forward current as a function of pulse duration.

Philips Semiconductors   Product specification

High-speed diodes   1N4148; 1N4446; 1N4448

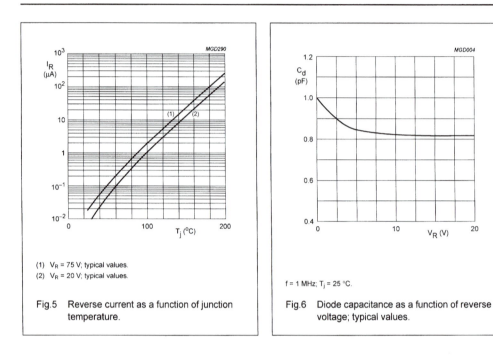

(1) V_R = 75 V; typical values.
(2) V_R = 20 V; typical values.

Fig.5   Reverse current as a function of junction
        temperature.

f = 1 MHz; T_j = 25 °C.

Fig.6   Diode capacitance as a function of reverse
        voltage; typical values.

Philips Semiconductors

Product specification

High-speed diodes

1N4148; 1N4446; 1N4448

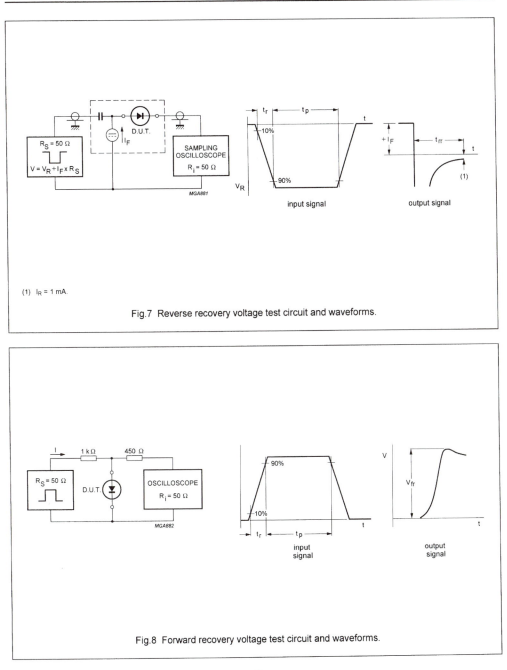

(1) $I_R = 1$ mA.

Fig.7 Reverse recovery voltage test circuit and waveforms.

Fig.8 Forward recovery voltage test circuit and waveforms.

High-speed diodes

1N4148; 1N4446; 1N4448

PACKAGE OUTLINE

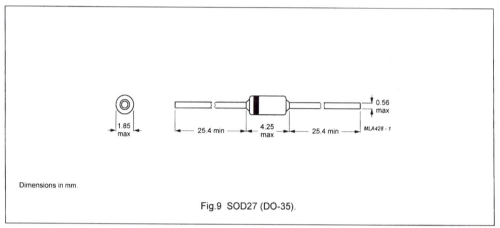

Dimensions in mm.

Fig.9 SOD27 (DO-35).

DEFINITIONS

Data Sheet Status	
Objective specification	This data sheet contains target or goal specifications for product development.
Preliminary specification	This data sheet contains preliminary data; supplementary data may be published later.
Product specification	This data sheet contains final product specifications.
Limiting values	
Limiting values given are in accordance with the Absolute Maximum Rating System (IEC 134). Stress above one or more of the limiting values may cause permanent damage to the device. These are stress ratings only and operation of the device at these or at any other conditions above those given in the Characteristics sections of the specification is not implied. Exposure to limiting values for extended periods may affect device reliability.	
Application information	
Where application information is given, it is advisory and does not form part of the specification.	

LIFE SUPPORT APPLICATIONS

These products are not designed for use in life support appliances, devices, or systems where malfunction of these products can reasonably be expected to result in personal injury. Philips customers using or selling these products for use in such applications do so at their own risk and agree to fully indemnify Philips for any damages resulting from such improper use or sale

RF COMMUNICATIONS PRODUCTS

DATA SHEET

SA602A
Double-balanced mixer and oscillator

Product specification
Replaces datasheet of April 17, 1990
IC17 Data Handbook

1997 Nov 07

Philips Semiconductors

PHILIPS

Philips Semiconductors

Product specification

Double-balanced mixer and oscillator

SA602A

DESCRIPTION
The SA602A is a low-power VHF monolithic double-balanced mixer with input amplifier, on-board oscillator, and voltage regulator. It is intended for high performance, low-power communication systems. The guaranteed parameters of the SA602A make this device particularly well suited for cellular radio applications. The mixer is a "Gilbert cell" multiplier configuration which typically provides 18dB of gain at 45MHz. The oscillator will operate to 200MHz. It can be configured as a crystal oscillator, a tuned tank oscillator, or a buffer for an external LO. For higher frequencies the LO input may be externally driven. The noise figure at 45MHz is typically less than 5dB. The gain, intercept performance, low-power and noise characteristics make the SA602A a superior choice for high-performance battery operated equipment. It is available in an 8-lead dual in-line plastic package and an 8-lead SO (surface-mount miniature package).

FEATURES
- Low current consumption: 2.4mA typical
- Excellent noise figure: <4.7dB typical at 45MHz
- High operating frequency
- Excellent gain, intercept and sensitivity
- Low external parts count; suitable for crystal/ceramic filters
- SA602A meets cellular radio specifications

PIN CONFIGURATION

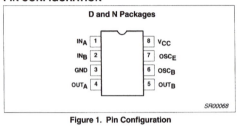

Figure 1. Pin Configuration

APPLICATIONS
- Cellular radio mixer/oscillator
- Portable radio
- VHF transceivers
- RF data links
- HF/VHF frequency conversion
- Instrumentation frequency conversion
- Broadband LANs

ORDERING INFORMATION

DESCRIPTION	TEMPERATURE RANGE	ORDER CODE	DWG #
8-Pin Plastic Dual In-Line Plastic (DIP)	-40 to +85°C	SA602AN	SOT97-1
8-Pin Plastic Small Outline (SO) package (Surface-mount)	-40 to +85°C	SA602AD	SOT96-1

ABSOLUTE MAXIMUM RATINGS

SYMBOL	PARAMETER		RATING	UNITS
V_{CC}	Maximum operating voltage		9	V
T_{STG}	Storage temperature range		-65 to +150	°C
T_A	Operating ambient temperature range SA602A		-40 to +85	°C
θ_{JA}	Thermal impedance	D package	90	°C/W
		N package	75	°C/W

Philips Semiconductors

Product specification

Double-balanced mixer and oscillator

SA602A

BLOCK DIAGRAM

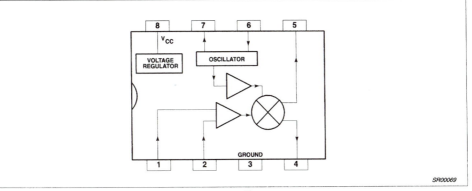

Figure 2. Block Diagram

AC/DC ELECTRICAL CHARACTERISTICS

$V_{CC} = +6V$, $T_A = 25°C$; unless otherwise stated.

SYMBOL	PARAMETER	TEST CONDITIONS	LIMITS SA602A			UNITS
			MIN	TYP	MAX	
V_{CC}	Power supply voltage range		4.5		8.0	V
	DC current drain			2.4	2.8	mA
f_{IN}	Input signal frequency			500		MHz
f_{OSC}	Oscillator frequency			200		MHz
	Noise figure at 45MHz			5.0	5.5	dB
	Third-order intercept point	$RF_{IN} = -45dBm$: $f_1 = 45.0MHz$ $f_2 = 45.06MHz$		-13	-15	dBm
	Conversion gain at 45MHz		14	17		dB
R_{IN}	RF input resistance		1.5			$k\Omega$
C_{IN}	RF input capacitance			3	3.5	pF
	Mixer output resistance	(Pin 4 or 5)		1.5		$k\Omega$

DESCRIPTION OF OPERATION

The SA602A is a Gilbert cell, an oscillator/buffer, and a temperature compensated bias network as shown in the equivalent circuit. The Gilbert cell is a differential amplifier (Pins 1 and 2) which drives a balanced switching cell. The differential input stage provides gain and determines the noise figure and signal handling performance of the system.

The SA602A is designed for optimum low-power performance. When used with the SA604 as a 45MHz cellular radio second IF and demodulator, the SA602A is capable of receiving -119dBm signals with a 12dB S/N ratio. Third-order intercept is typically -13dBm (that is approximately +5dBm output intercept because of the RF gain). The system designer must be cognizant of this large signal limitation. When designing LANs or other closed systems where transmission levels are high, and small-signal or signal-to-noise issues are not critical, the input to the SA602A should be appropriately scaled.

Besides excellent low-power performance well into VHF, the SA602A is designed to be flexible. The input, RF mixer output and oscillator ports can support a variety of configurations provided the designer understands certain constraints, which will be explained here.

The RF inputs (Pins 1 and 2) are biased internally. They are symmetrical. The equivalent AC input impedance is approximately 1.5k || 3pF through 50MHz. Pins 1 and 2 can be used interchangeably, but they should not be DC biased externally. Figure 5 shows three typical input configurations.

The mixer outputs (Pins 4 and 5) are also internally biased. Each output is connected to the internal positive supply by a 1.5kΩ resistor. This permits direct output termination yet allows for balanced output as well. Figure 6 shows three single-ended output configurations and a balanced output.

Philips Semiconductors   Product specification

Double-balanced mixer and oscillator   SA602A

The oscillator is capable of sustaining oscillation beyond 200MHz in crystal or tuned tank configurations. The upper limit of operation is determined by tank "Q" and required drive levels. The higher the "Q" of the tank or the smaller the required drive, the higher the permissible oscillation frequency. If the required LO is beyond oscillation limits, or the system calls for an external LO, the external signal can be injected at Pin 6 through a DC blocking capacitor. External LO should be at least 200mV$_{P-P}$.

Figure 7 shows several proven oscillator circuits. Figure 7a is appropriate for cellular radio. As shown, an overtone mode of operation is utilized. Capacitor C3 and inductor L1 suppress oscillation at the crystal fundamental frequency. In the fundamental mode, the suppression network is omitted.

Figure 8 shows a Colpitts varactor tuned tank oscillator suitable for synthesizer-controlled applications. It is important to buffer the

output of this circuit to assure that switching spikes from the first counter or prescaler do not end up in the oscillator spectrum. The dual-gate MOSFET provides optimum isolation with low current. The FET offers good isolation, simplicity, and low current, while the bipolar transistors provide the simple solution for noncritical applications. The resistive divider in the emitter-follower circuit should be chosen to provide the minimum input signal which will assure correct system operation.

When operated above 100MHz, the oscillator may not start if the Q of the tank is too low. A 22kΩ resistor from Pin 7 to ground will increase the DC bias current of the oscillator transistor. This improves the AC operating characteristic of the transistor and should help the oscillator to start. A 22kΩ resistor will not upset the other DC biasing internal to the device, but smaller resistance values should be avoided.

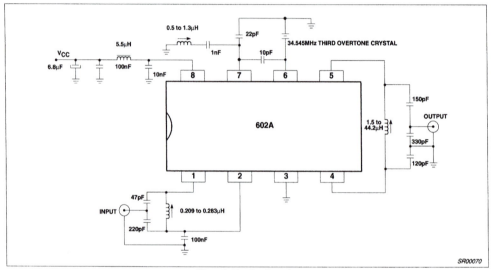

Figure 3. Test Configuration

Philips Semiconductors

Product specification

Double-balanced mixer and oscillator

SA602A

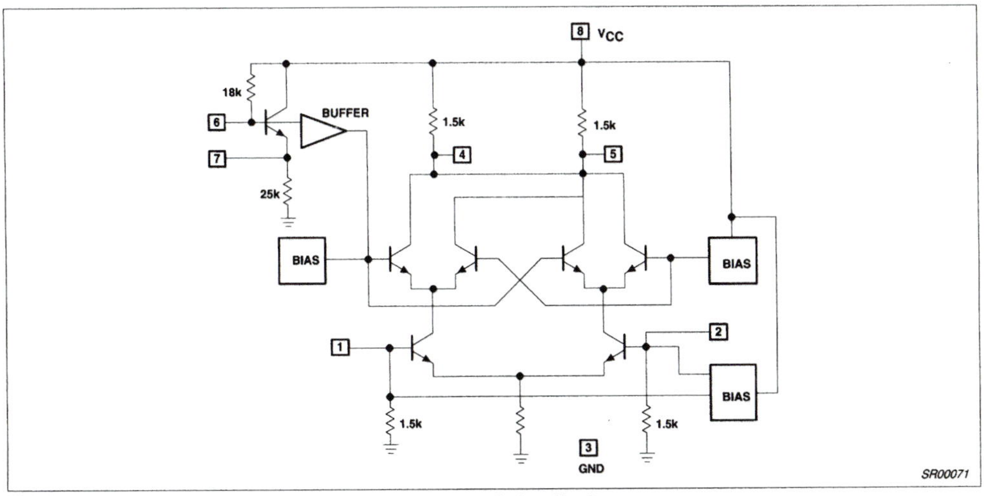

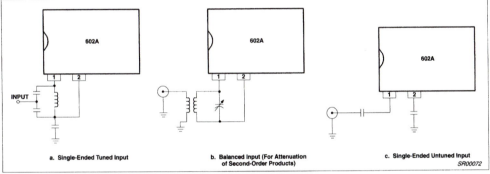

Figure 4. Equivalent Circuit

a. Single-Ended Tuned Input

b. Balanced Input (For Attenuation of Second-Order Products)

c. Single-Ended Untuned Input

Figure 5. Input Configuration

5

Philips Semiconductors

Product specification

Double-balanced mixer and oscillator

SA602A

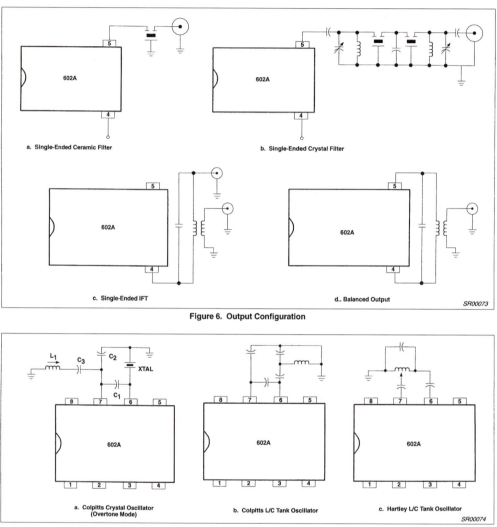

a. Single-Ended Ceramic Filter

b. Single-Ended Crystal Filter

c. Single-Ended IFT

d.. Balanced Output

SR00073

Figure 6. Output Configuration

a. Colpitts Crystal Oscillator
(Overtone Mode)

b. Colpitts L/C Tank Oscillator

c. Hartley L/C Tank Oscillator

SR00074

Figure 7. Oscillator Circuits

Philips Semiconductors

Product specification

Double-balanced mixer and oscillator

SA602A

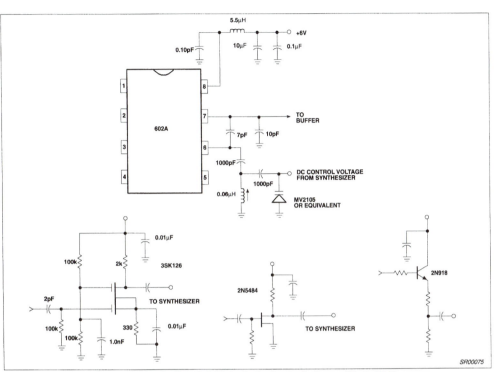

Figure 8. Colpitts Oscillator Suitable for Synthesizer Applications and Typical Buffers

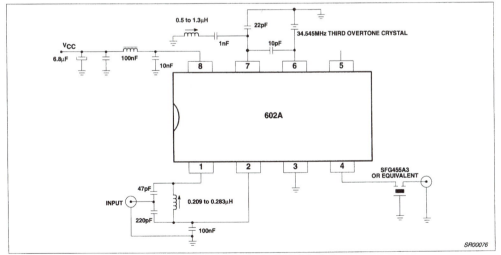

Figure 9. Typical Application for Cellular Radio

Philips Semiconductors

Product specification

Double-balanced mixer and oscillator

SA602A

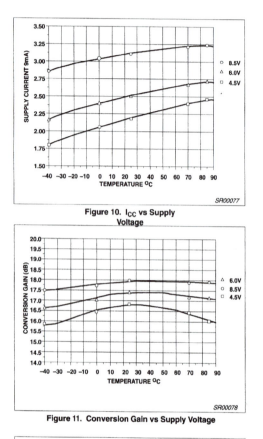

Figure 10. I_{CC} vs Supply Voltage

SR00077

Figure 11. Conversion Gain vs Supply Voltage

SR00078

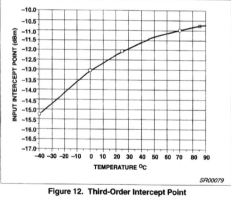

Figure 12. Third-Order Intercept Point

SR00079

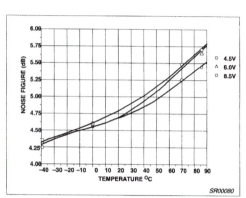

Figure 13. Noise Figure

SR00080

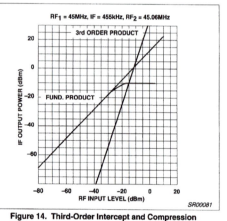

Figure 14. Third-Order Intercept and Compression

SR00081

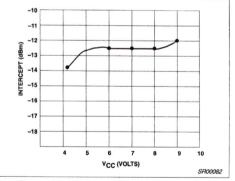

Figure 15. Input Third-Order Intermod Point vs V_{CC}

SR00082

Philips Semiconductors

Product specification

Double-balanced mixer and oscillator

SA602A

SO8:   plastic small outline package; 8 leads; body width 3.9mm   **SOT96-1**

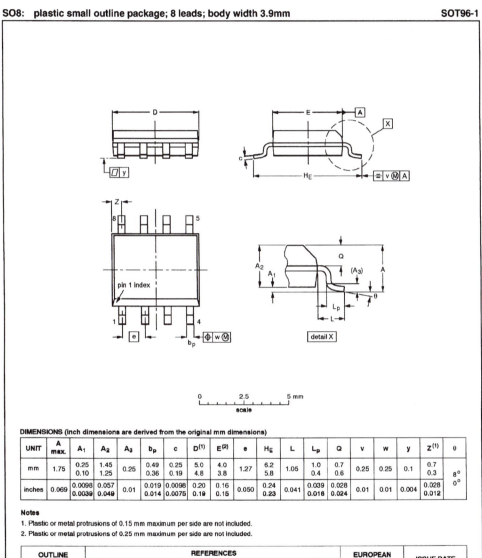

DIMENSIONS (inch dimensions are derived from the original mm dimensions)

UNIT	A max.	A₁	A₂	A₃	bₚ	c	D(1)	E(2)	e	Hₑ	L	Lₚ	Q	v	w	y	z(1)	θ
mm	1.75	0.25 / 0.10	1.45 / 1.25	0.25	0.49 / 0.36	0.25 / 0.19	5.0 / 4.8	4.0 / 3.8	1.27	6.2 / 5.8	1.05	1.0 / 0.4	0.7 / 0.6	0.25	0.25	0.1	0.7 / 0.3	8° 0°
inches	0.069	0.0098 / 0.0039	0.057 / 0.049	0.01	0.019 / 0.014	0.0098 / 0.0075	0.20 / 0.19	0.16 / 0.15	0.050	0.24 / 0.23	0.041	0.039 / 0.016	0.028 / 0.024	0.01	0.01	0.004	0.028 / 0.012	

Notes

1. Plastic or metal protrusions of 0.15 mm maximum per side are not included.
2. Plastic or metal protrusions of 0.25 mm maximum per side are not included.

OUTLINE VERSION	REFERENCES			EUROPEAN PROJECTION	ISSUE DATE
	IEC	JEDEC	EIAJ		
SOT96-1	076E03S	MS-012AA			92-11-17 95-02-04

Philips Semiconductors

Product specification

Double-balanced mixer and oscillator

SA602A

DIP8: plastic dual in-line package; 8 leads (300 mil)

SOT97-1

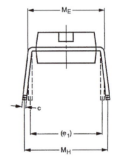

DIMENSIONS (inch dimensions are derived from the original mm dimensions)

UNIT	A max.	A$_1$ min.	A$_2$ max.	b	b$_1$	b$_2$	c	D$^{(1)}$	E$^{(1)}$	e	e$_1$	L	M$_E$	M$_H$	w	z$^{(1)}$ max.
mm	4.2	0.51	3.2	1.73 1.14	0.53 0.38	1.07 0.89	0.36 0.23	9.8 9.2	6.48 6.20	2.54	7.62	3.60 3.05	8.25 7.80	10.0 8.3	0.254	1.15
inches	0.17	0.020	0.13	0.068 0.045	0.021 0.015	0.042 0.035	0.014 0.009	0.39 0.36	0.26 0.24	0.10	0.30	0.14 0.12	0.32 0.31	0.39 0.33	0.01	0.045

Note

1. Plastic or metal protrusions of 0.25 mm maximum per side are not included.

OUTLINE VERSION	REFERENCES				EUROPEAN PROJECTION	ISSUE DATE
	IEC	JEDEC	EIAJ			
SOT97-1	050G01	MO-001AN				92-11-17 95-02-04

Philips Semiconductors   Product specification

Double-balanced mixer and oscillator   SA602A

DEFINITIONS

Data Sheet Identification	Product Status	Definition
Objective Specification	**Formative or In Design**	This data sheet contains the design target or goal specifications for product development. Specifications may change in any manner without notice.
Preliminary Specification	**Preproduction Product**	This data sheet contains preliminary data, and supplementary data will be published at a later date. Philips Semiconductors reserves the right to make changes at any time without notice in order to improve design and supply the best possible product.
Product Specification	**Full Production**	This data sheet contains Final Specifications. Philips Semiconductors reserves the right to make changes at any time without notice, in order to improve design and supply the best possible product.

Philips Semiconductors
811 East Arques Avenue
P.O. Box 3409
Sunnyvale, California 94088–3409
Telephone 800-234-7381

Let's make things better.

Philips
Semiconductors

PHILIPS

Index